AF360718

Drought Stress in Horticultural Plants

Drought Stress in Horticultural Plants

Editors

Stefania Toscano
Giulia Franzoni
Sara Álvarez

MDPI • Basel • Beijing • Wuhan • Barcelona • Belgrade • Manchester • Tokyo • Cluj • Tianjin

Editors

Stefania Toscano
University of Catania
Italy

Giulia Franzoni
University of Milan
Italy

Sara Álvarez
Instituto
Tecnológico
Agrario de Castilla
y León (ITACYL)
Spain

Editorial Office
MDPI
St. Alban-Anlage 66
4052 Basel, Switzerland

This is a reprint of articles from the Special Issue published online in the open access journal *Horticulturae* (ISSN 2311-7524) (available at: https://www.mdpi.com/journal/horticulturae/special_issues/Drought_Horticultural_Plants).

For citation purposes, cite each article independently as indicated on the article page online and as indicated below:

LastName, A.A.; LastName, B.B.; LastName, C.C. Article Title. *Journal Name* **Year**, *Volume Number*, Page Range.

ISBN 978-3-0365-6517-0 (Hbk)
ISBN 978-3-0365-6518-7 (PDF)

Contents

About the Editors

Stefania Toscano

Stefania Toscano Interest: drought stress, abiotic stress; functional characteristics of ornamental plants; product innovation; post-harvest; urban green areas and related functions; safeguarding biodiversity. Regarding my scientific production: 58 indexed papers in different journals appearing as the first author in 27 of them, in 3 as the last author and in 4 as the corresponding author; Citations according to Scopus: 569; h index Scopus: 13. In addition, 4 book chapters, 37 participations in international and national congresses; participations in courses, seminars, webinars, etc. related to my research; referee for many journals; guest editor in 4 Special Issues. Participation in 15 projects to date. Member of the "International Society for Horticultural Science" and of the "Italian Society of Horticulture (SOI)".

Giulia Franzoni

Giulia Franzoni Interest: abiotic stresses, biostimulants, vegetable crops, sustainable agriculture, horticulture, postharvest physiology, fruit quality. Regarding my scientific production: 11 indexed papers in different journals, citations according to Scopus: 312; h index Scopus: 6. In addition, Agronomy best paper awards 2019, participations in international and national congresses; participations in courses, seminars, webinars, etc. related to my research; referee for scientific journals. Member of the "Italian Society of Horticulture (SOI)".

Sara Álvarez

Sara Álvarez Interests: deficit irrigation; plant physiology; ornamental plants; stress physiology; evapotranspiration; salinity; water relations; tree nut crops; intrinsic water use efficiency Regarding my scientific production: 42 indexed papers in 17 different journals (19 Q1) appearing as first author in 15 of them, in 6 as last author and in 8 as corresponding author; Citations according to Scopus: 1340; h index Scopus: 24. In addition, 1 book chapters, 16 scientific-technical reports, 63 participations in international and national congresses, 23 oral communications, 3 invited speakers; 35 participations in courses, seminars, webinars, etc. related to my research; member of a jury of 2 doctoral Thesis, referee of 90 papers in SCI journal; guest editor in 5 Special Issues. Participation in 20 projects to date, 5 of which have been financed by European funds. Principal investigator of a FEADER project and three contracts of special relevance with companies. Finally, award for the best communication in the VII Congreso Ibérico de Agroingeniería y Ciencias Horticolas in 2013.

Preface to "Drought Stress in Horticultural Plants"

This Special Issue has been focused on "Drought stress in Horticultural plants". The authors throughout the duration of the Special Issue have contributed with original research and reviews covering a large number of topics associated with the effects of drought stress on different horticultural plants in open fields or in controlled environments. In addition, a good part of the articles is framed within the analysis of the response mechanism to drought stress. Water deficit, stress adaptation, drought responsive genes, gas exchange, enzyme activity are some of the keywords most used in the accepted manuscripts.

Stefania Toscano, Giulia Franzoni, and Sara Álvarez
Editors

 *horticulturae*

Editorial

Drought Stress in Horticultural Plants

Stefania Toscano [1],*, Giulia Franzoni [2] and Sara Álvarez [3]

[1] Department of Agriculture, Food and Environment, Università degli Studi di Catania, 95123 Catania, Italy
[2] Department Agricultural and Environmental Sciences, Università degli Studi di Milano, Via Celoria 2, 20133 Milano, Italy
[3] Unit of Woody and Horticultural Crops, Instituto Tecnológico Agrario de Castilla y Leon, 47071 Valladolid, Spain
* Correspondence: stefania.toscano@unict.it; Tel.: +39-095-4783303

Citation: Toscano, S.; Franzoni, G.; Álvarez, S. Drought Stress in Horticultural Plants. *Horticulturae* **2023**, *9*, 7. https://doi.org/10.3390/horticulturae9010007

Received: 5 December 2022
Revised: 9 December 2022
Accepted: 12 December 2022
Published: 21 December 2022

Drought stress is one of the main factors limiting horticultural crops, especially in environments such as the Mediterranean basin, which is often characterized by sub-optimal water availability. Global changes will determine the increase in semi-arid conditions, so all horticultural crops will have to cope with water scarcity. Appropriate plant selection and new cultivation methods, especially strategies of deficit irrigation, are crucial in improving crop cultivation performance.

This Special Issue, entitled "Drought Stress in Horticultural Plants", comprises 11 innovative publications, which could enrich our knowledge about the mechanisms of plant to drought stress.

The plants that overcome drought stress develop different morphological, physiological, and biochemical mechanisms. Yang et al. [1] provide a review that focuses on the molecular mechanisms, and in particular on the main drought stress signals and signal transduction pathways in plants, as well as the functional and regulatory genes related to drought stress. The authors summarized the above aspects to provide valuable background knowledge and a theoretical basis for future agriculture, forestry breeding, and cultivation.

Climate change is often cited as one of the future challenges facing the agricultural sector because it significantly impacts both the agricultural sector and food security. The incidence of extreme weather conditions, such as flooding, drought, heat or frost, among others, is becoming more frequent, placing plants under stressful conditions. Another review article written by Giordano et al. [2], focused its attention on the plants' defense mechanisms and the involved morpho-physiological, biochemical, and molecular changes in the responses of ornamental plants to deficit irrigation. Drought stress tolerance can be reached through the selection of species tolerant to drought stress or by increasing the tolerance of sensitive species. In particular, this study analyzed the little-known response of ornamental plants to water stress conditions, as this is an agricultural sector that is constantly growing. From the different species analyzed in this manuscript, it was shown that both sensitive and tolerant plants have innate defense mechanisms, which include morphological changes (increase in leaf thickness, stomata density reduction and reduction in plant growth), physiological changes (restoration of osmotic balance and the closure of stomata), and synthesis of antioxidant molecules and enzymes. The drought-stress response also includes hormonal activity, transcription factors, and the activation of specific genes.

Among the experimental articles, Ferreira et al. [3] evaluated the ecophysiological parameters and the productivity of *Vigna unguiculata* (L.) Walp in response to water deficiency during the reproductive phase. Over the course of this study, the limit value could be established as a threshold water potential (−0.88 MPa) from which the water scarcity has negative effects on the cowpea grown under the climatic conditions of northeastern Pará.

Duarte et al. [4] analyzed how water stress level estimation is possible with good accuracy on the whole canopy, using majority voting at the tuber differentiation and maximum rate of tuberization phenological stages; the use of machine learning algorithms allows us to determine which regions in the spectral signature of the leaves are more influential to better estimate water stress from remote sensing using images in the visible (400–700 nm) and near-infrared (NIR) (700–1000 nm) bands. The results could lead to the use of more specific normalized water indices for water stress detection and estimation in potato crops by using these machine learning algorithms. This work is an important base for further research considering actual potato crop field conditions and cultural practices.

Bedding plants play an important role in public parks and private gardens. Very frequently, particularly in the Mediterranean area, these plants can suffer from drought stress because they are not always adequately watered. Romano and Toscano [5] analyzed the morphological, physiological, and enzymatic responses of *Zinnia elegans* L. subjected to different drought-stress levels. This study showed that the species response was different in relation to their stress levels. With light deficit irrigation, the plants could perform as well as fully irrigated plants. With medium deficit irrigation, the mechanisms were not always suitable to overcome drought stress. With severe deficit irrigation, the strategies adopted by the plants were not able to resist drought stress (e.g., stomatal closure, photosynthesis reduction, increase in water use efficiency and increase in enzyme activity and proline content).

The Mediterranean environment is characterized by high summer temperatures often associated with a shortage and poor quality of irrigation water. The presence of simultaneous stresses (drought and saline aerosol) has a cumulative negative effect on plant growth and survival. Toscano et al. [6] wanted to analyze the effects of drought and aerosol stresses on two species (*Callistemon citrinus* (Curtis) Skeels and *Viburnum tinus* L. 'Lucidum'). The interaction between the two stress conditions was found to be additive for almost all the physiological parameters, resulting in enhanced damage on plants under stress combination. The overall data suggested that *Viburnum* was more tolerant compared to *Callistemon* under the experimental conditions studied.

Hessini et al. [7] investigated the limits of tolerance to water deficit of the Damask rose and identified the main physiological and biochemical mechanisms that are linked to drought resistance. An integrated approach combining biochemical and physiological studies revealed new insights into the mechanisms and processes involved in *Rosa damascena* Mill. var. *trigentipetala* drought adaptation. In particular, under water-deficit conditions, the contents of biomolecules were positively correlated with antioxidant and inhibitory enzyme activities (LOX, AChE). These results suggest adequate protection against oxidative damage, and, thus, adaptation to water limitation. Water deficit can successfully enhance health-promoting phytochemicals in roses, which could be manipulated through agricultural techniques and screening programs to develop drought-tolerant genotypes.

In Asia, the mung bean is a nutritionally and economically important legume crop. This crop is sensitive to water scarcity in the different stages of development of its growth period, but the information regarding this is rather scarce. Ariharasutharsan et al. [8] imposed water stress on two mung bean cultivars, VRM(Gg)1 and CO6 during the flowering stage, evaluating physiological, biochemical, and transcriptional changes. Transcriptional analysis of photosynthesis, antioxidants, and drought-sensitive genes showed that VRM(Gg)1 increased transcripts more than CO6 under drought stress. Increased transcripts of drought-responsive genes indicate that VRM(Gg)1 showed a better genetic basis against drought stress than CO6. These results help us understand the mung bean response to drought stress and will contribute to the development of genotypes with increased drought tolerance using naturally occurring genetic variants.

The exploration and conservation of genetic diversity among plant accessions are important for the management of plant genetic resources. Forty-six tomato accessions from Jordan were collected by Makhadmeh et al. [9] and evaluated for their performance and morphophysiology, as well as undergoing molecular characterization to detect genetic diversity. The accessions were also subjected to two levels of water stress. Drought stress negatively affected several traits, revealing a wide range of variations among tomato accessions. The results provide new insight into the use of informative molecular markers to elucidate such a large genetic variation discovered in this collection.

Nemeskéri et al. [10], analyzed the use of plant growth-promoting rhizobacteria (PGPR) to promote tolerance to drought stress in tomato cultivars. Drought-tolerant PGPR may support plant development under limited water supply conditions when the plant's water demand is not completely satisfied under rain-fed conditions or when the availability of irrigation water is limited. The authors analyzed the effects of two inoculation treatments compared to control plants without artificial inoculation and three irrigation levels. In particular, they measured the chlorophyll a fluorescence, leaf chlorophyll content (SPAD value), canopy temperature, and yield. Different effects resulted according to the growth stage and in relation to the irrigation levels. Based on the results, the authors recommend the application of the PGPR given the positive effects observes on physiological processes, leading to a higher marketable yield, particularly under water shortage.

Li et al. [11] analyzed the difference in drought tolerance of seven common lily varieties based on morphological and physiological markers. The results showed differences in the morphological indices of leaves and anatomical structures in these varieties. Drought reduced chlorophyll content, inhibited net photosynthesis and increased enzyme activity, malondialdehyde, proline, soluble sugar, and soluble protein. The structure of lily leaves can therefore be used as one of the indices for classifying drought resistance. Genetic diversity analysis and functional annotations of genes associated with the SSR information obtained in this study provide valuable information on the most suitable genotype that can be implemented in plant breeding programs and future molecular analysis to differentiate the drought resistance of flower varieties.

Drought stress must be continuously investigated through multidisciplinary approaches in order to obtain valuable information about the mechanisms involved in plant stress responses, leading to the identification of agronomic solutions to mitigate stressful effects. Moreover, the study of plant response mechanisms, and particularly the genes and the signal transduction involved in stress tolerance, will help the breeder in the selection of drought-tolerant plants, the identification of suitable genotypes, and the adoption of specific management strategies in drought-prone environments.

Overall, these papers provide insights into new research directions within this field, covering the horticultural and ornamental species of interest. In the future, further research will be focused on finding new approaches and species, with a particular interest in maintaining plant productivity and food quality.

Author Contributions: Conceptualization, S.T., G.F. and S.Á.; writing—original draft preparation, S.T.; writing—review and editing, S.T., G.F. and S.Á. All authors have read and agreed to the published version of the manuscript.

Funding: This research received no external funding.

Conflicts of Interest: The authors declare no conflict of interest.

References

1. Yang, X.; Lu, M.; Wang, Y.; Wang, Y.; Liu, Z.; Chen, S. Response Mechanism of Plants to Drought Stress. *Horticulturae* **2021**, *7*, 50. [CrossRef]
2. Giordano, M.; Petropoulos, S.; Cirillo, C.; Rouphael, Y. Biochemical, Physiological, and Molecular Aspects of Ornamental Plants Adaptation to Deficit Irrigation. *Horticulturae* **2021**, *7*, 107. [CrossRef]
3. Ferreira, D.; Sousa, D.; Nunes, H.; Pinto, J.; Farias, V.; Costa, D.; Moura, V.; Teixeira, E.; Sousa, A.; Pinheiro, H.; et al. Cowpea Ecophysiological Responses to Accumulated Water Deficiency during the Reproductive Phase in Northeastern Pará, Brazil. *Horticulturae* **2021**, *7*, 116. [CrossRef]
4. Duarte-Carvajalino, J.; Silva-Arero, E.; Góez-Vinasco, G.; Torres-Delgado, L.; Ocampo-Paez, O.; Castaño-Marín, A. Estimation of Water Stress in Potato Plants Using Hyperspectral Imagery and Machine Learning Algorithms. *Horticulturae* **2021**, *7*, 176. [CrossRef]
5. Toscano, S.; Romano, D. Morphological, Physiological, and Biochemical Responses of Zinnia to Drought Stress. *Horticulturae* **2021**, *7*, 362. [CrossRef]
6. Toscano, S.; Ferrante, A.; Romano, D.; Tribulato, A. Interactive Effects of Drought and Saline Aerosol Stress on Morphological and Physiological Characteristics of Two Ornamental Shrub Species. *Horticulturae* **2021**, *7*, 517. [CrossRef]

7.   Hessini, K.; Wasli, H.; Al-Yasi, H.; Ali, E.; Issa, A.; Hassan, F.; Siddique, K. Graded Moisture Deficit Effect on Secondary Metabolites, Antioxidant, and Inhibitory Enzyme Activities in Leaf Extracts of *Rosa damascena Mill. var. trigentipetala*. *Horticulturae* **2022**, *8*, 177. [CrossRef]

8.   Ariharasutharsan, G.; Karthikeyan, A.; Renganathan, V.; Marthandan, V.; Dhasarathan, M.; Ambigapathi, A.; Akilan, M.; Palaniyappan, S.; Mariyammal, I.; Pandiyan, M.; et al. Distinctive Physio-Biochemical Properties and Transcriptional Changes Unfold the Mungbean Cultivars Differing by Their Response to Drought Stress at Flowering Stage. *Horticulturae* **2022**, *8*, 424. [CrossRef]

9.   Makhadmeh, I.; Albalasmeh, A.; Ali, M.; Thabet, S.; Darabseh, W.; Jaradat, S.; Alqudah, A. Molecular Characterization of Tomato (*Solanum lycopersicum* L.) Accessions under Drought Stress. *Horticulturae* **2022**, *8*, 600. [CrossRef]

10.   Nemeskéri, E.; Horváth, K.; Andryei, B.; Ilahy, R.; Takács, S.; Neményi, A.; Pék, Z.; Helyes, L. Impact of Plant Growth-Promoting Rhizobacteria Inoculation on the Physiological Response and Productivity Traits of Field-Grown Tomatoes in Hungary. *Horticulturae* **2022**, *8*, 641. [CrossRef]

11.   Li, X.; Jia, W.; Zheng, J.; Ma, L.; Duan, Q.; Du, W.; Cui, G.; Wang, X.; Wang, J. Application of Morphological and Physiological Markers for Study of Drought Tolerance in Lilium Varieties. *Horticulturae* **2022**, *8*, 786. [CrossRef]

horticulturae

Article

Application of Morphological and Physiological Markers for Study of Drought Tolerance in *Lilium* Varieties

Xiang Li [1], Wenjie Jia [1,*], Jie Zheng [2], Lulin Ma [1], Qing Duan [1], Wenwen Du [1], Guangfen Cui [1], Xiangning Wang [1] and Jihua Wang [1,*]

[1] Flower Research Institute, Yunnan Academy of Agriculture Sciences, Kunming 650000, China
[2] Forestry Management Agency of Mount Emei, Leshan 614200, China
* Correspondence: jiawenjie917917@163.com (W.J.); wjh0505@gmail.com (J.W.)

Abstract: The shortage of water resources is an unfavourable factor that restricts the production of flowers. The use of drought-resistant morphological markers is of great significance to distinguish the drought resistance of flower varieties. In this paper, we study the difference in drought tolerance of seven common lily varieties in the flower market by morphological and physiological markers. The results showed that there were differences in leaf morphological indices and anatomical structures among the seven varieties. Drought reduced the chlorophyll content, inhibited the photosynthetic rate, and increased catalase (CAT), peroxidase (POD), superoxide dismutase (SOD), malondialdehyde (MDA), proline, soluble sugar, and soluble protein. After rewatering, the activities of CAT, POD, and SOD of 'Lyon', 'Royal Sunset', and 'Robina' varieties decreased, which was opposite to the varieties of 'Immaculate', 'Elena', 'Siberia', and 'Gelria'. According to the membership function value of physiological indices, the drought resistance of seven lily varieties from weak to strong was 'Immaculate', 'Elena', 'Siberia', 'Gelria', 'Robina', 'Royal Sunset', and 'Lyon'. Drought resistance is related to the thickness of leaves, palisade tissue, sponge tissue, and specific leaf area. Lily leaf structure can be used as one of the indices to judge drought resistance.

Keywords: *Lilium*; drought tolerance; leaf morphology and structure; photosynthetic capacity; antioxidant; membership function

Citation: Li, X.; Jia, W.; Zheng, J.; Ma, L.; Duan, Q.; Du, W.; Cui, G.; Wang, X.; Wang, J. Application of Morphological and Physiological Markers for Study of Drought Tolerance in *Lilium* Varieties. *Horticulturae* **2022**, *8*, 786. https://doi.org/10.3390/horticulturae8090786

Academic Editors: Stefania Toscano, Giulia Franzoni and Sara Álvarez

Received: 19 July 2022
Accepted: 16 August 2022
Published: 30 August 2022

Publisher's Note: MDPI stays neutral with regard to jurisdictional claims in published maps and institutional affiliations.

1. Introduction

Global climate change intensifies the problem of drought and could expand the scope of drought in subtropical arid areas [1]. The loss caused by drought is equivalent to 60% of the loss associated with all climate disasters. Due to water shortages, the growth and development of plants in arid and semiarid areas are restricted [2], and the physiological and biochemical characteristics of plants are changed [3]. Lilies (*Lilium* spp.) are perennial herbaceous bulbous flowers of the family *Liliaceae* (*Lilium*). Lilies are not only famous due to their beautiful flowers but also their wide use as potted flowers and in landscaping all over the world [4]. Abiotic stress is an important factor affecting the growth and development of the lily. Abiotic stress affects the yield and ornamental quality and restricts the application of lilies in open fields [5,6]. Drought and water shortages often occur in the main lily-producing areas. Water shortage directly affects plant growth and development and then affects the quality of cut flowers [7]. The shortage of water resources has become a serious ecological problem, restricting the development of global flower production.

In response to drought stress, there could be differences between interspecific or even intra-specific differences. Understanding the response of different species to water deficit conditions will help to identify drought-resistant morphological markers and facilitate the identification of drought-sensitive and drought-tolerant species [8]. Drought-induced morphological and physiological changes can be used to find drought-resistant genotypes or develop new flower varieties to improve drought productivity [9]. Leaves are the main organs of photosynthesis and transpiration in plants, and they are also the most sensitive organs of plants to drought stress. In the process of adapting to the external environment, the anatomical characteristics of the leaf structure are related to the drought resistance ability, such as the thickness of the cuticle on the epidermis, the developed transport tissue, and the smaller stomata [10]. Therefore, drought resistance among plants has a certain relationship with leaf morphology and anatomical structure [11]. Plants use the morphological characteristics of drought adaptation to ensure maximum water absorption under drought conditions. The leaf phenotype can be used to preliminarily judge the drought resistance adaptability of the variety [12].

When the plant is under drought stress, the root system will quickly generate chemical signals to transmit to the above-ground part, causing the stomata to close to reduce water loss. The plant shrinks the stomata to prevent more water loss. Stomatal closure will reduce the transpiration rate, limit CO_2 absorption and transmission, and reduce the net photosynthetic rate of the plant [13]. The change in chlorophyll content indicates the process of photooxidation and degradation and reflects the effect of drought stress on photosynthesis. The degree of inhibition of photosynthesis in plant leaves by drought stress reflects the strength of drought resistance [14]. When photosynthesis is inhibited, the supply of organic matter required for the development of plant reproductive organs will be reduced, which will hinder reproductive development [15]. Under the influence of drought stress, plant cells, tissues and organs accumulate reactive oxygen species (ROS), and the activities of catalase (CAT), superoxide dismutase (SOD), and peroxidases (PODs) are enhanced, which helps to remove ROS and reduce electrolyte leakage and lipid peroxidation to maintain the vitality and integrity of organelles and cell membranes. An effective antioxidant system is important to provide drought tolerance [16]. The balance of water metabolism is broken, resulting in dehydration of the cell protoplasm and a decline in water potential, resulting in the oxidative modification of proteins, lipids, and DNA and damaging normal cell functions. Proline and soluble sugar, as widely distributed important osmotic regulatory substances, can prevent damage to plants caused by drought stress by maintaining cell turgor [6]. The concentration of protein and soluble sugar is considered a general indicator of drought tolerance [17].

Plant drought resistance strategies are important in morphology, physiology, and biochemistry. The evaluation of plant drought resistance mainly considers the molecular as well as epigenetic levels of drought resistance and drought resistance recovery [18]. The study of the stress/recovery response helps us to better understand the ability of plants to adapt to different environmental and climatic conditions [17]. Stress recovery is an important part of the plant response. In the special case of drought, the carbon balance depends not only on the rate and degree of photosynthesis decline under water depletion but also on the capacity supply of photosynthesis restoration after water depletion. After drought and rewatering, plants can quickly resume growth, eliminate the inhibition of drought on plant growth, and sometimes even have a super compensation effect to compensate for the loss of plants caused by drought [19]. It is necessary to have more direct indicators for evaluation of flower production. Only by better understanding the differences in the main adaptation mechanisms of these lily varieties to cope with drought stress can we determine the key adaptation traits to distinguish their drought resistance. However, the application of morphological and physiological markers to evaluate drought tolerance of lily is very few.

The drought resistance membership function method is a better comprehensive evaluation method of drought resistance. The greater the average value is, the stronger the drought resistance. Membership function analysis provides a way to comprehensively evaluate the drought resistance of plants based on multiple determination, which can avoid the one-sidedness of a single index and improve the reliability and accuracy of drought resistance identification. The drought resistance membership function method is widely used to evaluate the drought resistance of iris [20], citrus [21], and other flowers, as well as abiotic stresses such as heavy metals [22], temperature [23], and salinity [24,25]. At present, few studies have compared and evaluated the drought resistance of several lily varieties using leaf morphological structure and physiological indicators as markers [6]. We selected seven common lily varieties in the flower market as plant materials and used pot experiments to determine the differences in leaf morphology and structure of different lily varieties. According to the changes in photosynthesis, the antioxidant system and physiological indices of lily leaves from drought stress to rewatering, we determined the drought-resistance level of seven lily varieties. We hypothesize that the difference in the leaf structure of the lily can reflect the drought resistance of varieties, which was consistent with the level of drought resistance reflected by physiological indicators. So, the results of this study can provide a reference for the selection of lily varieties in areas with a relative lack of water resources and can also provide a reference for the evaluation of drought resistance of other plants.

2. Materials and Methods

2.1. Test Site and Plant Materials

The experiment was conducted in the lily breeding experimental base ($25°7'33''$ N, $102°45'48''$ E, 1951.1 m above sea level) of the flower Research Institute of Yunnan Academy of Agricultural Sciences, China, from May 2020 to August 2020. The Seven different lily varieties used in the experiment were: Oriental Hybrids 'Siberia', Pollen Abortion Cultivar 'Immaculate', Longiflorum Hybrid 'Gelria', Double Petal Cultivar 'Elena', Oriental Trumpet (OT) Hybrid 'Robina', Longiflorum Asiatic (LA) Hybrid 'Royal Sunset', and Asiatic Hybrid 'Lyon'. The test materials were healthy in appearance, no plant diseases and insect pests, and the specifications were consistent specifications (bulb perimeter diameter 12~14 cm) after vernalization treatment (Figure 1). Cultivated in a greenhouse and exposed to a natural photoperiod (from May to August), the cultivation soil was sandy soil, with field moisture capacity (FMC) of 23.20%, organic matter content of 5.34%, pH 5.66, total nitrogen of 0.18%, total phosphorus of 0.08%, total potassium of 1.93%, alkali hydrolysable nitrogen of 139.92 mg·kg^{-1}, available phosphorus of 81.28 mg·kg^{-1}, available potassium of 112.68 mg·kg^{-1}, available iron of 421.55 mg·kg^{-1}, exchangeable calcium of 988.41 mg·kg^{-1}, and exchangeable magnesium of 110.70 mg·kg^{-1}. The greenhouse temperature was 14~25 °C, the relative humidity was 55~70%, and the greenhouse CO_2 concentration was 389 µmol·mol^{-1}.

2.2. Experimental Design

Lily bulbs were cultivated in boxes (0.6 m × 0.4 m × 0.25 m, length × width × height). The row spacing was 12 cm, the plant spacing was 15 cm, and 9 plants were planted in each box. Nine boxes were planted for each variety. One box was used for leaf morphological analysis under natural conditions, and the other 8 boxes were used to analyse the changes in leaf physiological indices under drought stress. When the plant grew to the budding stage and the top growth stopped, the morphological indices were measured, and drought stress began. The samples were collected at 0, 5, 10, 15, 20, 25, and 30 days of drought stress and 5 days (35 days) of continuous rehydration after 30 days.

2.3. Soil Water Status

The gravimetric water content (GWC) was measured by the commonly used drying and weighing method [26]. The soil water content was monitored 0–10 cm away from the soil surface of each pot. See Table 1 for the GWC and relative water content (RWC) of each treatment.

$$GWC = \frac{M - M_S}{M_S} \times 100\% \tag{1}$$

$$RWC = \frac{GWC}{FMC} \times 100\% \tag{2}$$

where M: original soil weight; M_S: dry soil weight; FMC: field moisture capacity.

'Siberia'   'Immaculate'   'Gelria'

'Elena'   'Robina'

'Royal Sunset'   'Lyon'

Figure 1. Seven varieties of lily.

Table 1. Changes in gravimetric water content (GWC) and relative water content (RWC).

	0 d	5 d	10 d	15 d	20 d	25 d	30 d	35 d
GWC (%)	19.67	14.86	13.11	10.12	8.98	7.06	5.67	20.34
RWC (%)	84.78	64.05	56.51	43.62	38.71	30.43	24.44	87.67

2.4. Determination of Leaf Morphological Indices

In the first box of lily varieties used for morphological index measurement, all leaves of 9 plants of each variety were collected for total leaf area calculation. The sixth fully expanded leaf from the top branch point of 9 plants was selected for single leaf length, width, area, dry weight, and specific leaf area (SLA).

$$\text{SLA} = \frac{\text{Leaf area}}{\text{Leaf dry weight}} \tag{3}$$

Five plants were selected to collect the seventh fully expanded leaf from the top branch point for paraffin section observation. The paraffin sections were placed into 1% safranine dye solution for staining for 1–2 h, washed with distilled water, and decolorized with 50%, 70%, and 80% gradient alcohol for 60 s. Then, the sections were placed into the 0.5% solid green dye solution for staining for 30–60 s, decolorized with anhydrous ethanol for 90 s, placed into a 60°C oven for drying, and cleared with xylene for 5 min. The sections were observed and imaged under an optical microscope (YS100, Nikon, Tokyo, Japan).

2.5. Determination of Photosynthetic Index

Among the 8 drought stress treatment groups of each variety, the plants to be tested were induced by light at 8:30~11:30 a.m. under natural light conditions. Three plants with strong growth, relatively consistent growth, and small fluctuations in photosynthesis were selected to measure the net photosynthetic rate (Anet), transpiration rate (E), stomatal conductance (gs), and intercellular CO_2 concentration (Ci) of the sixth fully expanded leaf from the growth point by the Li-6400 photosynthetic determination system (LI-COR, Lincoln, NE, USA). The readings were repeated 3 times, and the average value was taken (air chamber temperature 14~22 °C, external CO_2 concentration 389 μmol CO_2 mol^{-1}, air velocity in air chamber 500 m·s^{-1}) [27].

2.6. Determination of Physiological Index

In the 8 drought stress treatment groups of each variety, samples were taken from 9:00~10:00 a.m., and the seventh to twelfth leaves from the growth point were sampled completely, and each 3 leaves were mixed into a sample. The content of chlorophyll a (Chla) and b (Chlb) in leaves was determined according to the Arnon method [28]. The activity of SOD was determined by the nitrogen blue tetrazole method, with 50% inhibition of photochemical reduction in nitroblue tetrazolium as an enzyme activity unit (U); the activity of POD was measured by guaiacol colorimetry, and the change in optical density (d470 nm) at 470 nm per minute was 0.1, which is an enzyme activity unit (U); the activity of CAT was determined by the ultraviolet (UV) absorption method, and the decrease of 240 by 0.1 within 1 min was taken as an enzyme activity unit (U) [29]. The content of MDA was determined by two-component spectrophotometry [30]. The content of soluble sugar was determined by anthrone colorimetry [31]. The content of free proline was determined by the acid ninhydrin method [32]. Determination of protein content was performed by the Bradford method [33]. A UV-visible spectrophotometer (UV-5800, Shanghai Metash instruments Co., Ltd, Shanghai, China) was used to complete the index determination.

2.7. Data Processing and Statistical Analysis

The data were sorted with Excel 2021 and plotted with Origin 9.0. The Duncan test was used with the statistical software SPSS 26 (SPSS Inc., Chicago, IL, USA) to test the significance of differences between treatments ($p < 0.05$ level). Canonical discriminant analysis (CDA) and correlation analysis were carried out. The results of morphological and physiological indices were expressed as the mean $\pm$ standard deviation (SD). Based on determining several drought resistance evaluation indices, the photosynthetic, physiological, and antioxidant index data of lily leaves were quantitatively converted through subordination function analysis (SFA), and the drought resistance of seven lily varieties was comprehensively evaluated:

$$X(\mu) = (X - X_{min})/(X_{max} - X_{min}) \tag{4}$$

$$X(\mu) = 1 - (X - X_{min})/(X_{max} - X_{min}) \tag{5}$$

where $X(\mu)$ is the subordination function value; X is the relative value of an index under drought treatment; X_{max} is the maximum value in the index; and X_{min} is the minimum value in this indicator. If a certain index is positively correlated with drought resistance, it shall be calculated with Formula (3); if there is a negative correlation, Formula (4) will be used to calculate.

3. Results

3.1. Leaf Morphology and Anatomical Structure of Seven Lily Species

There were differences in leaf length, leaf width, single leaf area, and total leaf area among the different lily varieties (Figure 2, Table 2). The maximum difference multiples of leaf length, leaf width, single leaf area, and total leaf area among varieties were 1.5, 3.1, 4.0 and 2.2 times, respectively. 'Robina' had significantly higher leaf length, leaf width, single leaf area, and total leaf area than the other varieties, 'Lyon' had the lowest leaf length, leaf width, and single leaf area, and 'Gelria' had the lowest total leaf area. The specific leaf area of 'Lyon' was significantly lower than that of other lily varieties.

Figure 2. Leaf morphology of seven lily varieties.

Table 2. Leaf characteristics of seven lily varieties.

Varieties	Leaf Length (cm)	Leaf Width (cm)	Single Leaf Area (cm^2)	Specific Leaf Area ($cm^2 \cdot g^{-1}$)	Total Leaf Area (cm^2)
'Siberia'	10.71 ± 0.50 d	2.27 ± 0.05 d	16.33 ± 0.53 c	114.46 ± 20.63 ab	1148.30 ± 54.22 c
'Immaculate'	10.79 ± 0.58 c	2.49 ± 0.06 c	18.56 ± 0.50 b	123.01 ± 9.82 ab	1236.83 ± 43.93 b
'Gelria'	10.55 ± 0.62 d	1.41 ± 0.03 e	11.58 ± 0.80 d	129.48 ± 38.72 a	817.86 ± 27.39 e
'Elena'	10.79 ± 0.58 d	2.55 ± 0.07 b	17.55 ± 0.68 c	117.04 ± 18.87 ab	1209.21 ± 49.55 b
'Robina'	14.28 ± 0.65 a	2.61 ± 0.09 a	25.61 ± 0.50 a	103.72 ± 22.06 c	1826.84 ± 49.09 a
'Royal Sunset'	12.44 ± 0.67 b	1.22 ± 0.04 f	11.30 ± 0.46 d	101.58 ± 15.75 c	1138.22 ± 60.03 c
'Lyon'	9.37 ± 0.52 e	0.85 ± 0.05 g	6.38 ± 0.33 e	81.31 ± 15.03 d	992.79 ± 21.23 d

Values are the average ± standard error. Different letters within the same column indicate significant differences between treatments ($p < 0.05$, $n = 10$).

The leaf thickness of the 7 lily species was 330.04–638.25 μm (Figure 3, Table 3). The leaf thickness of 'Lyon' and 'Royal Sunset' was significantly higher than the leaf thickness of the other varieties, and 'Gelria' had the largest thickness of palisade tissue. A very interesting observation for us is that the mesophyll thickness is not in line with leaf dimensions for 'Lyon'. The thickness of spongy tissue varied from 182.57 to 453.48 μm. 'Lyon' and 'Royal Sunset' sponges have significantly higher tissue thicknesses than other varieties, and 'Gelria' sponges have the smallest tissue thickness. The maximum palisade tissue/spongy tissue is 'Gelria'. 'Lyon' and 'Robina' have the largest thickness of the upper epidermis. The thickness of the upper epidermis of 'Gelria' was significantly lower than the thickness of the upper epidermis of other varieties. Except for 'Gelria', the thickness of the upper epidermis was greater than the thickness of the lower epidermis.

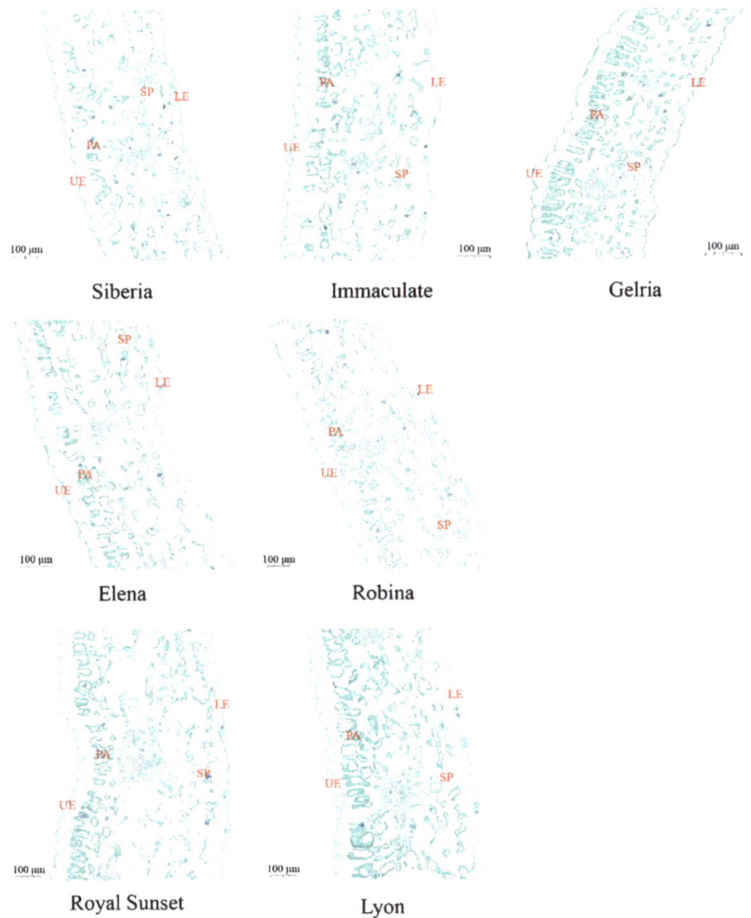

Figure 3. Anatomical structure of leaves of seven lily varieties. UE: upper epidermis; LE: lower epidermis; PA: palisade tissue; SP: spongy tissue.

Table 3. Anatomical structure parameters of leaves of seven lily varieties.

Varieties	Blade Thickness (μm)	Palisade Tissue Thickness (μm)	Spongy Tissue Thickness (μm)	Palisade Tissue/Spongy Tissue	Upper Epidermis Thickness (μm)	Lower Epidermis Thickness (μm)
'Siberia'	452.65 ± 11.24 c	61.4 ± 4.29 c	306.61 ± 10.53 b	0.20 ± 0.01 cd	66.13 ± 2.39 b	46.23 ± 3.05 bc
'Immaculate'	446.88 ± 11.38 c	46.44 ± 3.67 d	298.15 ± 12.54 b	0.16 ± 0.01 e	65.15 ± 3.36 b	49.57 ± 2.32 ab
'Gelria'	330.04 ± 9.86 e	61.5 ± 4.19 c	182.57 ± 6.91 d	0.34 ± 0.04 a	23.87 ± 1.29 c	36.08 ± 2.81 d
'Elena'	407.86 ± 9.69 d	53.3 ± 4.48 cd	254.80 ± 8.14 c	0.21 ± 0.02 c	63.27 ± 3.68 b	34.98 ± 1.87 d
'Robina'	484.25 ± 8.47 b	86.14 ± 3.88 a	280.81 ± 11.17 bc	0.31 ± 0.02 b	77.10 ± 3.57 a	42.74 ± 1.92 c
'Royal Sunset'	615.74 ± 12.28 a	77.80 ± 5.41 ab	437.49 ± 12.23 a	0.18 ± 0.02 de	64.46 ± 3.18 b	37.97 ± 3.15 d
'Lyon'	638.25 ± 11.97 a	75.83 ± 3.96 b	453.48 ± 11.96 a	0.17 ± 0.00 e	77.98 ± 2.38 a	51.68 ± 4.85 a

Values are the average $\pm$ standard error. Different letters within the same column indicate significant differences between treatments ($p < 0.05$, $n = 5$).

3.2. Effects of Drought Stress on Photosynthesis in Lily Leaves

With the extension of drought stress treatment time, the content of Chla and Chlb showed a downward trend (Figure 4). After 30 days of drought stress, the Chla of the seven lily varieties decreased by 16.6%, 16.4%, 13.7%, 16.7%, 13.2%, 12.7%, and 12.0%, and the Chlb content decreased by 27.5%, 32.2%, 24.1%, 32.0%, 26.3%, 21.5%, and 19.8%. The chlorophyll content of 'Lyon' was the lowest under drought stress. The chlorophyll content of 'Elena', 'Siberia', and 'Immaculate' was greatly affected by drought stress. Gs, E, and Anet in leaves of seven varieties decreased gradually with the extension of stress time; Ci increased gradually with the extension of stress time, and showed the opposite trend after rewatering (Figure 5). In general, 10 days of drought stress had little effect on the photosynthetic parameters of lily leaves. The variation ranges of the above parameters varied with different varieties, and they changed significantly on the 15th day. The photosynthetic parameters of 'Royal Sunset' and 'Lyon' changed significantly after 20 days of drought stress. On the 30th day, the gs of the leaves of the seven lily varieties decreased by 74.3%, 78.3%, 66.0%, 76.7%, 60.1%, 50.3%, and 21.8%, respectively. After re-watering, 'Robina', 'Royal Sunset', and 'Lyon' had the best recovery effects on the net photosynthetic rate, increasing by 56.0%, 43.9%, and 42.6%, respectively. A very interesting phenomenon is that the net photosynthetic rate of 'Royal Sunset', 'Lyon', and 'Robina' recovers faster than other varieties after rehydration.

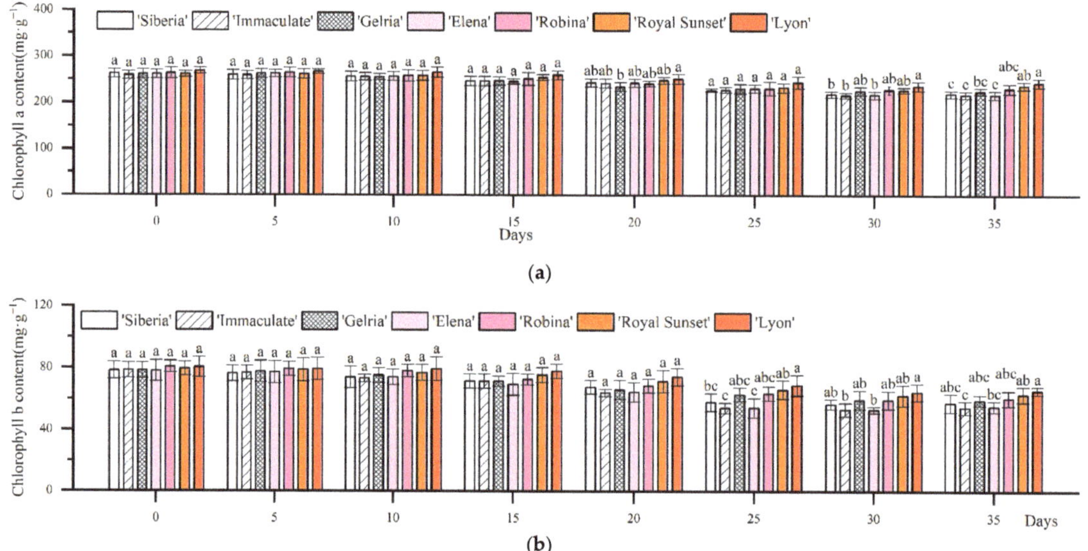

Figure 4. Effects of drought stress on chlorophyll content of lily leaves: (**a**) Effects of drought stress on chlorophyll a content; (**b**) Effects of drought stress on chlorophyll b content. Different letters above bar graphs indicate significant difference among lily varieties ($p < 0.05$, $n = 3$).

3.3. Effects of Drought Stress on Antioxidant Enzyme Activity in Lily Leaves

Under long-term drought stress, the activities of CAT, POD, and SOD in lily leaves showed an increasing trend (Figure 6). After 10 days of stress treatment, there was no significant difference in the activity of CAT and SOD in lily leaves of all varieties ($p > 0.05$). After 15 days of treatment, the activity of CAT, POD, and SOD in seven lily varieties showed an increasing trend. With the extension of stress time, the activity of CAT, POD, and SOD in 'Royal Sunset' and 'Lyon' continued to increase, and the content of MDA did not change significantly. On the 20th day, the activity of antioxidant enzymes in lily leaves of 'Lyon' was significantly higher than the activity of other varieties. The CAT, POD, and SOD activities of 'Siberia', 'Immaculate', 'Gelria', and 'Elena' decreased significantly on the 25th day, and the MDA content increased by 141%, 176%, 136%, and 161%, respectively, on the 25th day. After 30 days of stress treatment, the antioxidant enzyme activity of 'Robina', 'Royal Sunset', and 'Lyon' leaves were significantly higher than the antioxidant enzyme activity of other varieties. The CAT, POD, and SOD activity of 'Siberia', 'Immaculate',

'Gelria', and 'Elena' decreased compared with the CAT, POD, and SOD activity of the 25th day. After 35 days of re-watering treatment, the CAT, POD, and SOD activity of these four varieties increased, and the MDA content decreased. In contrast, 'Robina', 'Royal Sunset', and 'Lyon' decreased their antioxidant enzyme activity after stress recovery.

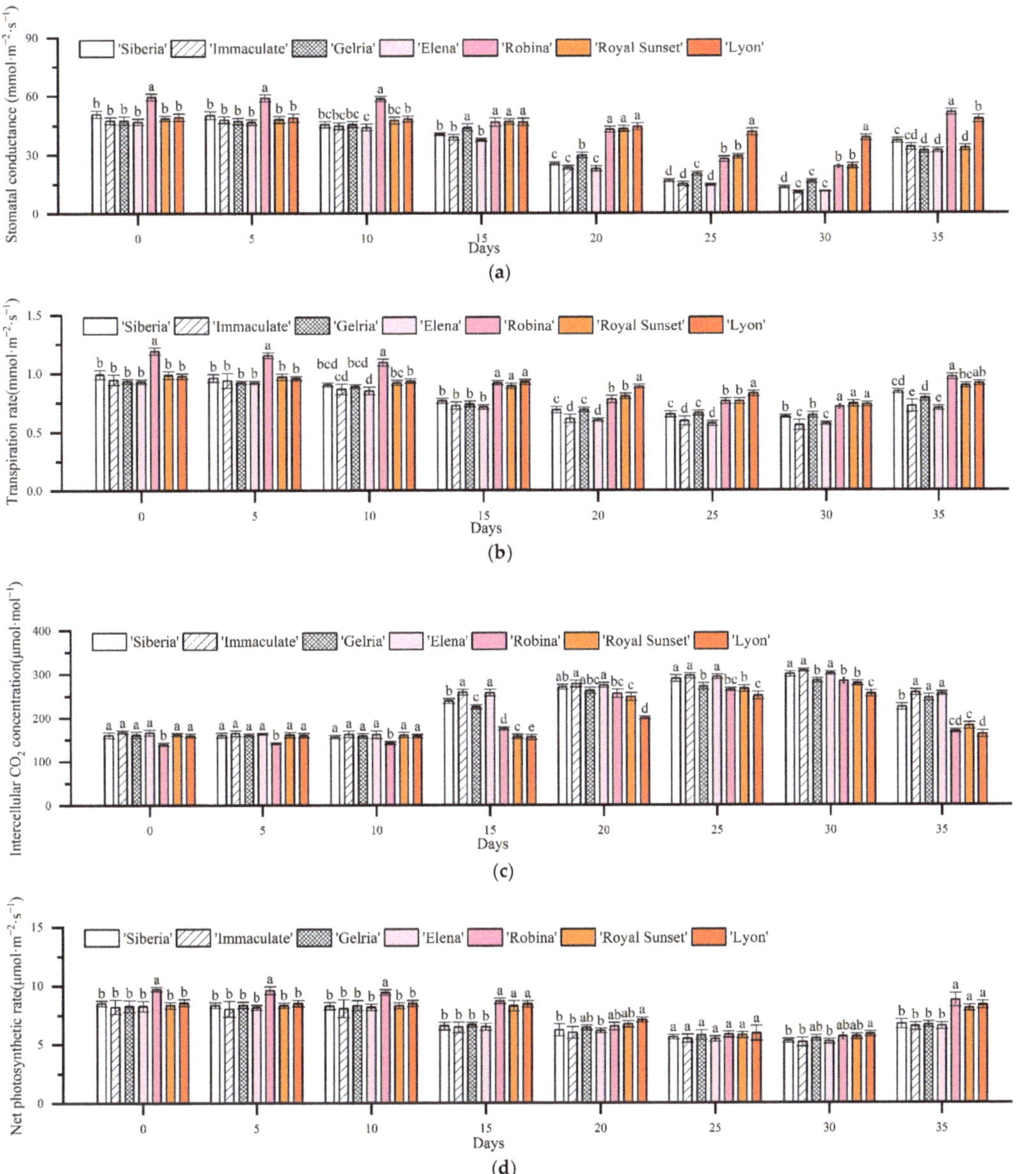

Figure 5. Effects of drought stress on photosynthetic parameters of lily leaves: (**a**) Effects of drought stress on stomatal conductance; (**b**) Effects of drought stress on transpiration rate; (**c**) Effects of drought stress on intercellular CO_2 concentration; (**d**) Effects of drought stress on net photosynthetic rate. Different letters above bar graphs indicate significant difference among lily varieties ($p < 0.05$, $n = 3$).

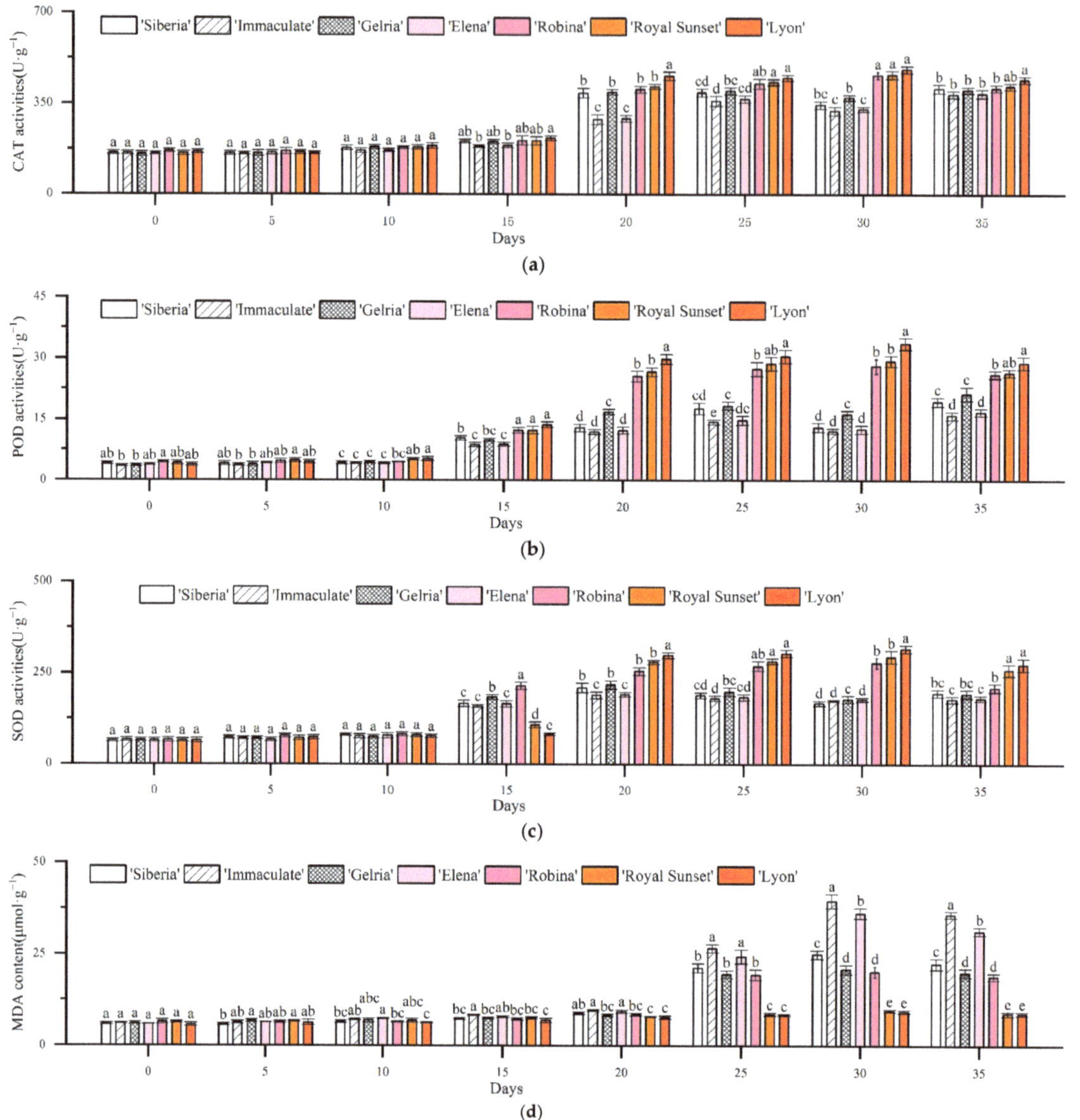

Figure 6. Effects of drought stress on antioxidant enzyme activity of lily leaves: (**a**) Effects of drought stress on CAT enzyme activity; (**b**) Effects of drought stress on POD enzyme activity; (**c**) Effects of drought stress on SOD enzyme activity; (**d**) Effects of drought stress on MDA content. Different letters above bar graphs indicate significant difference among lily varieties ($p < 0.05$, $n = 3$).

3.4. Effects of Drought Stress on the Physiological Metabolism of Lily Leaves

After drought stress, the content of free proline, soluble sugar, and soluble protein in lily leaves increased to varying degrees (Figure 7). Within 10 days of drought stress treatment, the content of soluble sugar in 'Lyon' leaves was significantly higher than the content of soluble sugar of other varieties. The content of free proline in 'Siberia', 'Immaculate', 'Gelria', 'Elena', and 'Robina' increased by 49.9%, 45.2%, 37.2%, 46.3%, and

35.9%, respectively, after 15 days of stress treatment. The content of free proline and soluble protein in 'Siberia', 'Immaculate', 'Gelria', and 'Elena' was significantly higher than the content of free proline and soluble protein in 'Robina', 'Royal Sunset', and 'Lyon'. The content of free proline in 'Lyon' leaves was significantly higher than the content of free proline of other varieties after 30 days of drought stress treatment. After stress recovery, the content of free proline, soluble sugar, and soluble protein in the leaves of the seven lily varieties decreased to varying degrees. After rehydration, the content of proline and soluble protein decreased significantly.

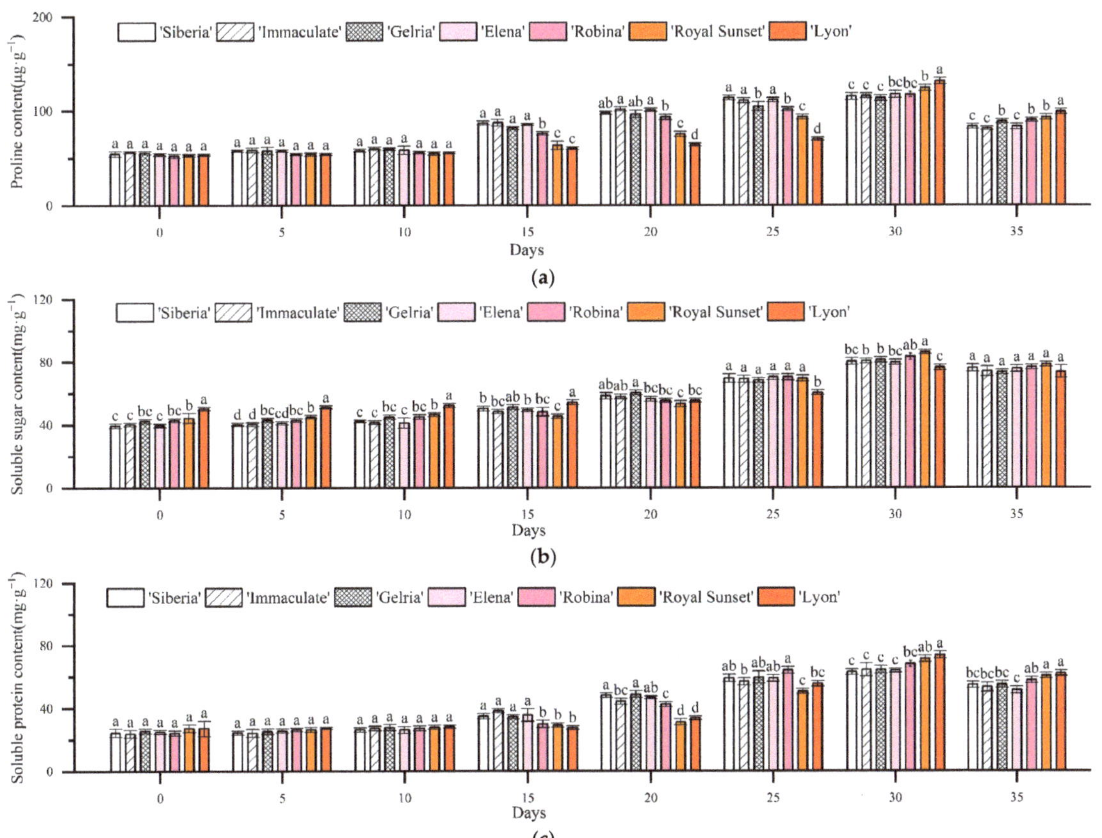

Figure 7. Effects of drought stress on the physiological metabolism of lily leaves: (**a**) Effects of drought stress on proline content; (**b**) Effects of drought stress on soluble sugar content; (**c**) Effects of drought stress on soluble protein content. Different letters above bar graphs indicate significant difference among lily varieties ($p < 0.05$, $n = 3$).

3.5. Evaluation of Drought Resistance in Lily Varieties

After 30 days of drought stress treatment, the physiological, photosynthetic, and antioxidant systems were affected to the greatest extent. These time point data can reflect the drought tolerance of lily. Rewatering treatment (35 d) could reflect the recovery ability of lily under drought stress. First, the specific subordinate values of each drought resistance index in each variety were obtained, and then the subordinate drought resistance values of each index of the specified variety were accumulated. The average value was obtained to evaluate the stress resistance, and the drought resistance strength was determined according to the average value of each variety. The drought resistance of lily leaves at

30 d and 35 d was evaluated by the indices CAT, POD, SOD, MDA, proline, soluble sugar, soluble protein, Chla, Chlb, gs, E, Ci, and Anet. According to the SFA value, the drought resistance of seven lily varieties was ranked from strong to weak as 'Lyon', 'Royal Sunset', 'Robina', 'Gelria', 'Siberia', 'Elena', and 'Immaculate' (Table 4).

Table 4. Membership function value and drought resistance ranking of lily.

	Days	Siberia	Immaculate	Gelria	Elena	Robina	Royal Sunset	Lyon
Chla	30	0.209	0.148	0.401	0.177	0.479	0.464	0.697
	35	0.256	0.213	0.390	0.210	0.524	0.675	0.815
Chlb	30	0.366	0.202	0.497	0.198	0.495	0.633	0.731
	35	0.396	0.225	0.494	0.266	0.566	0.703	0.846
CAT	30	0.220	0.104	0.362	0.139	0.801	0.811	0.918
	35	0.434	0.186	0.373	0.219	0.452	0.533	0.820
SOD	30	0.042	0.087	0.108	0.107	0.707	0.806	0.941
	35	0.216	0.085	0.198	0.108	0.324	0.695	0.811
POD	30	0.066	0.030	0.200	0.048	0.691	0.741	0.918
	35	0.280	0.075	0.401	0.124	0.693	0.717	0.866
MDA	30	0.508	0.067	0.632	0.168	0.652	0.977	0.984
	35	0.503	0.035	0.588	0.193	0.625	0.978	0.978
gs	30	0.117	0.029	0.219	0.048	0.465	0.478	0.943
	35	0.302	0.169	0.088	0.089	0.924	0.144	0.778
E	30	0.444	0.171	0.496	0.217	0.757	0.853	0.847
	35	0.548	0.173	0.380	0.119	0.929	0.712	0.770
Ci	30	0.190	0.064	0.413	0.162	0.432	0.503	0.852
	35	0.358	0.069	0.176	0.084	0.850	0.744	0.909
Anet	30	0.353	0.269	0.527	0.311	0.619	0.624	0.794
	35	0.198	0.139	0.180	0.145	0.797	0.603	0.675
Proline	30	0.227	0.254	0.162	0.317	0.304	0.585	0.879
	35	0.178	0.085	0.408	0.174	0.469	0.590	0.845
Soluble protein	30	0.202	0.284	0.289	0.258	0.513	0.705	0.854
	35	0.384	0.290	0.399	0.172	0.580	0.767	0.868
Soluble sugar	30	0.432	0.454	0.503	0.397	0.645	0.869	0.159
	35	0.668	0.483	0.441	0.626	0.709	0.867	0.456
Average membership function		0.311	0.169	0.359	0.195	0.615	0.684	0.806
Rank		5	7	4	6	3	2	1

3.6. Canonical Discriminant Analysis

After treatment at different time points, all physiological levels of the tested lily varieties were analysed by canonical discriminant analysis (CDA) to verify the drought resistance of the seven lily varieties (Figure 8). According to the CDA results, a total of six discriminant functions (DFSs) were identified, and the first two DFSs (DF1 and DF2) could clearly identify the differences between them (Figure 8, DF1 and DF2 accounted for 55.1% and 31.8% of the independent variable information in discriminant analysis, respectively). In general, under drought stress treatment and rewatering treatment, combined with the levels of key traits, including CAT, SOD, POD, MDA, proline, soluble sugar, soluble protein, Chla, Chlb, gs, E, Ci, and Anet, the horizontal distribution position of drought tolerance of seven lily varieties analysed by CDA was consistent with the results of the SFA.

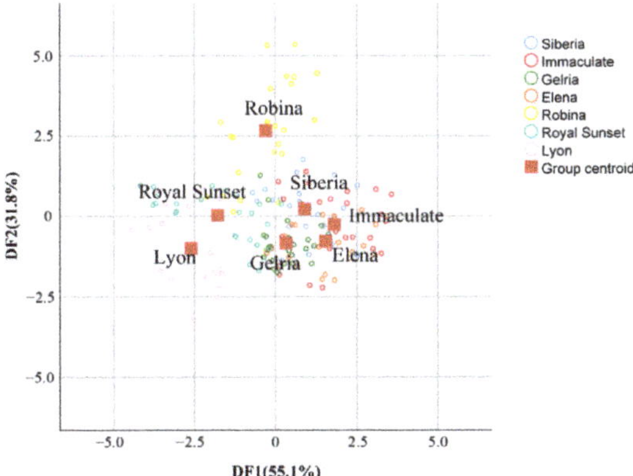

Figure 8. The first two discriminant function (DF) gradients were obtained by canonical discriminant analysis using the physiological indices of seven lily varieties from 0 to 35 days.

3.7. Correlation Analysis

The results showed that leaf thickness, palisade tissue, and sponge tissue thickness were positively correlated with antioxidant enzyme activity, photosynthesis, proline, and soluble protein and negatively correlated with MDA and Ci (Figure 9). The correlation between other leaf structure indices and drought resistance was low. MDA, Ci, and SLA were negatively correlated with antioxidant enzyme activity, photosynthesis, proline, and soluble protein.

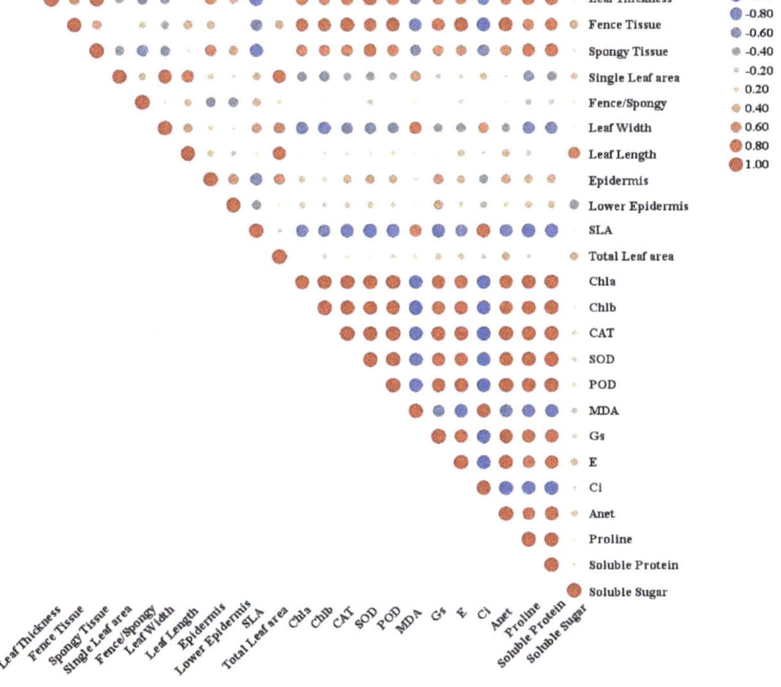

Figure 9. Heatmap of the correlation between leaf structure and physiological indices of lily under drought stress.

4. Discussion

Leaves are a manifestation of the long-traits adaptation of plants to the environment. Leaf morphology and structure are the organs with the greatest plasticity to adapt to environmental changes in the process of evolution, reflecting the characteristics of water use. It is necessary for plants to maintain a balance between water absorbed by roots and transpiration of leaves. Drought response and resistance is one of the main limiting factors of plant growth and development. Plant response and resistance to drought stress are a combination of complex biological processes at the cellular, physiological, and biochemical levels and the whole plant level [34,35]. Drought-resistant plants can adapt to water deficient environment through morphological adaptation, physiological regulation, and molecular signals. The response to drought stress is a process from "adaptation" to "injury" [13]. Rewatering after a certain degree of drought stress will have compensatory effects on photosynthesis, antioxidation, and physiology [36].

In the process of adapting to the arid environment, the plants growing in an arid area for a long time have formed special morphological structures, including the thickening of cuticle, the development of palisade tissue, the shrinking of cells, and the subsidence of stomata. The SLA reflects the ability of plants to obtain resources. The plants with low SLA value can better adapt to resource-poor and arid environments [37]. This study found that the change in the SLA of seven lily varieties is basically consistent with the drought resistance. 'Lyon' showed strong resistance among the seven lily varieties. 'Gelria' and 'Robina' have the largest ratio of palisade tissue/spongy tissue, indicating that the palisade tissue is developed, which is related to variety differences. However, due to their thin leaves, drought resistance is weakened, so the ratio of palisade tissue/spongy tissue cannot be used as an indicator to evaluate the drought resistance of the lily [38]. The leaf thicknesses of 'Royal Sunset' and 'Lyon' are significantly higher than the leaf thicknesses of other lily varieties, which is consistent with their drought resistance. Narrow and thick leaves and developed palisade tissue can prevent excessive transpiration and reduce leaf surface temperature, which is a sign of strong drought resistance. The fence organization ensures the transportation and maintenance of water and nutrients [39]. 'Lyon' has thick leaves, a small single leaf area, and well-developed palisade tissue and sponge tissue, showing stable characteristics of drought adaptation in morphology and anatomy. These morphological characteristics have a surface area that can effectively prevent evaporation and reduce evaporation and have important ecological significance for reducing water loss and maintaining water balance in plants.

Drought stress affects the normal physiological and metabolic processes of plants. Mesophyll is the main part of photosynthesis in leaves. This study found that drought stress causes the loss of chlorophyll in lily leaves. Water deficit caused the loss of water in cells and chloroplasts, and the activity of enzymes involved in carbon fixation in chloroplasts was inhibited, thus affecting the synthesis of chlorophyll. This change can also be used as an adaptive expression of the lily. By reducing the light absorption, reactive oxygen free radicals formed by excess excitation energy could be prevented from further damaging the photosynthetic system [40].

The decrease in chlorophyll in lily varieties with strong drought resistance was low, which may be related to leaf structure. Chlorophyll is distributed mainly in the chloroplasts of mesophyll cells. The mesophyll cells of the lily differentiate into palisade tissue and sponge tissue. The palisade tissue is close to the upper epidermis and mainly carries out photosynthesis. The thickness of palisade tissue and spongy tissue was positively correlated with chlorophyll content and photosynthetic rate. Under drought stress, the membrane system in plant cells, including the membrane structure related to photosynthesis, is destroyed due to the lack of water, nutrients, and energy, resulting in the interruption of physiological processes. These factors may directly or indirectly affect the chlorophyll content. Chlorophyll content is reported to decrease significantly with decreasing soil water content [41]. Plants that can maintain high chlorophyll content under drought stress are considered able to use light energy more effectively, so they are considered to have enhanced drought resistance [42]. The results showed that the chlorophyll content of 'Lyon' decreased the least and that photosynthesis was inhibited the least. Through the change range of Chla and Chlb under drought stress, the change range of Chlb under drought stress was found to be large. The change range of Chlb content can be used to evaluate the degree of inhibition of lily photosynthesis. The decrease in stomatal conductance is often the main reason for the decrease in photosynthesis under drought stress [43].

Drought stress can be divided into two types: stomatal limitation and nonstomatal limitation [44,45]. Under mild drought (0–10 d), all lily varieties showed a decrease in stomatal conductance, transpiration rate, and intercellular CO_2 concentration and no change in antioxidant enzyme system activity, cell osmotic substances, or malondialdehyde content, indicating that the lily was under mild drought stress at this stage (Figure 10). There are gaps between mesophyll cells to form a ventilation system, which exchanges gas with the outside world through pores on the epidermis [46]. The decrease in the photosynthetic rate was caused by the decrease in stomatal factors.

Under moderate drought (15–20 days), the antioxidant enzyme system activity of lily varieties began to increase, preventing the accumulation of ROS in cells and protecting cells from excessive ROS. Under normal circumstances, the production and clearance of ROS in cells are in a dynamic equilibrium state. At the same time, stimulated by the signal of reactive oxygen species, the content of cell osmotic substances began to increase, such as soluble sugars, Pro, and soluble protein, improving the drought resistance of plants [47]. The same results were also found in the study of walnut (*Juglans regia* L.) drought resistance. Walnut varieties with strong drought resistance contain less starch in the form of soluble sugar, and the content of free proline increases with the degree of drought stress [48]. At this stage, the stomatal conductance decreased, the transpiration rate decreased, and the Ci of 'Gelria', 'Siberia', 'Elena', and 'Immaculate' began to rise, indicating that the photosynthetic rate of the above varieties decreased and began to change from stomatal factors to nonstomatal factors. At the initial stage of drought stress, the photosynthetic rate of the leaves of the seven varieties of lily did not change significantly. The stomatal conductance of leaves decreased significantly, and Ci increased. At this time, the decline in photosynthesis was caused by the decrease in stomatal conductance.

Under severe drought (25–30 d), the antioxidant enzyme activities of 'Gelria', 'Siberia', 'Elena', and 'Immaculate' began to decrease, resulting in a reduction in the ability to remove ROS in cells, the accumulation of ROS, the oxidation of intracellular lipids, the disintegration of chloroplasts and an increase in MDA accumulation. Due to the imbalance between the production and utilization of electrons under drought stress, plants produce ROS in chloroplasts, peroxisomes, mitochondria, endoplasmic reticulum, plasma membrane, and the cell wall. When plants suffer from drought stress, the dynamic balance is broken, excessive accumulation of ROS damages cells, and oxidative deterioration may eventually lead to cell death [49]. Under ROS stress, the spatial configuration of various membrane proteins or enzymes is disturbed, resulting in increased membrane permeability and ion leakage, chlorophyll destruction, metabolic disorder, and even serious injury or death of plants [50]. ROS attack the most sensitive biological macromolecules in plant cells; induce lipid peroxidation, protein carbonylation, and DNA damage; and damage the plant's functions, leading to a disastrous series of events. To protect cells from the harmful effects of excessive ROS, plants have evolved a series of complex enzymatic and nonenzymatic antioxidant defence mechanisms to maintain the homeostasis of the intracellular redox state. At this stage, the reduction in the photosynthetic rate was transformed into nonstomatal factors. In comparison, 'Lyon', 'Royal Sunset', and 'Robina' continued to increase the activity of antioxidant enzymes in cells, with less ROS accumulation and slower chloroplast disintegration, and the intercellular CO_2 concentration began to rise, indicating that the photosynthetic rate of these varieties decreased at this stage and began to change from stomatal factors to nonstomatal factors. Under the stimulation of ROS, the content of osmotic substances in the cells of all varieties continued to increase, and the drought resistance of plants increased. Drought stress leads to damage to the cell membrane system, and the increase in MDA content reflects the degree of damage caused by drought to cells. Although 'Robina' has a large leaf area, its strong drought resistance may be related to its triploidity. Research shows that under moderate and severe drought stress, the utilization of water by triploids is better than the utilization of water by diploids [51]. 'Robina' and 'Immaculate' have similar leaf structures, but the increase in MDA content is low, indicating that the triploid population has less damage due to drought and strong tolerance to drought. This phenomenon is also found in polyploid *Lonicera japonica* [52]. Stomata can quickly and sensitively sense drought stress [53]. Under drought stress, the stomatal conductance of lily leaves decreased, and 'Lyon' stomatal conductance only decreased by 21.8% after 30 days of drought stress, which may be related to the lily having small and thick leaves and thicker palisade tissue. The same conclusion was reached in *Brassica napus* varieties [54]. The photosynthetic rate of leaves did not change significantly. At this time, the inhibition degree of stress on photosynthesis may be related to the role of the antioxidant system. In conclusion, Chlb is more sensitive to adverse drought environments, and the drought-resistant structure of leaves can protect the photosynthetic system of lily leaves under drought stress.

After 5 days of rewatering, except for 'Lyon', 'Royal Sunset', and 'Robina', the antioxidant enzyme activity of other varieties increased, the ability to clear intracellular ROS increased, ROS decreased, and the content of malondialdehyde decreased. The results showed that the arid soil environment was alleviated, and the membrane lipid oxidation caused by the accumulation of ROS was repaired after rehydration [55]. For varieties with weak drought tolerance, the physiological functions of plants are inhibited, and the activity of CAT, POD, and SOD cannot be maintained at a high level. After restoring water conditions, the protective mechanisms were enhanced by increasing enzyme activity [56]. The protective mechanisms of improving enzyme activity in drought-tolerant varieties will be relieved when the water condition is restored. Similar laws have been found in soybeans [57]. Chlorophyll synthesis began to recover, and stomatal conductance and the photosynthetic rate increased. Among these observations, the photosynthetic recovery rate of 'Lyon', 'Royal Sunset', and 'Robina' was higher than the photosynthetic recovery rate of 'Gelria', 'Siberia', 'Elena', and 'Immaculate'. At the same time, due to the decrease in the ROS signal, coupled with the increase in water absorption, the concentration of cell osmotic substances gradually decreased.

In this study, morphological and physiological markers were used to screen the drought resistance of different lily varieties. The relationship between drought resistance physiology and leaf morphology of lily was revealed. It provides a reference for the development of water-saving, efficient, and high-quality cut-flower production and the breeding of drought-resistant varieties. In the future, we can combine the water balance mechanism of different parts of lily and use transcriptomics, proteomics, and other biotechnology to further study the drought-resistance mechanism of lily.

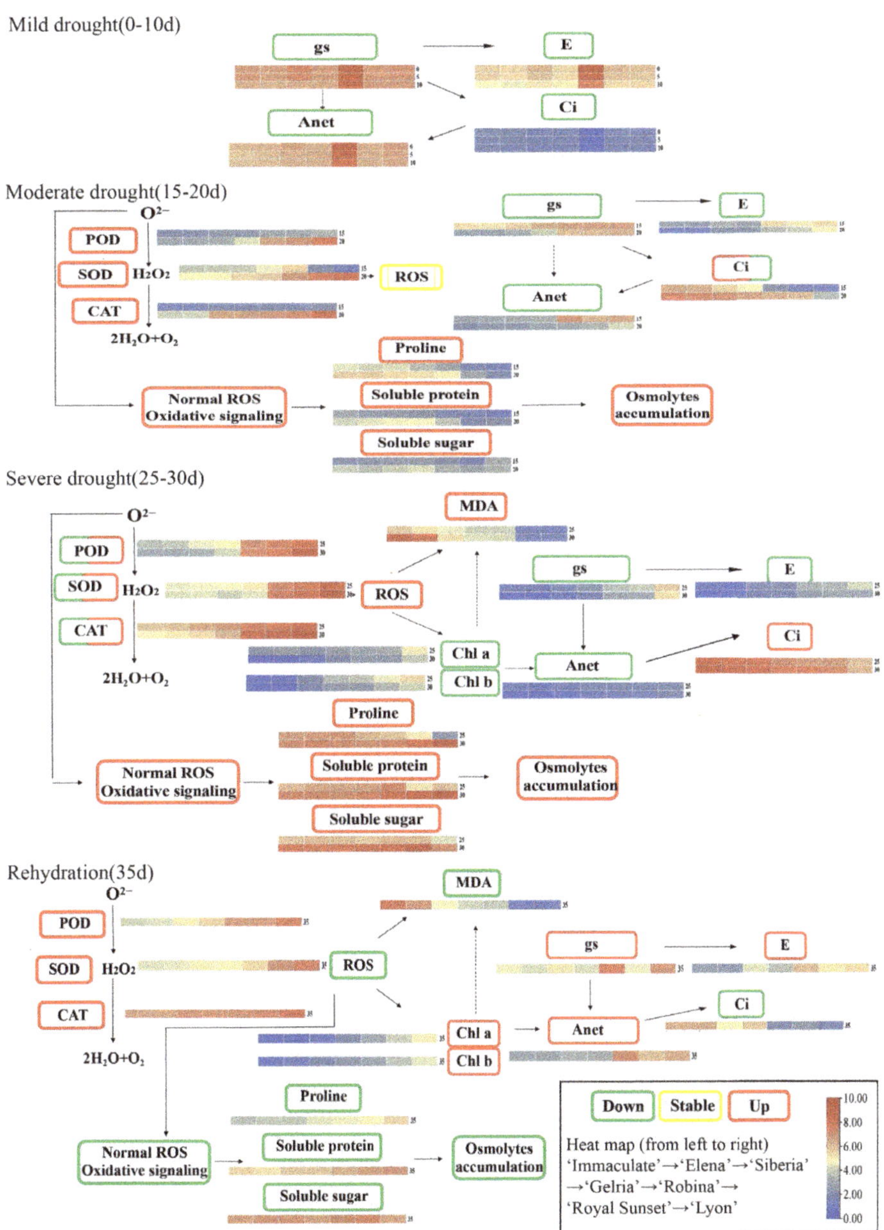

Figure 10. Physiological adaptation model of the lily under drought stress and rehydration. The green box indicates that the index decreases, the red box indicates that the index increases, and the yellow box indicates that the index does not change. gs, stomatal conductance; Ci, intercellular CO_2 concentration; Anet, net photosynthetic rate; E, transpiration; SOD, superoxide dismutase; POD, peroxidase; CAT, catalase; MDA, malondialdehyde; ROS, reactive oxygen species; Chl a, chlorophyll a; Chl b, chlorophyll b. The heatmap reflects the index changes of the seven lily varieties, and the drought resistance gradually increases from left to right, followed by 'Immaculate', 'Elena', 'Siberia', 'Gelria', 'Robina', 'Royal Sunset', 'Lyon'.

5. Conclusions

There were differences in leaf morphological indices and anatomical structures among the seven varieties, and the changes in the photosynthetic system, antioxidant enzyme system activity, and cell osmotic material system of the lily varieties with different leaf types were different under different drought conditions. From the analysis of the synergistic changes in the above systems, the drought resistance of 'Lyon', 'Royal Sunset', and 'Robina' is better than the drought resistance of 'Gelria', 'Siberia', 'Elena', and 'Immaculate'. Specific leaf area may be one of the apparent indices that affect the drought resistance of the lily.

Author Contributions: Conceptualization, X.L. and W.J.; methodology, X.L., J.W., L.M. and J.Z.; formal analysis, Q.D., W.D. and G.C.; investigation, X.L. and X.W.; resources, G.C.; data curation, X.L., J.W. and J.Z.; writing—original draft preparation, X.L. and W.J.; writing—review and editing, X.L. and W.J. All authors have read and agreed to the published version of the manuscript.

Funding: This research was funded by grants from the National Key R&D Program of China (No. 2018YFD1000400), the National Natural Science Foundation of China (31960614), the Yunnan Science and Technology Talents and Platform Program of China (2018HB083), and Open Fund of National Engineering Research Center for Ornamental Horticulture and Yunnan Key Laboratory of Flower Breeding (FKL-202103).

Institutional Review Board Statement: Not applicable.

Informed Consent Statement: Not applicable.

Data Availability Statement: Not applicable.

Conflicts of Interest: The authors declare no conflict of interest.

References

1. Trenberth, K.E.; Dai, A.; van der Schrier, G.; Jones, P.D.; Barichivich, J.; Briffa, K.R.; Sheffield, J. Global warming and changes in drought. *Nat. Clim. Chang.* **2014**, *4*, 17–22. [CrossRef]
2. Chaudhry, S.; Sidhu, G.P.S. Climate change regulated abiotic stress mechanisms in plants: A comprehensive review. *Plant Cell Rep.* **2022**, *41*, 1–31. [CrossRef]
3. Kapoor, D.; Bhardwaj, S.; Landi, M.; Sharma, A.; Ramakrishnan, M.; Sharma, A. The impact of drought in plant metabolism: How to exploit tolerance mechanisms to increase Crop Production. *Appl. Sci.* **2020**, *10*, 5692. [CrossRef]
4. Bakhshaie, M.; Khosravi, S.; Azadi, P.; Bagheri, H.; van Tuyl, J.M. Biotechnological advances in *Lilium*. *Plant Cell Rep.* **2016**, *35*, 1799–1826. [CrossRef]
5. Khoyerdi, F.F.; Shamshiri, M.H.; Estaji, A. Changes in some physiological and osmotic parameters of several pistachio genotypes under drought stress. *Sci. Hortic.* **2016**, *198*, 44–51. [CrossRef]
6. Li, W.; Wang, Y.; Zhang, Y.; Wang, R.; Guo, Z.; Xie, Z. Impacts of drought stress on the morphology, physiology, and sugar content of Lanzhou lily (*Lilium davidii* var. *unicolor*). *Acta Physiol. Plant.* **2020**, *42*, 127. [CrossRef]
7. Shi, L.; Wang, Z.; Kim, W.S. Effect of drought stress on shoot growth and physiological response in the cut rose 'charming black' at different developmental stages. *Hortic. Environ. Biotechnol.* **2019**, *60*, 1–8. [CrossRef]
8. Giordano, M.; Petropoulos, S.A.; Cirillo, C.; Rouphael, Y. Biochemical, physiological, and molecular aspects of ornamental plants adaptation to deficit irrigation. *Horticulturae* **2021**, *7*, 107. [CrossRef]
9. Kuppler, J.; Wieland, J.; Junker, R.R.; Ayasse, M. Drought-induced reduction in flower size and abundance correlates with reduced flower visits by bumble bees. *AoB Plants* **2021**, *13*, 1–8. [CrossRef]
10. Ennajeh, M.; Vadel, A.M.; Cochard, H.; Khemira, H. Comparative impacts of water stress on the leaf anatomy of a drought-resistant and a drought-sensitive olive cultivar. *J. Hortic. Sci. Biotechnol.* **2010**, *85*, 289–294. [CrossRef]
11. Zhang, F.-J.; Zhang, K.-K.; Du, C.-Z.; Li, J.; Xing, Y.-X.; Yang, L.-T.; Li, Y.-R. Effect of drought stress on anatomical structure and chloroplast ultrastructure in leaves of sugarcane. *Sugar Tech.* **2015**, *17*, 41–48. [CrossRef]
12. Westerband, A.C.; Bialic-Murphy, L.; Weisenberger, L.A.; Barton, K.E. Intraspecific variation in seedling drought tolerance and associated traits in a critically endangered, endemic Hawaiian shrub. *Plant Ecol. Divers.* **2020**, *13*, 159–174. [CrossRef]
13. Bhargava, S.; Sawant, K. Drought stress adaptation: Metabolic adjustment and regulation of gene expression. *Plant Breed.* **2013**, *132*, 21–32. [CrossRef]
14. Yang, P.M.; Huang, Q.C.; Qin, G.Y.; Zhao, S.P.; Zhou, J.G. Different drought-stress responses in photosynthesis and reactive oxygen metabolism between autotetraploid and diploid rice. *Photosynthetica* **2014**, *52*, 193–202. [CrossRef]
15. McLaughlin, J.E.; Boyer, J.S. Sugar-responsive gene expression, invertase activity, and senescence in aborting maize ovaries at low water potentials. *Ann. Bot.* **2004**, *94*, 675–689. [CrossRef]

16. Laxa, M.; Liebthal, M.; Telman, W.; Chibani, K.; Dietz, K.-J. The role of the plant antioxidant system in drought tolerance. *Antioxidants* **2019**, *8*, 94. [CrossRef]
17. Fang, Y.; Xiong, L. General mechanisms of drought response and their application in drought resistance improvement in plants. *Cell. Mol. Life Sci.* **2015**, *72*, 673–689. [CrossRef]
18. Zandalinas, S.I.; Mittler, R.; Balfagón, D.; Arbona, V.; Gómez-Cadenas, A. Plant adaptations to the combination of drought and high temperatures. *Physiol. Plant.* **2018**, *162*, 2–12. [CrossRef]
19. Xu, Z.; Zhou, G.; Shimizu, H. Plant responses to drought and rewatering. *Plant Signal. Behav.* **2010**, *5*, 649–654. [CrossRef]
20. Bo, W.; Fu, B.; Qin, G.; Xing, G.; Wang, Y. Evaluation of drought resistance in *Iris germanica* L. based on subordination function and principal component analysis. *Emir. J. Food Agric.* **2017**, *29*, 770–778. [CrossRef]
21. Yi-ling, Y.; Chun-hui, H.; Qing-qing, G.; Xue-yan, Q.; Xiao-biao, X. Evaluation of drought-resistance traits of citrus rootstock seedlings by multiple statistics analysis. *Acta Hortic.* **2015**, *1065*, 379–386. [CrossRef]
22. Liu, N.; Liu, S.; Gan, Y.; Zhang, Q.; Wang, X.; Liu, S.; Dai, J. Evaluation of mercury resistance and accumulation characteristics in wheat using a modified membership function. *Ecol. Indic.* **2017**, *78*, 292–300. [CrossRef]
23. Wassie, M.; Zhang, W.; Zhang, Q.; Ji, K.; Chen, L. Effect of heat stress on growth and physiological traits of Alfalfa (*Medicago sativa* L.) and a comprehensive evaluation for heat tolerance. *Agronomy* **2019**, *9*, 597. [CrossRef]
24. Gholizadeh, A.; Dehghani, H.; Akbarpour, O.; Amini, A.; Sadeghi, K.; Hanifei, M.; Sharifi-Zagheh, A. Assessment of Iranian wheat germplasm for salinity tolerance using analysis of the membership function value of salinity tolerance (MFVS). *J. Crop Sci. Biotechnol.* **2022**, 1–9. [CrossRef]
25. Ji, X.; Tang, J.; Fan, W.; Li, B.; Bai, Y.; He, J.; Pei, D.; Zhang, J. Phenotypic differences and physiological responses of salt resistance of walnut with four rootstock types. *Plants* **2022**, *11*, 1557. [CrossRef]
26. Bittelli, M. Measuring soil water content: A review. *HortTechnology* **2011**, *21*, 293–300. [CrossRef]
27. Kramer, D.M.; Johnson, G.; Kiirats, O.; Edwards, G.E. New Fluorescence Parameters for the Determination of QA Redox State and Excitation Energy Fluxes. *Photosynth. Res.* **2004**, *79*, 209. [CrossRef]
28. Arnon, D.I. Copper enzymes in isolated chloroplasts. polyphenoloxidase in *Beta vulgaris*. *Plant Physiol.* **1949**, *24*, 1–15. [CrossRef]
29. de Azevedo Neto, A.D.; Prisco, J.T.; Enéas-Filho, J.; Medeiros, J.-V.R.; Gomes-Filho, E. Hydrogen peroxide pre-treatment induces salt-stress acclimation in maize plants. *J. Plant Physiol.* **2005**, *162*, 1114–1122. [CrossRef]
30. Tsikas, D. Assessment of lipid peroxidation by measuring malondialdehyde (MDA) and relatives in biological samples: Analytical and biological challenges. *Anal. Biochem.* **2017**, *524*, 13–30. [CrossRef]
31. Buysse, J.; Merckx, R. An Improved Colorimetric Method to Quantify Sugar Content of Plant Tissue. *J. Exp. Bot.* **1993**, *44*, 1627–1629. [CrossRef]
32. Bates, L.S.; Waldren, R.P.; Teare, I.D. Rapid determination of free proline for water-stress studies. *Plant Soil* **1973**, *39*, 205–207. [CrossRef]
33. Bradford, M.M. A rapid and sensitive method for the quantitation of microgram quantities of protein utilizing the principle of protein-dye binding. *Anal. Biochem.* **1976**, *72*, 248–254. [CrossRef]
34. Harb, A.; Krishnan, A.; Ambavaram, M.M.R.; Pereira, A. Molecular and physiological analysis of drought stress in Arabidopsis reveals early responses leading to acclimation in plant growth. *Plant Physiol.* **2010**, *154*, 1254–1271. [CrossRef] [PubMed]
35. Deka, D.; Singh, A.K.; Singh, A.K. Effect of drought stress on crop plants with special reference to drought avoidance and tolerance mechanisms: A review. *Int. J. Curr. Microbiol. Appl. Sci.* **2018**, *7*, 2703–2721. [CrossRef]
36. Wang, G.; Yuan, Z.; Zhang, P.; Liu, Z.; Wang, T.; Wei, L. Genome-wide analysis of NAC transcription factor family in maize under drought stress and rewatering. *Physiol. Mol. Biol. Plants* **2020**, *26*, 705–717. [CrossRef] [PubMed]
37. Liu, M.; Wang, Z.; Li, S.; Lü, X.; Wang, X.; Han, X. Changes in specific leaf area of dominant plants in temperate grasslands along a 2500-km transect in northern China. *Sci. Rep.* **2017**, *7*, 10780. [CrossRef]
38. Luković, J.; Maksimović, I.; Zorić, L.; Nagl, N.; Perčić, M.; Polić, D.; Putnik-Delić, M. Histological characteristics of sugar beet leaves potentially linked to drought tolerance. *Ind. Crops Prod.* **2009**, *30*, 281–286. [CrossRef]
39. Guha, A.; Sengupta, D.; Kumar Rasineni, G.; Ramachandra Reddy, A. An integrated diagnostic approach to understand drought tolerance in mulberry (*Morus indica* L.). *Flora-Morphol. Distrib. Funct. Ecol. Plants* **2010**, *205*, 144–151. [CrossRef]
40. Dalal, V.K.; Tripathy, B.C. Water-stress induced downsizing of light-harvesting antenna complex protects developing rice seedlings from photo-oxidative damage. *Sci. Rep.* **2018**, *8*, 5955. [CrossRef]
41. Abdolahi, M.; Maleki Farahani, S. Seed quality, water use efficiency and eco physiological characteristics of Lallemantia (*Lallemantia* sp.) species as effected by soil moisture content. *Acta Agric. Slov.* **2019**, *113*, 307. [CrossRef]
42. Hossain, M.A.; Wani, S.H.; Bhattacharjee, S.; Burritt, D.J.; Tran, L.-S.P. (Eds.) *Drought Stress Tolerance in Plants*; Springer International Publishing: Cham, Switzerland, 2016; Volume 1, ISBN 978-3-319-28897-0.
43. Flexas, J.; Ribas-Carbó, M.; Bota, J.; Galmés, J.; Henkle, M.; Martínez-Cañellas, S.; Medrano, H. Decreased Rubisco activity during water stress is not induced by decreased relative water content but related to conditions of low stomatal conductance and chloroplast CO_2 concentration. *New Phytol.* **2006**, *172*, 73–82. [CrossRef]
44. Heilmeier, H.; Wartinger, A.; Erhard, M.; Zimmermann, R.; Horn, R.; Schulze, E.-D. Soil drought increases leaf and whole-plant water use of Prunus dulcis grown in the Negev Desert. *Oecologia* **2002**, *130*, 329–336. [CrossRef]
45. Wang, Y.; Yan, D.; Wang, J.; Ding, Y.; Song, X. Effects of elevated CO_2 and drought on plant physiology, soil carbon and soil enzyme activities. *Pedosphere* **2017**, *27*, 846–855. [CrossRef]

46. Lawson, T.; Simkin, A.J.; Kelly, G.; Granot, D. Mesophyll photosynthesis and guard cell metabolism impacts on stomatal behaviour. *New Phytol.* **2014**, *203*, 1064–1081. [CrossRef]

47. Naeem, M.; Naeem, M.S.; Ahmad, R.; Ahmad, R.; Ashraf, M.Y.; Ihsan, M.Z.; Nawaz, F.; Athar, H.-R.; Ashraf, M.; Abbas, H.T.; et al. Improving drought tolerance in maize by foliar application of boron: Water status, antioxidative defense and photosynthetic capacity. *Arch. Agron. Soil Sci.* **2018**, *64*, 626–639. [CrossRef]

48. Naser, L.; Kourosh, V.; Bahman, K.; Reza, A. Soluble sugars and proline accumulation play a role as effective indices for drought tolerance screening in *Persian walnut* (*Juglans regia* L.) during germination. *Fruits* **2010**, *65*, 97–112. [CrossRef]

49. García-Caparrós, P.; de Filippis, L.; Gul, A.; Hasanuzzaman, M.; Ozturk, M.; Altay, V.; Lao, M.T. Oxidative stress and antioxidant metabolism under adverse environmental conditions: A review. *Bot. Rev.* **2021**, *87*, 421–466. [CrossRef]

50. Petrov, V.; Hille, J.; Mueller-Roeber, B.; Gechev, T.S. ROS-mediated abiotic stress-induced programmed cell death in plants. *Front. Plant Sci.* **2015**, *6*, 69. [CrossRef]

51. Liao, T.; Wang, Y.; Xu, C.P.; Li, Y.; Kang, X.Y. Adaptive photosynthetic and physiological responses to drought and rewatering in triploid Populus populations. *Photosynthetica* **2018**, *56*, 578–590. [CrossRef]

52. Li, W.-D.; Biswas, D.K.; Xu, H.; Xu, C.-Q.; Wang, X.-Z.; Liu, J.-K.; Jiang, G.-M. Photosynthetic responses to chromosome doubling in relation to leaf anatomy in Lonicera japonica subjected to water stress. *Funct. Plant Biol.* **2009**, *36*, 783–792. [CrossRef] [PubMed]

53. Li, S.; Liu, J.; Liu, H.; Qiu, R.; Gao, Y.; Duan, A. Role of hydraulic signal and ABA in decrease of leaf stomatal and mesophyll conductance in Soil Drought-Stressed Tomato. *Front. Plant Sci.* **2021**, *12*, 653186. [CrossRef] [PubMed]

54. Batool, M.; El-Badri, A.M.; Hassan, M.U.; Haiyun, Y.; Chunyun, W.; Zhenkun, Y.; Jie, K.; Wang, B.; Zhou, G. Drought stress in *Brassica napus*: Effects, tolerance mechanisms, and management strategies. *J. Plant Growth Regul.* **2022**, 1–25. [CrossRef]

55. Osório, M.L.; Osório, J.; Vieira, A.C.; Gonçalves, S.; Romano, A. Influence of enhanced temperature on photosynthesis, photooxidative damage, and antioxidant strategies in *Ceratonia siliqua* L. seedlings subjected to water deficit and rewatering. *Photosynthetica* **2011**, *49*, 3–12. [CrossRef]

56. do Rosário Rosa, V.; Farias Dos Santos, A.L.; Da Alves Silva, A.; Peduti Vicentini Sab, M.; Germino, G.H.; Barcellos Cardoso, F.; de Almeida Silva, M. Increased soybean tolerance to water deficiency through biostimulant based on fulvic acids and *Ascophyllum nodosum* (L.) seaweed extract. *Plant Physiol. Biochem.* **2021**, *158*, 228–243. [CrossRef]

57. Akitha Devi, M.K.; Giridhar, P. Variations in physiological response, lipid peroxidation, antioxidant enzyme activities, proline and isoflavones content in soybean varieties subjected to drought stress. *Proc. Natl. Acad. Sci. India Sect. B Boil. Sci.* **2015**, *85*, 35–44. [CrossRef]

Article

Impact of Plant Growth-Promoting Rhizobacteria Inoculation on the Physiological Response and Productivity Traits of Field-Grown Tomatoes in Hungary

Eszter Nemeskéri [1], Kitti Zsuzsanna Horváth [1], Bulgan Andryei [1], Riadh Ilahy [2], Sándor Takács [1,*], András Neményi [1], Zoltán Pék [1] and Lajos Helyes [1]

[1] Horticulture Institute, Hungarian University of Agriculture and Life Sciences, Páter K. Street, 2100 Gödöllő, Hungary; nemeskeri.eszter@uni-mate.hu (E.N.); horvath.kitti.zsuzsanna@gmail.com (K.Z.H.); bulgan.mgl@muls.edu.mn (B.A.); nemenyi.andras.bela@uni-mate.hu (A.N.); pek.zoltan@uni-mate.hu (Z.P.); helyes.lajos@uni-mate.hu (L.H.)
[2] Laboratory of Horticultural Crops, National Agricultural Research Institute of Tunisia, University of Carthage, Carthage 1054, Tunisia; bn.riadh@gmail.com
* Correspondence: takacs.sandor@uni-mate.hu

Abstract: Drought-tolerant plant growth-promoting rhizobacteria (PGPR) may promote plant development under limited water supply conditions, when plant's water demand is not completely satisfied under rain-fed conditions or when irrigation water availability is limited. The aim of this study was to examine the effects of two inoculation treatments (B2: *Alcaligenes* sp. 3573, *Bacillus* sp. BAR16, and *Bacillus* sp. PAR11 strains and B3: *Pseudomonas* sp. MUS04, *Rhodococcus* sp. BAR03, and *Variovorax* sp. BAR04 strains) and compare those to a control (B0) without artificial inoculation on chlorophyll fluorescence, leaf chlorophyll content (SPAD value), canopy temperature, and the yield of the processing tomato cultivar H-1015 F1 grown under field conditions. The young seedlings of the hybrid tomato variety H-1015 F1 were immersed in 1% of B2 or B3 products (BAY-BIO, Szeged Hungary) for 5 min. Inoculated and untreated seedlings were grown under three irrigation treatments [regular irrigation (RI), deficit irrigation (DI), and no irrigation (I0)], to reveal the effect of PGPR under different levels of water stress. In the dry year (2018), higher canopy temperature and chlorophyll fluorescence (Fv/Fm) were measured during flowering in plants treated with bacteria than in untreated plants. In the stage of flowering and fruit setting, the B3 treatment led to a significant decrease in the Fv/Fm value, canopy temperature remained high, and the SPAD value was statistically the same in all treatments. Under limited water supply, in most cases, PGPR led to a significantly greater total yield but more unripe green berries compared to untreated plants. Under moderate water shortage (dry year + deficit irrigation), the B3 treatment resulted in 26% more ripe, marketable fruit and 49% less unripe fruit compared to the B2 treatment. On the other hand, in the wet year (2020), the bacterial treatments generally did not affect physiological properties, though the B2 treatment produced a higher marketable yield while the amount of green and diseased fruits did not differ statistically, compared to the B3 treatment under deficit irrigation. Based on our study, we recommend the application of the B3 PGPR product as it positively affected key physiological processes, leading to a higher marketable yield particularly under water shortage.

Keywords: canopy temperature; chlorophyll fluorescence; rhizobacteria; tomato; water deficit

Citation: Nemeskéri, E.; Horváth, K.Z.; Andryei, B.; Ilahy, R.; Takács, S.; Neményi, A.; Pék, Z.; Helyes, L. Impact of Plant Growth-Promoting Rhizobacteria Inoculation on the Physiological Response and Productivity Traits of Field-Grown Tomatoes in Hungary. *Horticulturae* **2022**, *8*, 641. https://doi.org/10.3390/horticulturae8070641

Academic Editors: Stefania Toscano, Giulia Franzoni, Sara Álvarez and Alessandra Francini

Received: 13 June 2022
Accepted: 12 July 2022
Published: 14 July 2022

Publisher's Note: MDPI stays neutral with regard to jurisdictional claims in published maps and institutional affiliations.

1. Introduction

Tomato (*Lycopersicon esculentum* Mill.) is a worldwide strategic crop. Recently, there has been an increase in the consumption and production of various tomato products. The growth and quality of tomato are influenced by several factors such as genotype, maturity stage, environmental factors and nutrient and water supply conditions. The greatest factor that limits yield in field-grown tomato is water shortage, which is increasingly prevalent due to ongoing climate change. The adverse effect of such factors on yield can be managed by adopting appropriate irrigation techniques, the use of bio-fertilizers, and the use of drought-resistant varieties. The application of different irrigation model systems provides the possibility to determine irrigation timing and irrigation water dose for crops [1–4], though this process requires knowledge of the response of the varieties to water-deficit stress. Under water shortage, plants respond by decreasing water loss as a result of stomata closing, which reduces transpiration as well as the inflow and delivery of atmospheric CO_2 to the cells [5] but increases stomata resistance, therefore influencing the efficacy of water consumption [6]. Transpiration efficiency (TE) is an important factor in water use efficiency (WUE) and is defined as the quantity of water used (WU) or the evaporation devoted to biomass production [7,8]. Under water shortage, a significant positive correlation was found between fruit yield per plant and TE in sesame plants [9] and between WU, yield, biomass, and WUE in green bean [6].

Under water shortage, plant growth is compromised following the decrease in transpiration, the increase in canopy temperature, low stomata conductance [10] and reduced chlorophyll fluorescence and photosynthesis activity [11]. Photosynthetic activity can be assessed through the effectiveness of photochemical systems (PSI and PSII) and by measuring leaf photosynthetic pigments (e.g., chlorophyll content). The maximum photochemical quantum quantity (Fv/Fm) of the PSII photochemical system, as chlorophyll fluorescence, is indicative of the degree to which photosynthesis is undisturbed [12]. Under dry conditions, the quantity of photosynthetic pigments and chlorophyll fluorescence (Fv/Fm) change, suggesting their importance for assessing drought tolerance in different tomato genotypes [13,14].

Tomato is generally considered to have a high water demand [15], though it shows different levels of sensitivity during development. If water shortage occurs in the early phases of flowering, it causes flower drop [16]; if it occurs during fruit setting, then smaller berries are formed [10]; and although the quantity of marketable fruit will be less, their soluble dry matter content would be higher under water abundance [17]. Yield loss associated with drought stress can reach 40–60% in cultivated crops [18]. Under such stress conditions, the distribution and extent of the root system in dry soil are severely compromised, thereby inhibiting water and nutrient uptake [9,19]. Genotypes with deep-reaching, developed root systems or that are able to develop an interaction with the microorganisms living in the soil may be able to withstand periods of dryness.

Recently, interest in soil microbes has significantly increased, with studies examining the role of various bacteria in plant development and the reactions to salt and water stress. In the ScienceDirect database (www.sciencedirect.com, accessed on 1 June 2022), using soil microbe as keywords, the number of published research items has increased sharply from 1271 to 7974 during the last decade (2012–2022), suggesting the importance of such factors for sustainable agriculture under changing climatic conditions. In the rhizosphere, fungi, bacteria form the largest group of soil microorganisms which interact with plants fine tuning, and therefore key plant physiological processes [20]. Low soil moisture levels (10–14%) stimulate increased root hair density and length compared to other soil moisture levels, depending mainly on genetic [21] and hormonal regulation [22–24]. The bacterial community alters the root's hormonal state, improving not only the plant's nutrient and water uptake and thereby its development, but also leading to the secretion of various biochemical substances on the root, which results in microbial colonization and a symbiotic relationship [25,26]. Under water shortage, these PGPRs increase the biosynthesis of indole-acetic acid (IAA) and abscisic acid (ABA), leading to a significant growth extension of roots [24,27]. Bacillus species specifically boosted the generation of antioxidant substances under salt stress [28]. Various microbes exhibited a significant role in mitigating the adverse effects of abiotic stress, which represents a new eco-friendly approach for sustainable agricultural production [29,30]. Rhizobacteria are used as bio-fertilizer, plant conditioners, plant stimulants, and bio-pesticides [30,31]. PGPR inoculation with four strains has been shown to help minimize fertilizer inputs by maintaining the photosynthetic efficiency and seed yield of durum wheat under especially dry conditions [32]. Tomato seedling inoculation with PGPR had a positive effect on the process of photosynthesis, resulting in enhanced chlorophyll fluorescence parameters due to increased evapotranspiration in the thylakoid membrane. Consequently, improvements in marketable fruit yield, biomass quantity, and water use efficiency were observed [33]. Additionally, other authors [34] noticed that certain PGPR products also increased the soluble dry matter (Brix°) and vitamin C content of tomato berries. Despite the fact that microbes have been successfully used in a number of plant species [35–37], to our knowledge, no bio-fertilizer provides a full range of nutrients or is suitable for all plant species under changing climatic conditions [38]. There is still a large gap between the potential of PGPR and its practical use as an organic fertilizer in plant production [39] as far as the assembly, composition, and structure of rhizospheric microbes; and plant–microbe interactions occurring under drought stress conditions remain mostly unknown [40]. There is also a considerable lack of knowledge not only about the physiological effects of PGPR preparations suitable for tomato production, but also suitable bacterial strain combinations and the ripening stage at which the treatment should be applied. Since the frequency of heatwaves and drought periods is increasing, and the availability of irrigation water is limited, stress-mitigating solutions should be researched and applied, such as the application of PGPRs, but its effect might vary under different water supply levels, so the research should cover that field as well.

The aim of the experiments was to examine the effect of new preparations containing different rhizobacterial strains on the physiological characteristics, productivity, and quality of processing tomatoes under different water supply conditions.

2. Materials and Methods

2.1. Research Materials and Planning

The tomato variety H-1015 F_1 hybrid (Heinz Seeds Company, Germany) was used in the 2018 and 2020 growing seasons. The experimental plot was located at the horticultural experimental farm of the Hungarian University of Agriculture and Life Sciences in Gödöllő, Hungary. The H-1015 F_1 tomato variety is a medium-sized plant with early ripeness (approx. 114 days), resistant to plant pathogens and characterized by long berries of 75–80 g as well as a high Brix° (6.2–6.8). The cultivation was performed in a slightly alkaline, clayey brown forest soil, containing 40% sand, 47.5% silt and 11.5% clay in the 0–60 cm soil layer, neutral in pH. Soil mineral content was: P_2O_5 281 mg kg^{-1}, K_2O 203 mg kg^{-1}, and NH_4 2.5 mg kg^{-1} with humus content of 1.6%. Sowing was performed in a greenhouse using a Klasmann TS3 (Klasmann-Deilmann GmbH, Geeste, Germany) substrate. Before being transplanted to the field, the four-week-old seedlings were inoculated with a PGPR product and then grown in the field.

2.2. Rhizobacteria Inoculation

The seedlings of the hybrid tomato variety H-1015 F_1 were treated with a mixture of two bacterial strains (colony-forming unit: 10^9 CFU/mL) provided by Zoltán Bay Research Institute (BAY-BIO, Szeged, Hungary) right before the transplantation to the field. Two bacterial treatments were used: the B2 treatment consisting of *Alcaligenes* sp. 3573, *Bacillus* sp. BAR16, and *Bacillus* sp. PAR11 strains, the B3 treatment consisting of *Pseudomonas* sp. MUS04, *Rhodococcus* sp. BAR03, and *Variovorax* sp. BAR04 strains and the untreated plants (control) were marked as B0. Before planting, the seedlings were immersed for 5 min in 20 litres of water containing 1% of bacterial solution. Seedlings were transplanted in a plant density of ~35,000 plants ha^{-1}, meaning single row distance of 150 cm and 19 cm distance between plants. The two factors (irrigation and PGPR inoculation) were designed in split-blocks. The size of each plot was 75 m^2. Samples were taken randomly from each plot in 4 replications. One replication consisted of 10 plants.

2.3. Climatic Data and Irrigation Treatments during Tomato Development

Three irrigation treatments were applied: regular irrigation (RI), where the evapotranspiration (ETc 100%) amount was replenished; deficit irrigation (DI), where only 50% of the ETc value was replenished to provide a moderate water shortage; and the I0 treatment, where ETc was not replenished. Under I0 treatment, the plants received natural precipitation during development along with nutrients after transplantation. The software AquaCrop 6.0 (FAO Rome, Italy) was used to calculate the RI, i.e., the plant evapotranspiration (ETc) value. A drip irrigation system was used to provide the necessary water.

Based on the precipitation quantities and temperatures of the years, 2018 can be considered to be moderately dry and 2020 to be wet (Table 1). In 2018, temperatures were relatively higher during tomato development than in 2020, with a significant difference in precipitation abundance. In 2018, a significant amount of rain fell prior to flowering; however, comparatively small amount of rain fell during flowering (ST1) and fruit setting and fruit development (ST2 and ST3, respectively) and the temperatures that rose above 30 °C impacted yield. In contrast, with the exception of a few days, 2020 was characterized by an even distribution of precipitation and temperatures of 30 °C during the development phase (ST1–ST3) (Figure 1).

Table 1. Temperature (°C), precipitation (mm) and irrigation (mm) during the 2018 and 2020 growing years under deficit (DI), regular (RI) and no-irrigation (I0) conditions.

Year	T_{max} Average °C	T_{min} Average °C	Precipitation (mm)	Irrigation (mm)			Total Water (mm)	
				DI	RI	I0	DI	RI
2018	27.5	15.7	304.6	80.2	160.3	349.9	384.8	464.9
2020	25.7	14.5	375.1	54.8	102.7	380.1	429.9	477.8

DI = deficit irrigation; RI = regular irrigation.

2.4. Field Measurements

Ten plants were selected in each plot to physiological properties measurement. Leaf chlorophyll content, chlorophyll fluorescence, and canopy temperature were measured on four occasions between 10:00 a.m. and 1:00 p.m. using selected plants, at the beginning of flowering (ST1), during flowering and fruit setting (ST2), during early fruit development (ST3), and during fruit ripening (ST4). Leaf chlorophyll content was measured with a portable chlorophyll meter (SPAD 502 Konica Minolta, Warington, UK). The instrument measures leaf photosynthetic light adsorption percentage of leaves in red and near-infrared range, with the specified calculated SPAD value correlating to the leaf chlorophyll content. Chlorophyll fluorescence was measured using a PAM 2500 (Heinz, Walz, GmbH, Effeltrich, Germany) portable chlorophyll fluorometer. The fluorometer measures initial (F0) and maximum (Fm) fluorescence in samples adjusted against the dark. Chlorophyll fluorescence, expressed as the percentage of Fv/Fm, was specified using the following formula:

$$\mathrm{Fv/Fm} = \frac{\mathrm{Fm} - \mathrm{F0}}{\mathrm{Fm}}$$

where F0 = initial fluorescence, Fm = maximum fluorescence, and Fv = variable fluorescence (Fm − F0).

Canopy temperature was measured using a Raytek MX4 (Raytek Corporation, Santa Cruz, CA, USA) portable infrared thermometer, simultaneously with the SPAD and chlorophyll fluorescence measurements.

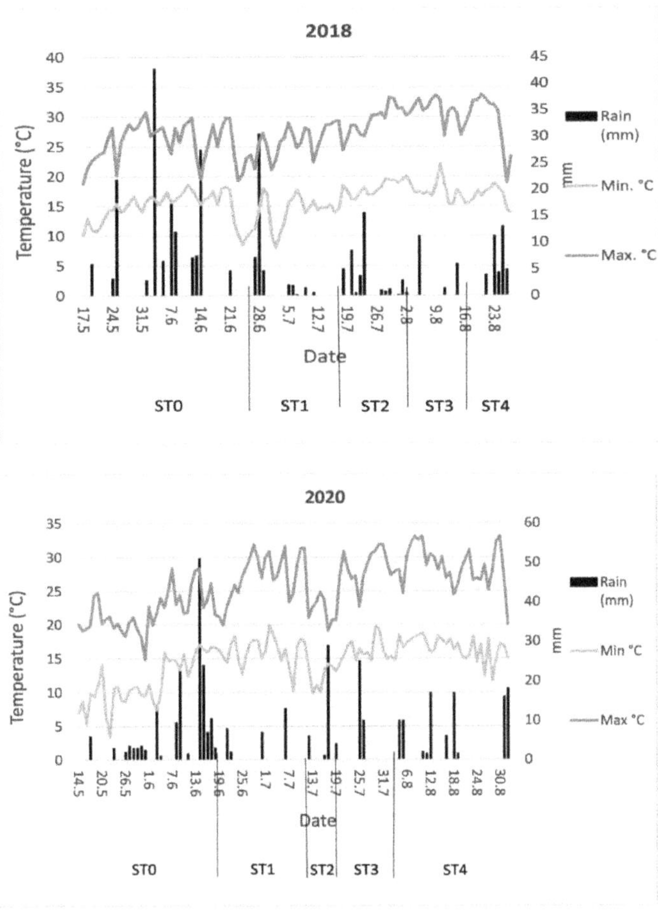

Figure 1. Temperature (°C) and precipitation (mm) data recorded during the growth stages of the hybrid tomato variety H-1015 F_1: ST0 = plantation–flowering, ST1 = flowering, ST2 = flowering and fruit setting, ST3 = early fruit development, and ST4 = fruit ripening.

2.5. Yield Analysis

The harvested fruits of the specified plants were weighed and sorted in different groups: group 1 (healthy, red-ripe marketable fruits), group 2 (healthy, green-unripe fruits, and group 3 (rotten, damaged and unmarketable fruits). The soluble dry matter content of the red-ripe, marketable fruits was measured using a Krüss DR201-95 handheld refractometer (A. Krüss Optronic GmbH, Hamburg, Germany) and expressed as degrees Brix (°Brix).

2.6. Statistical Analysis

The data were analyzed using one-way and two-way analysis of variance using SPSS 20 (IBM Hungary Ltd., Budapest, Hungary) Windows software. The homogeneity of the variance was assessed using Levene's test. Duncan's multi-range test was also used for means separation among treatment at the $p < 0.05$ level.

3. Results

Previously, we demonstrated that water shortage occurring during tomato plant's generative phase, from flowering to early berry ripening, negatively affected water consumption and photosynthesis-related processes [11] as well as fruit quality. In a moderately dry growing season (2018), canopy temperature and chlorophyll fluorescence were higher with respect to the wet and humid growing season (2020). Nevertheless, leaf relative chlorophyll content (SPAD values) (Figures 2–4) remained unaffected by the growing seasons. The effect of PGPR products varied with the development stages and climatic conditions of the growing years. Under dry conditions (2018 growing season), the canopy temperature of untreated plants was the lowest (26 °C) during flowering (ST1) but increased significantly followed by the B2 and B3 treatments, attaining values $\geq$30 °C at the early fruit development stage (ST3) particularly following the B3 treatment (Figure 2). In the wet growing year (2020), the low canopy temperature (22.5–23.7 °C) was recorded during flowering (ST1) increased by 13–15% during the flowering and fruit setting (ST2) as well as the early fruit development stages (ST3). Under these particular conditions, the effect of the bacterial treatment on canopy temperature was not revealed statistically (Figure 2).

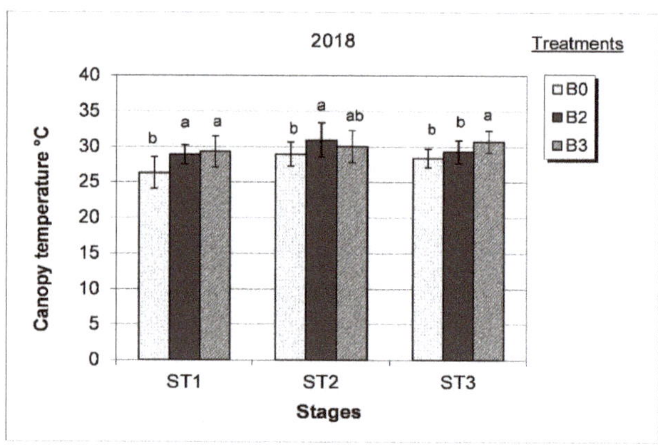

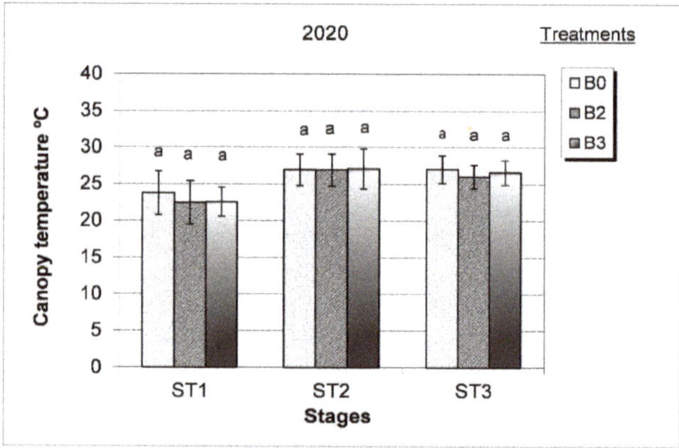

Figure 2. Effect of different bacterial treatments (B0 = control, B2 = *Alcaligenes* sp. 3573, *Bacillus* sp. BAR16, and *Bacillus* sp. PAR11 strains and B3 = *Pseudomonas* sp. MUS04, *Rhodococcus* sp. BAR03, and *Variovorax* sp. BAR04 strains) on canopy temperature during development stages (ST1 = during flowering, ST2 = flowering with fruit setting, and ST3 = early fruit development) in the dry (2018) and wet (2020) growing years. Values represent the mean $\pm$ SD of four replicates. Bars marked with different letters are significantly different at the $p < 0.05$ level.

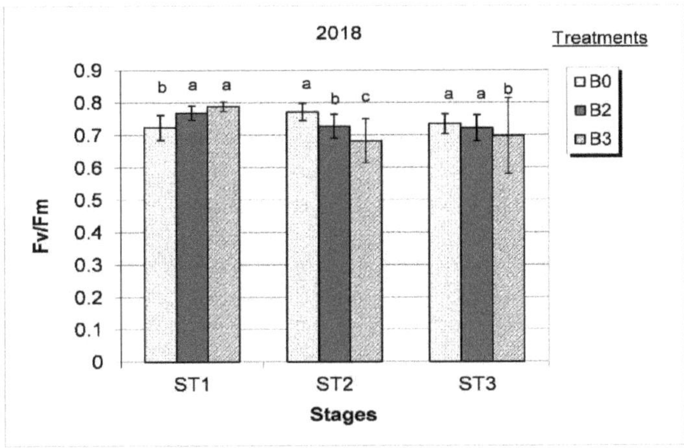

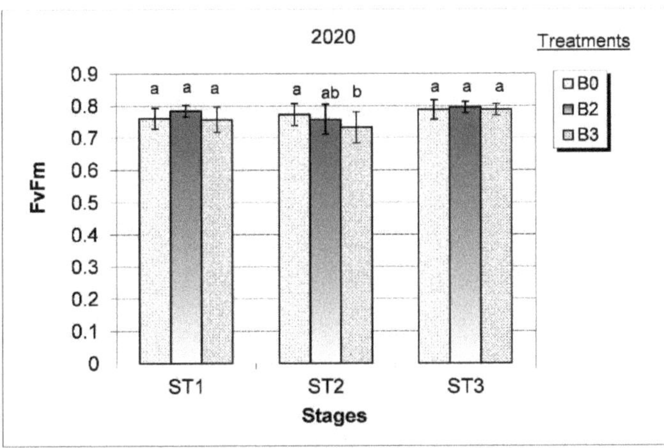

Figure 3. Effect of different bacterial treatments (B0 = control, B2 = *Alcaligenes* sp. 3573, *Bacillus* sp. BAR16, and *Bacillus* sp. PAR11 strains and B3 = *Pseudomonas* sp. MUS04, *Rhodococcus* sp. BAR03, and *Variovorax* sp. BAR04 strains) on chlorophyll fluorescence (Fv/Fm) during development stages (ST1 = during flowering, ST2 = flowering with fruit setting, and ST3 = early fruit development) in the dry (2018) and wet (2020) growing years. Values represent the mean ± SD of four replicates. Bars marked with different letters are significantly different at the $p < 0.05$ level.

Photosynthesis remains unchanged and plant development is not impeded in the case of high chlorophyll fluorescence (Fv/Fm) and low SPAD value. Under dry growing conditions, leaf chlorophyll content decreased, leading to reduced light absorption (i.e., the utilization of light energy), increased light reflection, and a decreased Fv/Fm ratio. In a moderately dry growing season (2018), B2- and B3-treated plants exhibited a significant increase in chlorophyll fluorescence (0.77–0.79 Fv/Fm) during tomato flowering (ST1), which subsequently decreased significantly during the flowering and fruit setting phase (ST2). The largest decrease in Fv/Fm values was measured in B3-treated plants with respect to untreated plants (Figure 3).

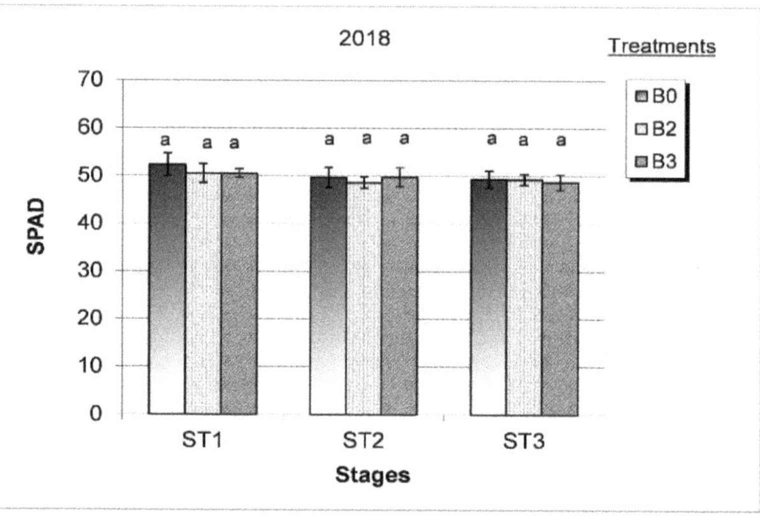

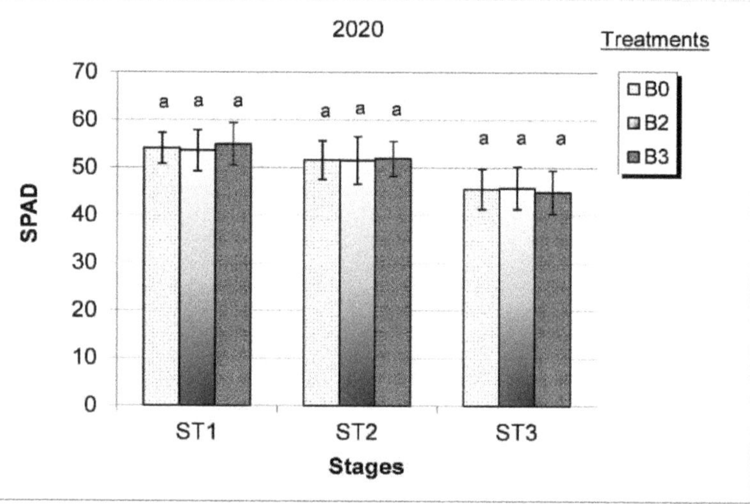

Figure 4. Effect of different bacterial treatments (B0 = control, B2 = *Alcaligenes* sp. 3573, *Bacillus* sp. BAR16, and *Bacillus* sp. PAR11 strains and B3 = *Pseudomonas* sp. MUS04, *Rhodococcus* sp. BAR03, and *Variovorax* sp. BAR04 strains) on SPAD values during development stages (ST1 = during flowering, ST2 = flowering with fruit setting, and ST3 = early fruit development) in the dry (2018) and wet (2020) growing years. Values represent the mean ± SD of four replicates. Bars marked with different letters are significantly different at the $p < 0.05$ level.

In the wet growing year (2020), photosynthesis was well balanced from flowering until early berry development (ST1–ST3) with a Fv/Fm value between 0.76 and 0.79 in untreated plants (Figure 3). Under these conditions, the Fv/Fm value was not significantly affected by bacterial treatments, except for B3-treated plants, where values decreased significantly during the flowering and fruit setting (ST2) stage. Regardless of the changes in climatic conditions between the two years, the highest chlorophyll fluorescence (Fv/Fm) decrease was measured during flowering and fruit setting (ST2) in the B3 treatment, thus influencing photosynthetic activity and the ripening process.

Under moderate dry growing conditions (2018), the leaf SPAD index ranged between 49 and 52 during the ST1–ST3 phase. The SPAD value measured in H 1015 F_1 tomato plants with bacterial treatment was only 4% less than values in untreated plants during flowering stage (ST1) and remained unchanged with advancing maturity (Figure 4). However, in the wet growing year (2020), the SPAD index gradually decreased during the ST1–ST3 phase. The SPAD value was the highest during flowering (54), 6% lower during flowering and fruit setting (ST2), and reached its lowest value (45 SPAD) during early fruit ripening (ST3); however, the effects of the bacteria treatment were not statistically significant (Figure 4).

Results indicate that during the wet growing year, the effect of PGPR was not obvious on physiological properties during plant's generative phase. However, the impact was with differing extent during the dry growing year and depending on water supply (Figure 5). Under water shortage (I0) associated with moderately dry growing conditions (2018), the canopy temperature of untreated plants (B0) was high and significantly decreased by irrigation. Without irrigation (I0), canopy temperature remained unchanged around 30 °C and was not significantly influenced by PGPR treatments but under irrigated conditions, (DI, RI) the canopy temperature of bacterial treatments significantly increased with respect to untreated plants (Figure 5a).

Regardless of water supply and in the dry growing season, chlorophyll fluorescence ranged from 0.73 to 0.75 Fv/Fm during the generative phase and was not significantly affected by bacterial treatments, except for B3, which caused a significant decrease in chlorophyll fluorescence (0.69 Fv/Fm) during deficit irrigation (Figure 5b). A similar result was observed during flowering and fruit setting (ST2), indicating that, at this stage, water shortage (DI) and bacterial treatments significantly affected photosynthetic activity. In the dry year growing year (2018), during flowering (ST1) and under non-irrigated (I0) conditions, the leaf SPAD value was slightly reduced in the treatments treated with PGPR, with no significant difference between PGPR-treated and untreated plants for leaf SPAD values during the generative stage or under irrigated (DI, RI) conditions (Figures 4 and 5c).

Under moderate water shortage (dry year + deficit irrigation), PGPR treatments (B2, B3) significantly affected plant water balance. Canopy temperature increased as transpiration decreased, and photosynthetic activity (Fv/Fm) was satisfactory with the exception of B3, as reflected in the crop's quality traits. Under water shortage, PGPR treatments significantly increased fruit quantity compared to untreated plants, though the percentage of healthy green fruits increased considerably among the distribution of the fruit types (Table 2). Depending on the degree of water shortage and different bacterial formulations was associated with different effects on the crop.

In the dry growing season and I0 treatment, B3-treated plants produced 32% less green berries than those B2-treated plants. A significant difference between the PGPR treatments under moderate water shortage (dry year + deficit irrigation) was noticed regarding green fruit yields. In fact, B3 treatment resulted in 26% more ripe marketable fruits and 49% less unripe fruits compared to the B2 treatment.

Soluble solid content was the highest under dry conditions (I0) and decreased significantly with irrigation. In the moderately dry growing year (2018), PGPR treatments had no effect on the Brix° value, though both water supply and PGPR treatments exhibited a negative impact under wet growing year (2020) (Table 2).

In the wet growing year (2020), there was a significant difference between the PGPR treatments. In fact, the B2 treatment resulted in a greater marketable yield and a higher Brix° value, but with higher percentage of diseased fruit compared to B3-treated plants and this effect was more pronounced under regular irrigation (RI) (Table 2).

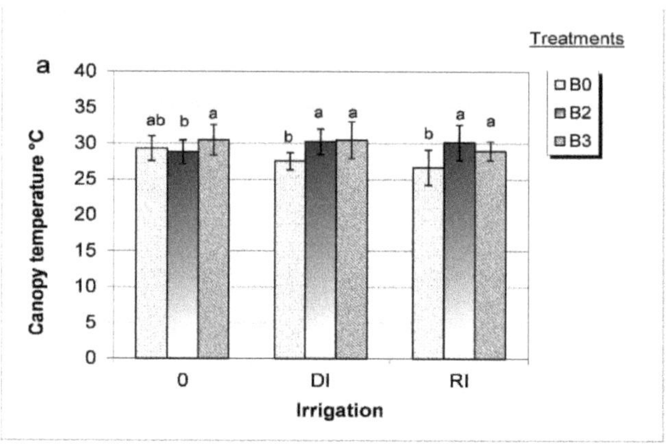
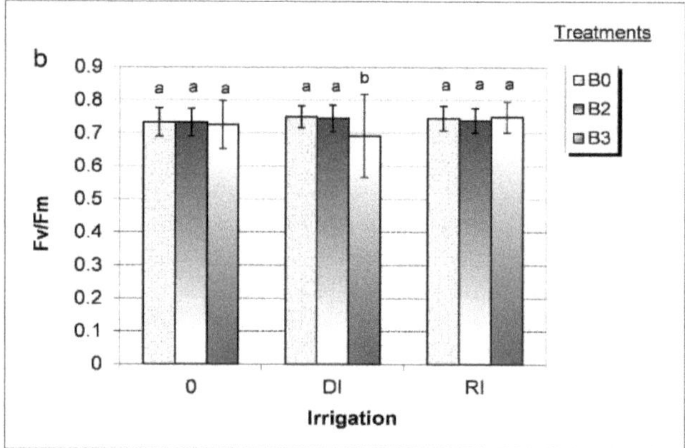
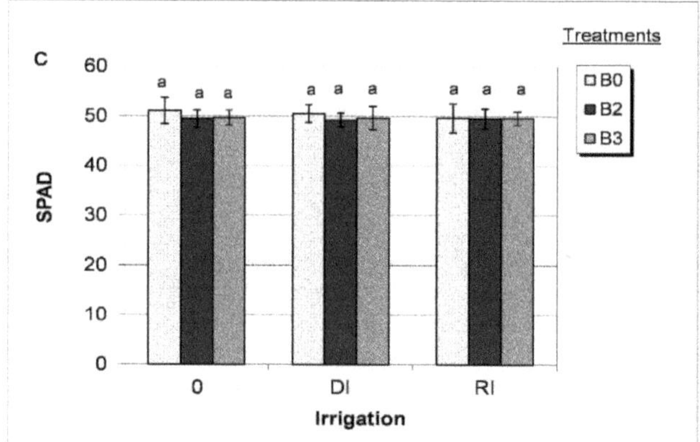

Figure 5. Effect of different bacterial treatments (B0 = control, B2 = Alcaligenes sp. 3573, *Bacillus* sp. BAR16, and *Bacillus* sp. PAR11 strains and B3 = *Pseudomonas* sp. MUS04, *Rhodococcus* sp. BAR03, and *Variovorax* sp. BAR04 strains) on canopy temperature (**a**), chlorophyll fluorescence (**b**) and SPAD (**c**) from flowering to early fruit development under different water supplies (I0 = non irrigation, DI = deficit irrigation and RI = regular irrigation) during the dry growing year (2018). Values represent the mean ± SD of four replicates. Bars marked with different letters are significantly different at the $p < 0.05$ level.

Table 2. Effects of different water supply (I0 = non irrigation, DI = deficit irrigation and RI = regular irrigation) and bacterial treatments (PGPR) (B0 = control, B2 = *Alcaligenes* sp. 3573, *Bacillus* sp. BAR16, and *Bacillus* sp. PAR11 strains and B3 = *Pseudomonas* sp. MUS04, *Rhodococcus* sp. BAR03, and *Variovorax* sp. BAR04 strains) on the total yield (TY, t.ha^{-1}), marketable yield (MY, t.ha^{-1}), green yield (GY, t.ha^{-1}), diseased/unmarketable yield (DY, t.ha^{-1}) and green biomass (GBm, t.ha^{-1}) as well as soluble solids (°Brix) of the processing tomato variety H-1015 F$_1$ grown during the dry (2018) and wet (2020) growing years.

Water Supply (WS) [a]	PGPR [b]	2018						2020					
		TY tha^{-1}	MY ha^{-1}	GY tha^{-1}	DY tha^{-1}	GBm tha^{-1}	Soluble Solids (°Brix)	TY tha^{-1}	MY ha^{-1}	GY tha^{-1}	DY tha^{-1}	GBm tha^{-1}	Soluble Solids (°Brix)
I0	B0	50.71 b	45.45 b	4.40 c	0.86 a	66.3 b	4.25 a	49.78 b	40.13 a	2.52 a	7.08 c	60.3 b	4.64 a
	B2	68.6 a	57.46 a	11.05 a	1.70 a	84.7 a	4.36 a	56.78 a	39.92 a	4.43 a	12.40 a	70.0 a	4.55 a
	B3	68.83 a	58.27 a	8.32 b	2.25 a	88.6 a	4.39 a	51.48 a	38.71 a	3.51 a	9.28 b	61.4 b	4.20 b
I0		62.72 B	53.72 B	7.92 A	1.60 A	79.87 B	4.33 A	52.68 A	39.59 A	3.49 A	9.59 B	63.9 A	4.46 A
DI	B0	64.97 b	59.26 c	4.65 d	1.06 a	81.1 b	3.65 a	51.10 b	40.33 a	2.05 a	8.71 b	62.7 ab	4.00 a
	B2	83.54 a	65.30 b	15.36 a	2.89 a	100.4 a	3.61 a	55.75 a	39.11 a	2.57 a	14.03 a	66.9 a	3.80 a
	B3	93.28 a	82.56 a	10.28 b	2.36 a	116.3 a	3.69 a	47.05 c	31.71 b	1.70 a	13.62 a	56.2 b	3.88 a
DI		80.57 A	69.04 A	10.10 A	2.10 A	99.3 A	3.65 B	51.30 A	37.05 A	2.11 B	12.12 A	61.9 A	3.89 B
RI	B0	72.43 b	59.75 b	9.92 b	2.71 a	87.5 b	3.38 a	48.40 b	34.51 a	1.71 a	12.21 b	64.5 a	3.40 a
	B2	76.37 b	62.99 a	10.41 b	14.44 a	89.8 b	3.34 a	56.18 a	35.71 a	2.23 a	18.22 a	67.1 a	3.42 a
	B3	85.01 a	69.72 a	13.16 a	2.14 a	98.1 b	3.22 a	43.58 c	29.36 b	2.83 a	11.34 b	54.7 b	3.04 b
RI		77.94 A	64.15 A	11.16 A	2.61 A	91.8 A	3.31 C	49.38 A	33.19 B	2.26 B	13.93 A	62.1 A	3.29 C
Significance	WS	*	*	ns	ns	*	***	ns	*	*	**	ns	***
	PGPR	*	*	*	ns	*	ns	***	*	Ns	***	**	***
	WS × PGPR	ns	ns	ns	ns	ns	ns	ns	ns	ns	ns	ns	ns

*** $p < 0.001$; ** $p < 0.01$; * $p < 0.05$; ns = non-significant. Capital letters indicate the mean separation among water supply conditions, lower case letters indicate the mean separation among bacterial treatments. Averages followed by different letters are significantly different at the $p < 0.05$ level using Duncan's test.

4. Discussion

The frequency of high temperatures and dry periods caused by climate change negatively affects the yield of major field-grown horticultural crops. It is widely recognized that yield loss can be mitigated by irrigation management and bio-fertilizer application. However, limited bacterial strains establish a symbiotic relationship with horticultural crops. PGPR excretes various phytohormones and osmolytes which contribute to improve plant stress tolerance by altering key physiological processes such as stomata closing, transpiration and photosynthesis [10,22,41–44]. Nevertheless, the determination of the suitable application period of the PGPR treatment during plant development might be crucial for efficient output.

During tomato development stages, the PGPR treatments indirectly affected canopy temperature and chlorophyll fluorescence. In a moderate dry growing year, canopy temperature and chlorophyll fluorescence (Fv/Fm) significantly increased prior to flowering in the B2 and B3 treatments, compared to the untreated control plants. However, during flowering and fruit setting stages, chlorophyll fluorescence (Fv/Fm) underwent the greatest decrease in B3-treated plants, regardless of the considered growing year (dry or humid). Similar results were reported previously as a result of mycorrhiza inoculation, where chlorophyll fluorescence (Fv/Fm) increased in tomato until flowering and fruit setting but decreased with advancing maturity stages, especially under dry growing conditions [45,46]. Some PGPR strains (*Bacillus safensis* (FV46), and *Brevibacillus lateosporus* (C9F)) reduced the Fv/Fm value compared to control according to the finding of another study [47], which is similar to our experiment in 2018 (ST2 and ST3) and in 2020 (ST2).

Tahir et al. [48] found that PGPR strains increased relative chlorophyll content (SPAD) in leaves under good water supply and dry growing conditions compared to untreated plants. *Bacillus* strains were reported to improve plant growth by affecting physiological processes, such as increasing plant relative water and chlorophyll content under drought stress conditions [49,50] and increasing chlorophyll production in maize during salt stress [51].

Phylazonit bio-fertilizer containing *Bacillus* strains resulted in significant increases in SPAD values in Uno Rosso F_1 tomato cultivar with respect to untreated plants [45]. In contrast, our results showed that the chlorophyll content of PGPR-treated H-1015 F_1 tomato plants remained unchanged during the phenological stages under different water supply conditions. The different values are likely attributed to the different interactions of tomato cultivars X PGPR strains.

Altered photosynthesis-related traits such as leaf chlorophyll content and chlorophyll fluorescence arising from water shortage during flowering and fruit setting affected plant development, tomato berry weight, and consequently marketable yield [11]. The physiological processes underlying plant tolerance to water shortage and the deleterious effects on yield depend on various factors such as the considered plant species, the PGPR strain [51,52] and the developmental stage at which their application is more efficient.

Pseudomonas species were found to be particularly suitable during tomato flowering and fruit setting stages [53] while the beneficial effects of *Bacillus* bacteria were mostly reported for soybean [54]. *Bacillus*-based bio-fertilizer increased tomato berry weight under moderate water shortage [45]. Results indicated that under moderate water shortage (dry year + deficit irrigation) occurring during the tomato flowering to early fruit development stages, PGPR treatments resulted in improved water balance in H-1015 F_1 tomato while photosynthetic activity remained unchanged. The application of the B3 treatment under these conditions led to a significant decrease in chlorophyll fluorescence and thus a reduction in photosynthetic activity. The B2 product contains *Bacillus* bacteria strains, leading to the highest photosynthetic activity, greater green mass, and an extended ripening process. This is in accordance with the findings of others, who reported that *Bacillus*-based products successfully forms colonies in tomato rhizosphere and contribute to improved growth and yield [55,56]. Root colonization and tomato plant growth were observed after 15 and 60 days, respectively [57].

The intense photosynthetic activity in leaves of B2-treated plants may have delayed the translocation of assimilates into the berries, which was reflected in the distribution of the fruit quality fractions and in a higher proportion of immature green berry formation. Under moderate water shortage (dry year + deficit irrigation), the B3 treatment resulted in greater quantities of marketable fruit and less green berries than the B2 treatment. This proves that under water shortage, plant physiological processes, yield quantities, and qualitative distribution are not only influenced by the bacterial composition of the applied products but also by their interaction with the host plant. Based on our results, under water shortage, both B2- and B3-treated plants produce better yields and more healthy green berries than untreated plants (Table 2). A significant difference could be detected between the two bacterial treatments under moderate water shortage (dry year + deficit irrigation). In fact, the B3 treatment resulted in 26% more marketable and 49% less unripe fruit than the B2 treatment. Contrary to expectations, the PGPR treatments did not impact either the number of diseased berries or their soluble dry matter content (Brix°) under moderate water shortage. Other authors [34] reported that certain PGPR bio-fertilizers decreased the quantity of diseased tomato berries and improved the Brix° value.

In the humid growing year, plants treated with B2 produced a greater marketable yield and improved the Brix° value but also lead to more rotten berries than in B3-treated plants. The results showed that the B2 bio-fertilizer, containing *Alcaligenes* strains in addition to *Bacillus*, resulted in prolonged ripening and a greater quantity of unripe green fruits via regulation of key physiological processes. However, the bio-fertilizer B3, containing three different rhizobacteria strains, resulted in more concentrated ripening stages and less unripe green fruits under water shortage conditions.

5. Conclusions

Our results show that in the case of water shortage, PGPR products influence plant physiological processes and increase yield quantities. Under water shortage occurring during tomato flowering and fruit setting, PGPR products significantly affected canopy temperature, chlorophyll fluorescence (Fv/Fm), and plant water balance and photosynthesis. The quali-quantitative distribution of fruit depends on the composition of the applied products, variety X bacterial strain interaction, and the water supply regime. Under water shortage, the *Bacillus*-species-containing B2 product resulted in undisturbed photosynthesis, prolonged ripening, a higher percentage of unripe green fruit, and more diseased fruit in good water supply. The B3 product, which has a different composition, was found to be more effective than the B2 product because the decrease in chlorophyll fluorescence led to a drop in photosynthetic activity and accelerated ripening processes, which contributed to the generation of a greater marketable yield and less unripe green berries. Based on our data, we recommend immersing young tomato seedlings, prior to transplantation, in a 1% solution of B3 combination product consisting of Pseudomonas sp. MUS04, Rhodococcus sp. BAR03, and Variovorax sp. BAR04 strains to improve the efficiency of the treatment. In the dry year, the examined PGPR products did not influence the Brix° value of the tomato fruit but had a negative effect in the wet year.

Author Contributions: Conceptualization, E.N. and L.H.; investigations and analysis, K.Z.H., B.A. and A.N.; methodology, Z.P.; writing—draft preparation, E.N.; write—review and editing, L.H., R.I. and S.T. All authors have read and agreed to the published version of the manuscript.

Funding: This research was supported by the EFOP-3.6.3-VEKOP-16-2017-00008 project. This work was supported by the ÚNKP-21-4 New National Excellence Program of the Ministry for Innovation and Technology from the source of the National Research, Development and Innovation Fund.

Acknowledgments: The authors would like to thank the staff of the Horticultural research farm at Gödöllő and also to everyone who contributed to this research.

Conflicts of Interest: The authors declare no conflict of interest.

References

1. Agbna, G.H.D.; Donglia, S.; Zhipeng, L.; Elshaikha, N.A.; Guangchenga, S.; Timm, L.C. Effects of deficit irrigation and biochar addition on the growth, yield, and quality of tomato. *Sci. Hortic.* **2017**, *222*, 90–101. [CrossRef]
2. Du, Y.D.; Cao, H.-X.; Liu, S.-Q.; Gu, X.-B.; Cao, Y.-X. Response of yield, quality, water and nitrogen use efficiency of tomato to different levels of water and nitrogen under drip irrigation in Northwestern China. *J. Integr. Agric.* **2017**, *16*, 1153–1161. [CrossRef]
3. Lorenzi, F.; Alfieri, S.M.; Monaco, E.; Bonfante, A.; Basile, A.; Patanè, C.; Menenti, M. Adaptability to future climate of irrigated crops: The interplay of water management and cultivars responses. A case study on tomato. *Biosyst. Eng.* **2017**, *157*, 45–62. [CrossRef]
4. Takács, S.; Csengeri, E.; Pék, Z.; Bíró, T.; Szuvandzisev, P.; Palotás, G.; Helyes, L. Performance evaluation of AquaCrop model in processing tomato biomass, fruit yield and water stress indicator modelling. *Water* **2021**, *13*, 3587. [CrossRef]
5. Sing, S.K.; Reddy, K.R. Regulation of photosynthesis, fluorescence, stomatal conductance and water-use efficiency of cowpea (*Vigna unguiculata* [L.] Walp.) under drought. *J. Photoch. Photobiol. B Biol.* **2011**, *105*, 40–50. [CrossRef]
6. Nemeskéri, E.; Molnár, K.; Pék, Z.; Helyes, L. Effect of water supply on water use related physiological traits and yield of snap beans in dry seasons. *Irrig. Sci.* **2018**, *36*, 143–158. [CrossRef]
7. Vadez, V. Root hydraulics: The forgotten side of roots in drought adaptation. *Field Crops Res.* **2014**, *165*, 15–24. [CrossRef]
8. Ratnakumar, P.; Vadez, V. Tolerant groundnut (*A. hypogaea* L.) genotypes to intermittent drought maintains high harvest index and has small leaf canopy under stress. *Funct. Plant Biol.* **2011**, *38*, 1016–1023. [CrossRef]
9. Pasala, R.; ·Pandey, B.B.; Gandi, S.L.; Kulasekaran, R.; Guhey, A.; Reddy, A.V. An insight into the mechanisms of intermittent drought adaptation in sesame (*Sesamum indicum* L.): Linking transpiration efficiency and root architecture to seed yield. *Acta Physiol. Plant* **2021**, *43*, 148. [CrossRef]
10. Helyes, L.; Bőcs, A.; Pék, Z. Effect of water supply on canopy temperature, stomatal conductance and yield quantity of processing tomato (*Lycopersicon esculentum* Mill.). *Int. J. Hortic. Sci.* **2010**, *16*, 13–15. [CrossRef]
11. Nemeskéri, E.; Neményi, A.; Bőcs, A.; Pék, Z.; Helyes, L. Physiological factors and their relationship with the productivity of processing tomato under different water supplies. *Water* **2019**, *11*, 586. [CrossRef]
12. Lobos, G.; Retamales, J.B.; Hancock, J.F.; Flore, J.A.; Cobo, N.; del Pozo, A. Spectral irradiance, gas exchange characteristics and leaf traits of *Vaccinium corymbosum* L. 'Elliott' grown under photo-selective nets. *Environ. Exp. Bot.* **2012**, *75*, 142–149. [CrossRef]
13. Ogweno, J.O.; Song, X.S.; Hu, W.H.; Shi, K.; Zhou, Y.H.; Yu, J.Q. Detached leaves of tomato differ in their photosynthetic physiological response to moderate high and low temperature stress. *Sci. Hortic.* **2009**, *123*, 17–22. [CrossRef]
14. Mishra, K.B.; Iannacone, R.; Petrozza, A.; Mishra, A.; Armentano, N.; La Vecchia, G.; Trtilek, M.; Cellini, F.; Nedbal, L. Engineered drought tolerance in tomato plants is reflected in chlorophyll fluorescence emission. *Plant Sci.* **2012**, *182*, 79–86. [CrossRef] [PubMed]

15. Zheng, J.; Huang, G.; Jia, D.; Wang, J.; Mota, M.; Pereira, L.S.; Huang, Q.; Xu, X.; Liu, H. Responses of drip irrigated tomato (*Solanum lycopersicum* L.) yield, quality and water productivity to various soil matric potential thresholds in an arid region of Northwest China. *Agric. Water Manag.* **2013**, *129*, 181–193. [CrossRef]

16. Bahadur, A.; Chatterjee, A.; Kumar, R.; Singh, M.; Naik, P.S. Physiological and biochemical basis of drought tolerance in vegetables. *Veg. Sci.* **2011**, *38*, 1–16.

17. Pék, Z.; Szuvandzsiev, P.; Neményi, A.; Helyes, L. Effect of season and irrigation on yield parameters and soluble solids content of processing cherry tomato. *Acta Hortic.* **2015**, *1081*, 197–202. [CrossRef]

18. Shao, H.B.; Chu, L.Y.; Jaleel, A.; Zhao, C.X. Water-deficit stress-induced anatomical changes in higher plants. *CR Biol.* **2008**, *331*, 215–225. [CrossRef]

19. Zhang, X.; Lu, G.; Long, W.; Zou, X.; Li, F.; Nishio, T. Recent progress in drought and salt tolerance studies in *Brassica* crops. *Breed. Sci.* **2014**, *64*, 60–73. [CrossRef]

20. Barriuso, J.; Solano, B.R.; Mañero, F.J.G. Protection against pathogen and salt stress by four plant growth-promoting rhizobacteria isolated from *Pinus* sp. on *Arabidopsis thaliana*. *Phytopathology* **2008**, *98*, 666–672. [CrossRef]

21. Liu, T.; Yang, X.G.; Batchelor, W.D.; Liu, Z.J.; Zhang, Z.T.; Wan, N.H.; Sun, S.; He, B.; Gao, J.Q.; Bai, F.; et al. A case study of climate-smart management in foxtail millet (*Setaria italica*) production under future climate change in Lishu county of Jilin, China. *Agric. For. Meteorol.* **2020**, *292–293*, 108131. [CrossRef]

22. Sauter, A.; Dietz, K.J.; Hartung, W. A possible stress physiological role of abscisic acid conjugates in root-to-shoot signalling. *Plant Cell Environ.* **2002**, *25*, 223–228. [CrossRef] [PubMed]

23. Cohen, A.C.; Travaglia, C.N.; Bottini, R.; Piccoli, P.N. Participation of abscisic acid and gibberellins produced by endophytic *Azospirillum* in the alleviation of drought effects in maize. *Botany* **2009**, *87*, 455–462. [CrossRef]

24. Sharifi, R.; Ryu, C.M. Chatting with a Tiny Belowground Member of Holobiome: Communication Between Plants and Gowth-Promoting Rhizobacteria. *Adv. Bot. Res.* **2017**, *82*, 135–160.

25. Kamilova, F.; Kravchenko, L.V.; Shaposhnikov, A.I.; Azarova, T.; Makarova, N.; Lugtenberg, B. Organic acids, sugars and L-tryptophane in exudates of vegetables growing on stonewool and their effects on activities of rhizosphere bacteria. *Mol. Plant Microbe Interact.* **2006**, *9*, 250–256. [CrossRef]

26. Oku, S.; Komatsu, A.; Tajima, T.; Nakashimada, Y.; Kato, J. Identification of chemotaxis sensory proteins for amino acids in *Pseudomonas fluorescens* Pf0-1 and their involvement in chemotaxis to tomato root exudate and root colonization. *Microbes Environ.* **2012**, *27*, 462–469. [CrossRef]

27. Dodd, I.C.; Zinovkina, N.Y.; Safronova, V.I.; Belimov, A.A. Rhizobacterial mediation of plant hormone status. *Ann. Appl. Biol.* **2010**, *157*, 361–379. [CrossRef]

28. Zhang, Y.; Cui, G.; Zhang, W.; Lang, D.; Zhixian, L.; Zhang, X. *Bacillus* sp. G_2 improved the growth of *Glycyrrhiza uralensis* Fisch. related to antioxidant metabolism and osmotic adjustment. *Acta Physiol. Plant* **2021**, *43*, 152. [CrossRef]

29. Nadeem, S.M.; Ahmad, M.; Zahir, Z.A.; Javaid, A.; Ashraf, M. The role of mycorrhizae and plant growth promoting rhizobacteria (PGPR) in improving crop productivity under stressful environments. *Biotechnol. Adv.* **2014**, *32*, 429–448. [CrossRef]

30. Singh, V.K.; Singh, A.K.; Singh, P.P.; Kumar, A. Interaction of plant growth promoting bacteria with tomato under abiotic stress: A review. *Agric. Ecosyst. Environ.* **2018**, *267*, 129–140. [CrossRef]

31. Berg, G. Plant-microbe interactions promoting plant growth and health: Perspectives for controlled use of microorganisms in agriculture. *Appl. Microbiol. Biotechnol.* **2009**, *84*, 11–18. [CrossRef] [PubMed]

32. Khanghahi, M.Y.; Leoni, B.; Crecchio, C. Photosynthetic responses of durum wheat to chemical/microbiological fertilization management under salt and drought stresses. *Acta Physiol. Plant* **2021**, *43*, 123. [CrossRef]

33. Le, A.T.; Pék, Z.; Takács, S.; Neményi, A.; Helyes, L. The effect of plant growth-promoting rhizobacteria on yield, water use efficiency and Brix Degree of processing tomato. *Plant Soil Environ.* **2018**, *64*, 523–529. [CrossRef]

34. Andryei, B.; Horváth, K.Z.; Agyemang Duah, S.; Takács, S.; Égei, M.; Szuvandzsiev, P.; Neményi, A. Use of plant growth promoting rhizobacteria (PGPRs) in the mitigation of water deficiency of tomato plants (*Solanum lycopersicum* L.). *J. Cent. Eur. Agric.* **2021**, *22*, 167–177.

35. Zhou, Y.; Sang, T.; Tian, M.; Jahan, M.S.; Wang, J.; Li, X.; Guo, S.; Liu, H.; Wang, Y.; Shu, S. Effects of *Bacillus cereus* on photosynthesis and antioxidant metabolism of cucumber seedlings under salt stress. *Horticulturae* **2022**, *8*, 463. [CrossRef]

36. Papoui, E.; Bantis, F.; Kapoulas, N.; Ipsilantis, I.; Koukounaras, A. A sustainable intercropping system for organically produced lettuce and green onion with the use of arbuscular mycorrhizal inocula. *Horticulturae* **2022**, *8*, 466. [CrossRef]

37. Sneha, S.; Anitha, B.; Sahair, R.A.; Raghu, N.; Gopenath, T.S.; Chandrashekrappa, G.K.; Basalingappa, K.M. Biofertilizer for crop production and soil fertility. *Acad. J. Agric. Res.* **2018**, *6*, 299–306.

38. Kaur, R.; Kaur, S.; Kaur, G. Molecular and physiological manipulations in rhizospheric bacteria. *Acta Physiol. Plant* **2021**, *43*, 77. [CrossRef]

39. Vejan, P.; Abdullah, R.; Khadiran, T.; Ismail, S.; Boyce, A.N. Role of plant growth promoting rhizobacteria in agricultural sustainability: A review. *Molecules* **2016**, *21*, 573. [CrossRef]

40. Aslam, M.M.; Okal, E.J.; Idris, A.L.; Qian, Z.; Xu, W.; Karanja, J.K.; Wani, S.H.; Yuan, W. Rhizosphere microbiomes can regulate plant drought tolerance. *Pedosphere* **2022**, *32*, 61–74. [CrossRef]

41. Yang, J.; Kloepper, J.W.; Ryu, C.M. Rhizosphere bacteria help plants tolerate abiotic stress. *Trends Plant Sci.* **2009**, *14*, 1–4. [CrossRef] [PubMed]

42. Forni, C.; Duca, D.; Glick, B.R. Mechanisms of plant response to salt and drought stress and their alteration by rhizobacteria. *Plant Soil* **2017**, *410*, 335–356. [CrossRef]

43. Davies, W.J.; Zhang, J. Root signals and the regulation of growth and development of plants in drying soil. *Annu. Rev. Plant Physiol. Plant Mol. Biol.* **1991**, *42*, 55–76. [CrossRef]

44. Bresson, J.; Varoquaux, F.; Bontpart, T.; Touraine, B.; Vile, D. The PGPR strain *Phyllobacterium brassicacearum* STM196 induces a reproductive delay and physiological changes that result in improved drought tolerance in *Arabidopsis*. *N. Phytol.* **2013**, *200*, 558–569. [CrossRef]

45. Nemeskéri, E.; Horváth, K.; Pék, Z.; Helyes, L. Effect of mycorrhizal and bacterial products on the traits related to photosynthesis and fruit quality of tomato under water deficiency conditions. *Acta Hortic.* **2019**, *1233*, 61–66. [CrossRef]

46. Horvath, K.Z.; Andryei, B.; Helyes, L.; Pék, Z.; Neményi, A.; Nemeskéri, E. Effect of mycorrhizal inoculations on physiological traits and bioactive compounds of tomato under water scarcity in field conditions. *Not. Bot. Hortic. Agrobot Cluj Napoca* **2020**, *48*, 1233–1247. [CrossRef]

47. Costa-Santos, M.; Mariz-Ponte, N.; Dias, M.C.; Moura, L.; Marques, G.; Santos, C. Effect of *Bacillus* spp. and *Brevibacillus* sp. on the Photosynthesis and Redox Status of *Solanum lycopersicum*. *Horticulturae* **2021**, *7*, 24. [CrossRef]

48. Tahir, M.; Khalid, U.; Khan, M.B.; Shahid, M.; Ahmad, I.; Akram, M.; Ijaz, M.; Hussain, M.; Farooq, A.B.; Naeem, M.A.; et al. Auxin and 1-aminocyclopropane-1-carboxylate deaminase activity exhibiting rhizobacteria enhanced maize quality and productivity under water deficit conditions. *Int. J. Agric. Biol.* **2019**, *21*, 943–954.

49. Gagné-Bourque, F.; Bertrand, A.; Claessens, A.; Aliferis, K.A.; Jabaji, S. Alleviation of drought stress and metabolic changes in timothy (*Phleum pratense* L.) colonized with *Bacillus subtilis* B26. *Front. Plant Sci.* **2016**, *7*, 584. [CrossRef]

50. Zhou, C.; Ma, Z.; Zhu, L.; Xiao, X.; Xie, Y.; Zhu, J.; Wang, J. Rhizobacterial strain *Bacillus megaterium* BOFC15 induces cellular polyamine changes that improve plant growth and drought resistance. *Int. J. Mol. Sci.* **2016**, *17*, 976. [CrossRef]

51. Chen, L.; Liu, Y.; Wu, G.; Veronican, N.; Jeri, K.; Shen, Q.; Zhang, N.; Zhang, R. Induced maize salt tolerance by rhizosphere inoculation of *Bacillus amyloliquefaciens* SQR9. *Physiol. Plant* **2016**, *158*, 34–44. [CrossRef] [PubMed]

52. Kloepper, J.W.; Lifshitz, R.; Zablotowich, R.K. Free-living bacterial inocula for enhancing crop productivity. *Trends Biotechnol.* **1989**, *7*, 39–43. [CrossRef]

53. Bona, E.; Cantamessa, S.; Massa, N.; Manassero, P.; Marsano, F.; Copetta, A.; Lingua, G.; D'Agostino, G.; Gamalero, E.; Berta, G. Arbuscular mycorrhizal fungi and plant growth-promoting pseudomonads improve yield, quality and nutritional value of tomato: A field study. *Mycorrhiza* **2017**, *27*, 1–11. [CrossRef] [PubMed]

54. Mondani, F.; Khani, K.; Honarmand, S.L.; Saeidi, M. Evaluating effects of plant growth-promoting rhizobacteria on the radiation use efficiency and yield of soybean (*Glycine max*) under water deficit stress condition. *Agric. Water Manag.* **2019**, *213*, 707–713. [CrossRef]

55. Vaikuntapu, P.R.; Dutta, S.; Samudrala, R.B.; Rao, V.R.V.N.; Kalam, S.; Podile, A.R. Preferential promotion of *Lycopersicon esculentum* (tomato) growth by plant growth promoting bacteria associated with tomato. *Indian J. Microbiol.* **2014**, *54*, 403–412. [CrossRef]

56. Zhou, D.; Huang, X.F.; Chaparro, J.M.; Badri, D.V.; Manter, D.K.; Vivanco, J.M.; Guo, J. Root and bacterial secretions regulate the interaction between plants and PGPR leading to distinct plant growth promotion effects. *Plant Soil* **2016**, *401*, 259–272. [CrossRef]

57. Kalam, S.; Basu, A.; Podile, A.R. Functional and molecular characterization of plant growth promoting *Bacillus* isolates from tomato rhizosphere. *Heliyon* **2020**, *6*, e04734. [CrossRef]

 *horticulturae*

Article

Molecular Characterization of Tomato (*Solanum lycopersicum* L.) Accessions under Drought Stress

Ibrahim Makhadmeh [1,*], Ammar A. Albalasmeh [2], Mohammed Ali [3], Samar G. Thabet [4], Walaa Ali Darabseh [1], Saied Jaradat [5] and Ahmad M. Alqudah [6,*]

1 Department of Plant Production, Jordan University of Science and Technology, Irbid 22110, Jordan; walaa2017ww@yahoo.com
2 Department of Natural Resources and Environment, Faculty of Agriculture, Jordan University of Science and Technology, Irbid 22110, Jordan; aalbalasmeh@just.edu.jo
3 Egyptian Deserts Gene Bank, Desert Research Center, Department of Genetic Resources, Cairo 11753, Egypt; mohammedalidrc@gmail.com
4 Department of Botany, Faculty of Science, University of Fayoum, Fayoum 63514, Egypt; sgs03@fayoum.edu.eg
5 Princess Haya Biotechnology Center, Jordan University of Science and Technology, Irbid 22110, Jordan; sjaradat@just.edu.jo
6 Department of Agroecology, Aarhus University Flakkebjerg, 4200 Slagelse, Denmark
* Correspondence: ibrahimm@just.edu.jo (I.M.); ama@agro.au.dk or ahqudah@gmail.com (A.M.A.)

Citation: Makhadmeh, I.; Albalasmeh, A.A.; Ali, M.; Thabet, S.G.; Darabseh, W.A.; Jaradat, S.; Alqudah, A.M. Molecular Characterization of Tomato (*Solanum lycopersicum* L.) Accessions under Drought Stress. *Horticulturae* **2022**, *8*, 600. https://doi.org/10.3390/horticulturae8070600

Academic Editors: Stefania Toscano, Giulia Franzoni, Sara Álvarez and Aušra Brazaitytė

Received: 23 May 2022
Accepted: 29 June 2022
Published: 4 July 2022

Publisher's Note: MDPI stays neutral with regard to jurisdictional claims in published maps and institutional affiliations.

Abstract: Exploring the genetic diversity among plant accessions is important for conserving and managing plant genetic resources. In the current study, a collection of forty-six tomato accessions from Jordan were evaluated based on their performance and their morpho-physiological, in addition to molecularly characterizing to detect genetic diversity. Tomato accessions seedlings were exposed to drought stress with 70% field capacity and 40% field capacity under field conditions in Jordan. Drought stress had significantly negatively influenced the dry root weight, fresh root weight, root growth rate, fresh shoot weight, dry shoot weight, and shoot growth rate. Moreover, proline content showed a highly significant increase of 304.2% in response to drought stress. The analysis of twenty morphological characters revealed a wide range of variations among tomato accessions. Accessions were screened with fourteen SSR primers; six primers were informative to explain the genetic diversity. Based on resolving power, primers LEct004 and LEat018 were most significant with all 46 accessions. Interestingly, polymorphic information content (PIC) values ranged from 0.00 (Asr2 marker) to 0.499 (LEct004), which confirms that the SSR markers are highly informative. Our findings provide new insights into using informative molecular markers to elucidate such wide genetic variation discovered in our collections from Afraa and Abeel (the southern part of Jordan). Interestingly, the SSR markers were associated with genes, e.g., LEat018 with *ACTIN_RELATED PROTEIN* gene, the LEct004 with the *HOMEOBOX PROTEIN TRANSCRIPTION FACTORS* gene, and Asr2 with ABA/WDS. Moreover, the *AUXIN RESPONSE FACTOR8* gene was associated with the LEta014 SSR marker and the LEta020 with the *THIOREDOXIN FAMILY TRP26* gene. Therefore, the genetic diversity analysis and functional annotations of the genes associated with SSR information obtained in this study provide valuable information about the most suitable genotype that can be implemented in plant breeding programs and future molecular analysis. Furthermore, evaluating the performance of the collection under different water regimes is essential to produce new tomato varieties coping with drought stress conditions.

Keywords: tomato; SSRs; breeding; gene-associated SSRs; genetic diversity; drought stress

1. Introduction

Tomato (*Solanum lycopersicum* L.) is an annual herbaceous plant that belongs to the Solanaceae Family [1–3]. Tomato is the second most economically important vegetable grown worldwide, it forms a significant part of the agricultural industry and is also the second most consumed vegetable. It is known that tomato production is increased considerably worldwide [2–5]. In Jordan, a wide range of tomato cultivars and accessions were gathered from farmers and stored in the seed bank at the National Agricultural Research Center (NARC) [6–8]. In Jordan, the production of tomatoes increases year by year with high ability of export. Accession tomatoes are highly adapted to the Jordanian conditions with large hereditary types used in breeding for tomato productivity and adaptation improvement. Therefore, they have been used for some time in breeding to make a character of adaptation to abiotic stresses [6,7]. Accessions are an excellent source for improving the tomato crop for drought and have existed in Jordan for several years. The collections of accession frequently exist in the thousands, which can usually be subject to pre-screening under different stress conditions to identify the most promising genotype [9]. Tomato is known to be sensitive to environmental stresses, including drought stress which impacts seed germination and plant development and performance [10]. Cattivelli et al. [11] reported the effect of drought on stages of plant development to understand the drought tolerance from germination to reproduction through conventional and molecular approaches. Drought stress during the vegetative or early reproductive phase usually reduces yield. It often induces physiological and molecular changes in plant water relation parameters such as cell turgor pressure, osmotic pressure, and water potential [12,13]. Drought stress at different developmental stages causes several morphological and biochemical alternations in various plant species. Water deficit at the early seedling stage might lead to higher dry root weights, longer roots, coleoptiles, and higher root-to-shoot ratios [14–16], all of these changes are parameters of interest and have been widely used as reliable markers toward drought stress tolerance for evaluating various crop plants. Reduction in water potential induces stomatal closure resulting in a decrease in photosynthesis, leaf expansion and orientation, stomatal behavior, photosynthesis, respiration rate, solute translocation, and ultimately yields [13,17]. Toxic substances generated during stress, such as reactive oxygen species (ROS), cause oxidative damage to the cellular organization. Tomato plants have developed an antioxidant system that scavenges toxic elements and accumulates osmoprotectants, including proline, glycine betaine, and other osmoprotectants to keep osmotic balance [13,17]. Drought stress tolerance was evaluated according to different tolerance indices to characterize tomatoes' physiological and genetic basis, including plant development, fruit set, fruit weight, shoot and root morphology, water use efficiency (WUE), and other physiological parameters [18]. Researchers have evaluated drought-tolerant tomato breeding cultivars in response to drought conditions drought stress [18]. Pakmore VF and the breeding line L03306 showed better performance in several deficit irrigation regimes. These genotypes are considered a resource for the drought tolerance breeding program.

In recent years, crop physiology and agronomics have led to new insights into drought tolerance. These insights have provided the breeders with new knowledge and tools for plant improvement and the ability to detect the variation between species, varieties, and accessions; for example, using several types of DNA molecular markers can be used such as random amplified polymorphic DNA (RAPD), restriction fragment length polymorphism (RFLP), amplified fragment length polymorphism (AFLP), minisatellites, variable number of tandem repeats (VNTRs), and simple sequence repeats (SSRs) [11,19–21]. SSRs are sections of DNA consisting of (1–6) or (2–7) base pair units tandemly repeated throughout the genome [20,22–24], to detect the variation degree [9,21,24] or phylogenetic [25–27]. The hypervariability character of SSR markers, based on microsatellite DNA loci, enables this method to be the main major in studying plant population genetics [8,20,28–31]. SSRs can be used in gene mapping studies, for example, using 65 SSR primer pairs by Liu et al. [32] in the cotton genome mapping [33]. Additionally, Shiri [34] used 38 maize hybrids and 12 SSR pairs to investigate the genetic diversity and then identify the informative SSRs for drought tolerance. The amplified bands were 40, the number of alleles ranged from 2 to 6, and the Polymorphism Information Content (PIC) ranged from 0.23 to 0.79. Tam et al. [35] detected the genetic diversity within the 34 lines of tomato and 35 lines of pepper using 29 SSR primers (16 for tomato and 13 for pepper). Our study noticed that the genetic variations within the tomato and pepper collections were similar because all the bands for both collections were polymorphic and the polymorphism percentages were equal (100%).

The specific goals of this research were to evaluate the genetic diversity of the tomato accessions grown in Jordan and to determine (1) the extent of morphological variations among tomato accessions, (2) the allele distribution in relation to the gene pool origins and probable drought tolerance based on geographic origin, (3) evaluate the effect of drought stress on tomato accessions at the seedling stage, (4) predict the function of candidate genes that are associated with our SSR primers in tomato, (5) determine the expression pattern of our target genes in plant tissues-specific, and (6) by combining phenotypes, SSR marker genotypes, and putative expression pattern, we can understand the functional roles of our genes which are related to the local adaptation to drought stress.

2. Materials and Methods

2.1. Plant Materials and Effect of Drought on Tomato Accessions at Germination Stage

Forty-six tomato accessions originating from different geographical regions in Jordan were ordered from the Genebank of the National Center for Agricultural Research and Extension (NCARE), Amman, Jordan, and were used in this study (Table 1). The seeds were grown in growth chambers at the plant production department laboratories as well as glasshouse of Jordan University of Science and Technology (JUST).At seedling stage, the leaf sampleswere collected for DNA analysis at the Princess Haya Biotechnology Center (PHBC) at the King Abdullah University Hospital (KAUH) and the University of Jordan.

Table 1. Tomato accession collection regions used in this study.

No.	Accession No.	Region of Location	No.	Accession No.	Region of Location
1	JOR111	Kharja/Irbid	24	JOR972	Rhaba/Irbid
2	JOR112	Kharja/Irbid	25	JOR973	Rhaba/Irbid
3	JOR950	Kharja/Irbid	26	JOR974	Rhaba/Irbid
4	JOR951	Kharja/Irbid	27	JOR975	Rhaba/Irbid
5	JOR952	Al-al/Irbid	28	JOR976	Rhaba/Irbid
6	JOR953	Al-al/Irbid	29	JOR977	Rhaba/Irbid
7	JOR954	Al-al/Irbid	30	JOR978	Rhaba/Irbid
8	JOR955	Kharja/Irbid	31	JOR979	Rhaba/Irbid
9	JOR956	Qasfa	32	JOR980	Rhaba/Irbid
10	JOR957	Hebras/Irbid	33	JOR981	Rhaba/Irbid
11	JOR958	Hebras/Irbid	34	JOR982	Afra/Tafileh
12	JOR959	Sakib/Jarash	35	JOR984	Afra/Tafileh
13	JOR960	Anjara/Ajloun	36	JOR985	Abel/Tafileh
14	JOR961	Shtafina/Ajloun	37	JOR986	Abel/Tafileh
15	JOR962	AAfna-Ain Jannah/Ajloun	38	JOR987	Abel//Tafileh
16	JOR963	Afna-Ain Jannah/Ajloun	39	JOR988	Ain Al-Baida/TafilehTafileh
17	JOR964	Rhaba/Irbid	40	JOR989	Ain Al-Baida/Tafileh
18	JOR965	Rhaba/Irbid	41	JOR990	Ain Al-Baida/Tafileh
19	JOR966	Rhaba/Irbid	42	JOR991	Ain Al-Baida/Tafileh
20	JOR967	Rhaba/Irbid	43	JOR992	Ain Al-Baida/Tafileh
21	JOR968	Rhaba/Irbid	44	JOR993	Ain Al-Baida/Tafileh
22	JOR970	Rhaba/Irbid	45	JOR994	Ain Al-Baida/Tafileh
23	JOR971	Rhaba/Irbid	46	JOR995	Rashadeyeh/Tafileh

2.2. Field Trial

Ten seeds of each tomato accession were sown at 8 cm depth (in 8 L pots containing a mixture of soil: sand: peat moss in a volume ratio of 2:1:1) under greenhouse conditions. Field capacity (FC) was determined by saturating the soil with water and recording the weight of the soil after drainage had stopped. Soil moisture content was measured gravimetrically by weighting soil samples before and after oven-drying at 105 °C for 24 h divided by the weight of the dry soil. Di-ammonium phosphate (DAP) fertilizer was added to the pots, the seeds were sown, and then the pots were covered with plastic to reduce evaporation during development. When plumule started to emerge, small holes were made carefully in the covers to enable the plants to grow, and 50 mL NPK (30:10:10) was added per pot (60 g per 20 L) to avoid the appearance of mineral deficiency in plants. Plants were exposed to drought stress at the beginning of the early seedling stage with 70% field capacity and 40% field capacity under field conditions, and three randomly selected accessions were used as a reference. Before irrigation, three reference pots were weighed and watered to adjust the corresponding FC. The experiment was carried out using a completely randomized design with three replications (ten seeds per replicate).

2.3. Vegetative Traits

After 60 days of growing, tomato plants were harvested. Three plants from each replicate of each treatment were randomly selected to be used for further analysis. Shoots and roots were separated manually to measure the fresh and dry weights after drying them at (65 °C for 72 h) in the oven. The relative shoot or root weight was calculated as follows:

Relative of shoot or root weight = [shoot or root weight at drought treatment/shoot or root weight at control] ×100.

The growth rate of shoot or root = shoot or root fresh weight/45 days.

Relative shoot or root growth rate = [shoot or root growth rate at drought treatment/shoot or root fresh growth rate at control] ×100.

2.4. Determination of Proline Content

The free content of proline was estimated and extracted according to the protocol of Bates et al. [36] _ENREF_32, and more measurement steps are discussed in the protocol by Senthilkumar et al. [37].

2.5. Morphological Characteristics

The following 20 morphological characters related to the tomato accessions (Table S4) during the seedling, immature, mature, and ripening stages were measured by a Tomato descriptor (IPGRI, 1999). Additionally, fruit shape-related perimeter traits and fruit shape index for the external shape and other characters were measured by a Tomato Analyzer (TA) software program version 3 (Rodríguez et al., 2010).

2.6. Statistical Analysis

2.6.1. Drought Data Analysis

The phenotypic data were statistically analyzed by the SPSS software (version 17). Analysis of variance (ANOVA) and means separation at LSD (0.05) in addition to T-Test were also calculated using SPSS software to compare the treatments.

2.6.2. Morphological Data Analysis

Data were statistically analyzed by the SPSS software (version 17). Means, range, maximum value, minimum value, standard deviation, standard error of the mean, and sum were measured. Hierarchical cluster analysis was used to calculate morphological similarity values between accessions using the Euclidean distance interval option and then classified them by dendrogram using average linkage (within groups).

2.7. DNA Extraction and Simple Sequence Repeat (SSR) Assays

Total genomic DNA was extracted from young leaves for five different plants per line using a DNA Plant Kit (Qiagen). DNA quantification was performed with an ND-1000 spectrophotometer (Nanodrop Technologies, Wilmington, DE, USA). The DNA quality was assessed using the absorbance ratio at 260 to that at 280 nm wavelengths (A260/A280). DNA quantity was calculated as DNA (μg/μL) = A260 $\times$ 50, where A260 is the absorbance at 260 nm. Thus, the concentration of DNA in μg/mL was calculated as DNA (μg/mL) = [A260 $\times$ 50] $\times$ DF where DF is the dilution factor. Fourteen SSR primers designed for tomato DNA fingerprinting were used. These SSR primer sequences were obtained from [38]. Six primers were selected for the next analysis to determine genetic diversity in tomato collection (LEat018, LEct004, LEta014, LEta020, CT114, and Asr2) based on the screening of fourteen SSR primers. Information about these primer sequences and their information are presented in Table 2.

Table 2. Primers and their sequence, expected fragment, and melting temperature.

SSR Name	Primer Sequence (5′~3′)	Expected Fragment Size (bp)	Melting Temperature Tm (°C)
LEat018	F: CGG CGT ATT CAA ACT CTT GG R: GCG GAC CTT TGT TTT GGT GA	120	46.7
LEct 004	F: AGC CAC CCA TCA CAA AGA TT R: GTC GCA CTA TCG GTC ACG TA	354	44.6
LEta 014	F: ACA AAC TCA AGA TAA GTA AGA GC R: GTG AAT TGT GTT TTA ACA	120	44.8
LEta020	F: AAC GGT GGA AAC TAT TGAAAG G R: CAC CAC CAA ACC CAT CGT C	175	46
CT114	F: ATA TTG CTT AGG CGT CAT CCA R: TTG AAA CCA GCC GTT GC	1125	58
Asr2	F: AGA GAA GCA ATA CAA TAGGC T R: TAT TAG ACA AAA CAT AGAGTC C	520	52

2.8. PCR Amplification and Product Electrophoresis

PCR amplification was performed for all the SSR markers and the best performing conditions were identified. During the primer testing, a fraction of the total number of plants was used for the polymerase chain reaction. PCR reactions were performed in 96-well plates using either the Perkin Elmer GeneAmp PCR system 9600 (PE Biosystems) or the TECHNE Genius thermal cycler (Techne Ltd., Cambridge, UK) with the same amplification program. Six SSR primer pairs were used for amplification reaction (Table 2). The DNA from the 46 tomato samples was amplified using SSR markers following the PCR amplification protocol by Promega (Madison, WI, USA). The PCR amplification conditions were set up for one cycle of denaturation for 2 min at 94 °C, followed by 33-cycle amplification with a 25 s denaturing at 94 °C, a 25 s annealing at the Tm (Tm varies for the individual primers), a 25 s extension at 68 °C, and a final extension cycle at 68 °C for 5 min.

The PCR products of SSRs were separated using 3% Metaphor Agarose gel (FM-CBioProducts) that was recommended to separate small-sized bands from SSRs [39] and electrophoretic apparatus (MS Major Science, UK) and BIO-RAD (Criterion TM cassettes) 100 V for 1 h was applied. DNA loading dye of 1 X was added to the PCR products for visual capture of DNA migration during electrophoresis. Five μL of 1000 bp and 100 bp DNA ladder were used as a reference to estimate the size of specific DNA bands. Finally, Gel Works ID advanced software analyzed the amplified DNA banding patterns. The size of the allele fragments that SSR amplified was measured. Polymorphism information content (PIC) was calculated as PIC = 1 − Spi2, where pi is the allele frequency [40].

2.9. Data Scoring

The gel for each primer was analyzed separately by scoring the bands and coded by (0) and (1) for the absent and present amplification bands for all test markers, respectively, using SAGA 6 software). Genetic similarity values between accessions were calculated using the Dice coefficient according to Dice [41]:

$$2 \times |X \cap Y|/(|X| + |Y|)$$

From the NTSYS-pc, version 2.0 software [42]. The genetic similarity matrix was used to generate a dendrogram using the Unweighted Pair Group Method of Arithmetic Averages (UPGMA).

2.10. Functional Assignments for Gene-Associated SSRs in Tomato

The sequence of SSR markers was used as a query to search against the *Solanum lycopersicum* genomics that we already downloaded from NCBI genomics (https://www.ncbi.nlm.nih.gov/genome/?term=tomato, accessed on 12 April 2022). Then, the alignment sequence was compared using various databases, such as National Center for Biotechnology Information (NCBI) gene bank, Phytozome, InterPro, and KEGG databases to predict the candidate genes associated with our SSR primers in tomatoes. For the potential functions of these genes, Phytozome v13 was used to obtain the annotations by KOG (Eukaryotic Orthologous Groups), KEGG (Kyoto Encyclopedia of Genes and Genomes), ENZYME, Pathway, and the InterPro family of protein analysis (Classification of protein families) tools. In Phytozome, we made the blast sequence against five tomato genomics such as *Solanum lycopersicum* ITAG2.4, *Solanum lycopersicum* ITAG3.2, *Solanum lycopersicum* ITAG4.0, *Solanum tuberosum* v4.03, and *Solanum tuberosum* v6.1.

2.11. Putative Tissue Expression Pattern, Subcellular Localization, Root Cell Types and Tissues of Our Target Genes

Putative tissue-specific expression profiles of *Solyc11g005330*, *Solyc02g089940*, *Solyc03g031970*, *Solyc01g097450*, *Solyc10g011690*, and *Solyc04g071580* genes that are associated with our SSR markers were extracted based on *Solanum lycopersicum* transcript expression database from nineteen tissues and organs including flowers, leaves, roots, and fruit from different developmental stages. Expression profiles were built using the tomato plant Electronic Fluorescent Pictograph Browsers (Tomato eFP browsers) (http://bar.utoronto.ca/eplant_tomato/) accessed on 12 April 2022 [43]. Moreover, the putative subcellular localizations of our previous gene from *Solanum lycopersicum* were examined based on tomato protein localization of fourteen different cell organs to recognize possible synthesis sites using the tomato Cell eFP browsers (Tomato eFP browsers) (http://bar.utoronto.ca/eplant_tomato/) accessed on 12 April 2022. Furthermore, the putative root cell types and tissues specific to our genes were examined using different root cell types and tissues under various promoter toolboxes, such as AtWER, SlPEP, AtPEP, SlCO2, SlSCR, SlSHR, AtS32, AtS18, SlWOX5, SlRPL11C, and 35S promoters to determine the putative function of our genes at specific root cell types http://bar.utoronto.ca/eplant_tomato/ accessed on 12 April 2022.

3. Results

3.1. The Effects of Drought on Tomato Accessions at the Seedling Stage

Several parameters such as fresh root weight, dry root weight, root growth rate, fresh shoot weight, dry shoot weight, and shoot growth rate were measured on the 46 tomato accessions. Seedlings were grown under three levels of water stress (control, 70% FC, 40% FC). However, all the parameters showed significant differences at 40% FC compared to the control treatment, which was also significant at 70%FC (Table S1). As expected, all morphological traits had a lower mean performance under drought stress (70% FC and

40% FC) than under normal conditions. On average, all parameters had a reduction due to drought stress at 70% FC and 40% FC compared with the control treatment (Table S3).

Drought stress is significantly affected by the proline concentrations in the leaf tissue for 46 tomato accessions (Table S1). A highly significant increment in proline content was detected at 40% FC compared to the control treatment (100% FC) by 304.2%.

3.2. Morphological Characterization among Tomato Accessions

In this study, Tomato Analyzer (TA) was used to assess the fruit shape variation in the accessions by measuring the morphological characterizations rapidly and accurately and quantifying traits that are impossible to quantify manually. The analysis of variance for all the morphological characters indicates a wide range of variability among tomato accessions, including fruit shape-related perimeter traits and fruit shape index for the external shape fruit shape index internal (Table S4).

3.3. Genetic Variation among Tomato Accessions Revealed by SSRs

Our investigation tested 14 SSR primers; of these, six yielding polymorphic amplification products were used, and the remaining 8 SSR primers either yielded no amplification product or no polymorphic. The banding patterns of SSRs are shown in Supplementary Figures S1–S6. Of the 46 tomato accessions, the genetic relationship among thirty-six tomato accessions was analyzed using six SSR primer pairs. Two hundred and forty-seven amplified bands were produced for 12 loci; of them, (11) loci were polymorphic and (1) loci were monomorphic, shown in Table 3, indicating that there is high allelic variation. The molecular weights ranged from 128 to 1170 bp. The number of alleles per locus varied from 1 for (CT114 and Asr2) markers to 3 for (LEat018 and LEat020) SSR markers. The percentage of polymorphic was 91.67% with a range between (zero to 0.49) (Table 4). The highest values of the effective number of alleles (Ne*) [44], the gene diversity (h*) [45], and the Shannon Index (I*) [46] were recorded for the LEct004 primer (350 bp loci) with values of 1.99, 0.4996, and 0.6927, respectively. While the lowest value (zero) of the effective number of alleles (Ne*), the gene diversity (h*), and the Shannon Index (I*) was shown by Asr2 primer (536 bp loci) (Tables 3 and 4).

Table 3. SSR names, the total number of bands/primers, loci, monomorphic, polymorphic loci, and percentage of polymorphism.

SSR Marker	Locus	Sample Size	Ne*	H*	Average H*	PIC	I*
LEat018	139	36	1.52	0.34	0.34	0.34	0.52
	132	36	1.26	0.21		0.20	0.36
	128	36	1.95	0.49		0.48	0.68
LEct004	364	34	1.60	0.38	0.43	0.37	0.56
	350	34	1.99	0.49		0.49	0.69
Lea014	175	36	1.38	0.27	0.27	0.28	0.45
	190	36	1.38	0.27		0.28	0.45
LEta020	223	36	1.75	0.43	0.36	0.42	0.62
	217	36	1.65	0.39		0.39	0.58
	206	36	1.34	0.25		0.25	0.42
CT114	1170	36	1.69	0.41		0.41	0.60
Asr2	536	36	1.00	0.00		0.00	0.00
Mean		36	1.55	0.33		0.33	0.49

Ne* = Effective number of alleles, H* = gene diversity, I* = Shannon's information index. The number of polymorphic loci is 11 and the percentage of polymorphic loci is 91.67.

Table 4. Diversity parameters of tomato accessions obtained from the analysis of SSR alleles.

SSR Name	Total Bands/Primers	No. of Loci	Monomorphic Loci	Polymorphic Loci	Polymorphism %
LEat018	46	3	0	3	100
LEct004	40	2	0	2	100
Lea014	46	2	0	2	100
LEta020	46	3	0	3	100
CT114	33	1	0	1	100
Asr2	36	1	1	0	0
Total	247	12	1	11	

3.4. UPGMA Dendrogram and Similarity

Genetic variation among tomatoes was evaluated based on bands obtained from SSR profiling using Nie genetic distance and Unweighted Pair Group with Arithmetic Averages (UPGMA). The coefficient of genetic similarity ranged from 0.30 between JOR956 and JOR966 accessions to 1 between JOR950 and JOR951, JOR964 and JOR965, JOR955 and JOR979, JOR955 and JOR980, JOR979 and JOR980, JOR955 and JOR988, JOR979 and JOR988, and JOR980 and JOR988. The most similar tomato accessions reported above are from Kharja, Rhaba, and Ain Al-Biada, whereas the most different are from Qasfa and Rhaba shown in Table 1.

Moreover, similarity values for all tomato accessions using the UPGMA dendrogram are shown in Figure 1. At a genetic similarity value of 0.62, the dendrogram is divided into two groups except for 25 (JOR978) accessions collected from Rhaba. The first group consists of three sub-groups: 1 (JOR111), 5 (JOR953), and 30 (JOR984) accessions with 80% similarity, 6 (JOR954), 7 (JOR955), 26 (JOR979), 27 (JOR980), 34 (JOR988), 22 (JOR972), 19 (JOR968), 28 (JOR981), 18 (JOR967), 23 (JOR973), 20 (JOR970), 33 (JOR987), 35 (JOR989), and 36 (JOR990) accessions with 76% similarity, and 17 (JOR966) and 21 (JOR971) accessions with 88% similarity. The tomato accessions reported in the first group were from kharja, Al-al, Rhaba, Afra, Abel, and Ain Al-Baida. The second group consists of three sub-groups: 2 (JOR950), 3 (JOR951), 8 (JOR956), and 12 (JOR961) accessions with 76.5% similarity, 4 (JOR952), 11 (JOR959), 10 (JOR958), 15 (JOR964), 16 (JOR965), 9 (JOR957), 13 (JOR962), and 32 (JOR986) accessions with 85% similarity, and 14 (JOR963), 29 (JOR982), and 31 (JOR985) accessions with 77% similarity. The tomato accessions reported in the second group were from kharja, Qasfa, Shtafina, Al-al, Sakib, Hebras, Ain-Jannah, Abel, Afra, and Rhaba. The number of alleles and the PIC value for each SSR marker are presented in Table 4.

Polymorphic information content (PIC) values ranged from 0.00 to 0.499 (mean 0.33), confirming that the SSR markers are highly informative. The SSR marker had the highest PIC value (LEct004), followed by LEat018 (0.488), while the (Asr2) marker had the lowest PIC value. In this study, we found no relationship between the number of nucleotides per repeat and PIC shown in Table 4. For example, LEat018 with the lower PIC (0.34) has 29 repeats compared to LEta020, which has PIC (0.36) with 11 repeats.

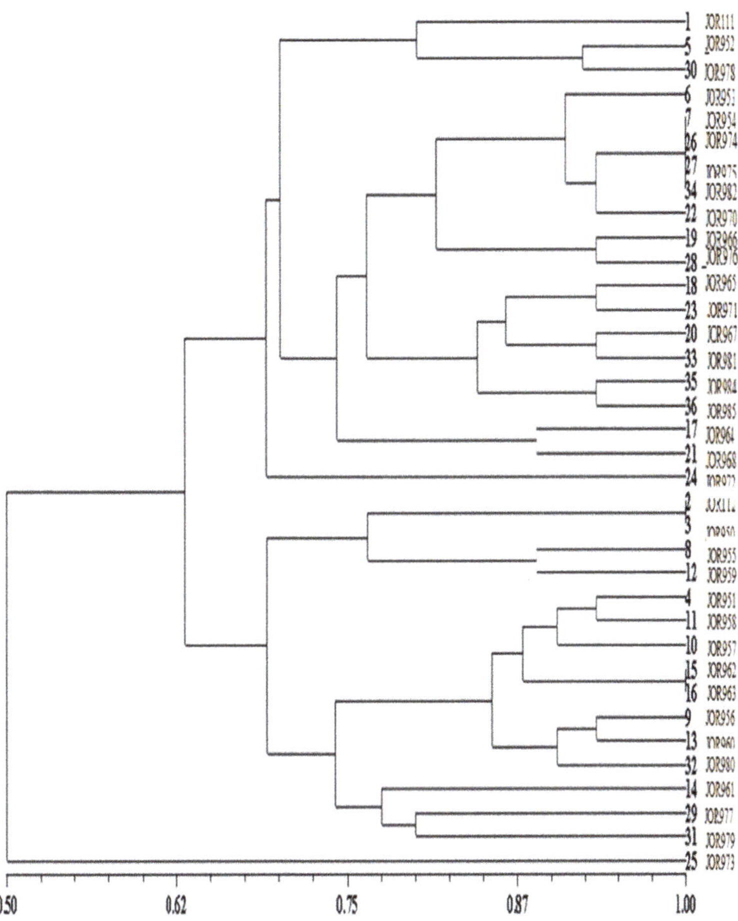

Figure 1. Dendrogram of tomato accessions generated by UPGMA cluster analysis of the dissimilarity values based on [45] coefficient.

3.5. The Functional Analysis of the Associated Genes with SSRs in Tomato

S. lycopersicum genomic sequence was used as a template for searching the sequence of SSR primers to predict the potential functions of our six genes. Then, databases such as Phytozome, NCBI, InterPro, and KEGG predicted more function annotations for these genes. In context, these genes were related to a wide range of functions, indicating that these gene-associated SSRs were potentially associated with essential biological functions, such as the LEat018 SSR marker associated with ACTIN_RELATED PROTEIN gene (*Solyc11g005330*) (Table 5). Additionally, the LEct004 SSR marker was associated with the *HOMEOBOX PROTEIN TRANSCRIPTION FACTORS* gene (*Solyc02g089940*). Moreover, the LEta014 SSR marker had an associated gene annotated as *AUXIN RESPONSE FACTOR 8* gene (*Solyc03g031970*), and the LEta020 SSR marker was found to be associated with the gene *THIOREDOXIN FAMILY TRP26* gene (*Solyc01g097450*). Furthermore, the CT114 marker was associated with the protein suppressor of the *PHYA-105 1* (*SPA1*) gene (*Solyc10g011690*). In addition, the ABA/WDS-induced protein (ABA_WDS) gene (*Solyc04g071580*) was associated with the Asr2 SSR marker (Table 5).

Table 5. BLAST corresponding Solyc (*Solanum lycopersicum*) gene sequences and annotation for SSR marker sequences.

SSR Name	Solyc Gene	Annotation Result
LEat018	*Solyc11g005330*	ACTIN_RELATED PROTEIN contains Interpro domain IPR004000 (actin family domain).
LEct004	*Solyc02g089940*	HOMEOBOX PROTEIN TRANSCRIPTION FACTORS contain Interpro domain(s) IPR009057(Homeodomain-like), IPR006563 (POX domain), IPR008422 (homeobox KN domain), IPR016039 (thiolase-like), and IPR001356 (homeobox domain).
LEta014	*Solyc03g031970*	AUXIN RESPONSE FACTOR 8 contains Interpro domain(s) IPR003340 (B3 DNA binding domain), IPR010525 (auxin response factor), IPR003311 (AUX/IAA protein), and IPR015300 (DNA-binding pseudobarrel domain).
LEta020	*Solyc01g097450*	THIOREDOXIN FAMILY TRP26 contains Interpro domain(s) IPR008979 (galactose-binding domain-like) and IPR010400 (PITH domain).
CT114	*Solyc10g011690*	Protein suppressor of PHYA-105 1 (SPA1) contains Interpro domain(s) IPR000719 (protein kinase domain), IPR017986 (WD40 repeat-containing domain), IPR002290 (serine/threonine/dual-specificity protein kinase, catalytic domain), IPR001680 (WD40 repeat), IPR015943 (WD40/YVTN repeat-like-containing domain), and IPR011009 (protein kinase-like domain).
Asr2	*Solyc04g071580*	ABA/WDS-induced protein (ABA_WDS) contains Interpro domain IPR003496 (ABA/WDS-induced protein).

3.6. Putative Tissue Expression Pattern of Genes S. lycopersicum Transcript Expression

The expression profile of the genes based on *S. lycopersicum* transcript expression was analyzed to understand their potential functions in different tissues (Figure 2). The results showed that the gene *Solyc11g005330* which is related to the ACTIN_RELATED PROTEIN was highly expressed in all tomato tissues, especially in Mature Green Fruit, Breaker Fruit, 3 cm Fruit, Root, and Breaker Fruit + 10 (Figure 2 and Table 5), while the highest expression levels for *Solyc02g089940* gene concerning HOMEOBOX PROTEIN TRANSCRIPTION FACTORS were observed in Fully Opened Flower, Leaves, Pimpinellifolium Leaf, Unopened Flower Bud, and Root (Figure 2 and Table 5). Additionally, the highest expression levels for the *Solyc03g031970* gene encoding to AUXIN RESPONSE FACTOR 8 were recorded at 1 cm Fruit, 3 cm Fruit, 2 cm Fruit, Unopened Flower Bud, Mature Green Fruit, and Leaves. The *Solyc01g097450* gene, which is related to THIOREDOXIN FAMILY TRP26, showed no clear expression level in any tomato tissue. On the other side, the highest expression levels for the *Solyc10g011690* gene concerning the protein suppressor of PHYA-105 1 (SPA1) were observed in Root, 3 cm Fruit, Pimpinellifolium Immature Green Fruit, Pimpinellifolium Breaker Fruit, Mature Green Fruit, and Breaker Fruit. In addition, the *Solyc04g071580* gene related to ABA/WDS-induced protein (ABA_WDS) was a high expression in Breaker Fruit + 10, Breaker Fruit, Mature Green Fruit, Pimpinellifolium Immature Green Fruit, 3 cm Fruit, and Root.

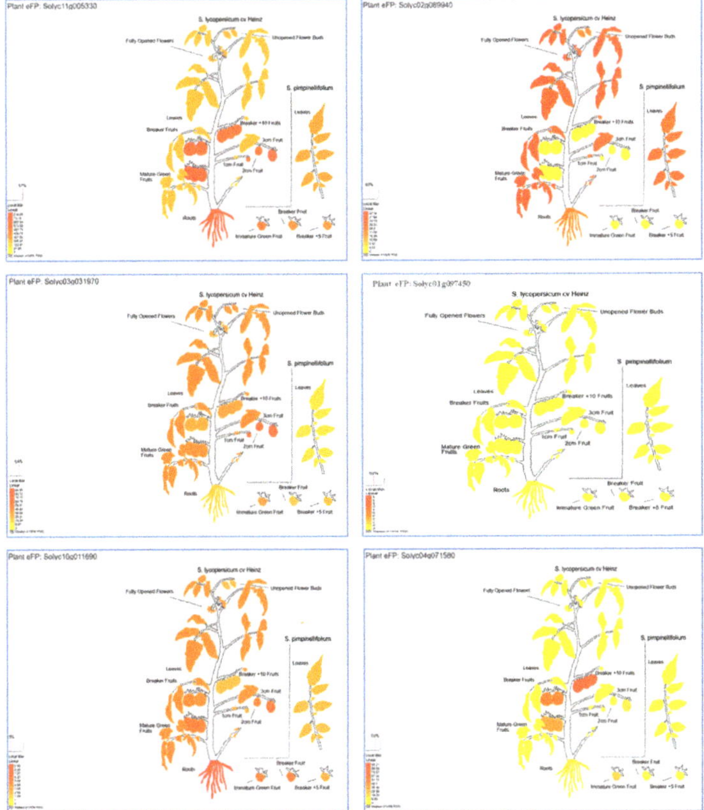

Figure 2. The putative "plant electronic fluorescent pictograph" tissue expression of *Solyc11g005330, Solyc02g089940, Solyc03g031970, Solyc01g097450, Solyc10g011690*, and *Solyc04g071580* genes of different tissues and developmental stages. The more intense the red color of the expression bar, the more gene expression detected [43].

3.7. Putative Subcellular Localizations of the Genes Based on S. lycopersicum Transcript Expression

Cell Electronic Fluorescent Pictograph tools were used to predict the putative subcellular localizations of the genes according to the protein localization of different cell organelles in the tomatoes. The subcellular localization profiles showed that *Solyc11g005330* and *Solyc01g097450* genes were highly expressed and presented in the cytosol (Figure 3), while the *Solyc02g089940*, *Solyc03g031970*, *Solyc10g011690*, and *Solyc04g071580* genes were highly expressed and presented in the nucleus (Figure 3).

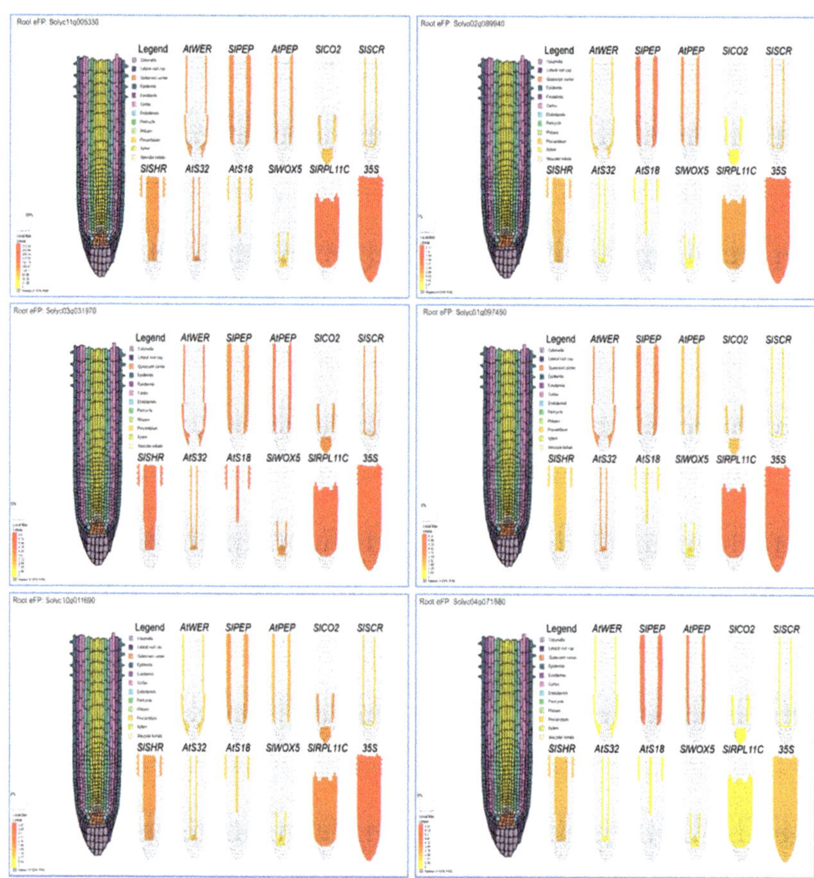

Figure 3. The putative expression of *Solyc11g005330*, *Solyc02g089940*, *Solyc03g031970*, *Solyc01g097450*, *Solyc10g011690*, and *Solyc04g071580* genes with different root cell types and/or tissues that related tovarious promoters. The more intense the red color of the expression bar, the more gene expression detected [43].

3.8. Putative Root Cell Types and Tissues Specific to the Genes Based on S. lycopersicum Transcript Expression

Root analysis using eplant_tomato tools observed the highest expression levels of the *Solyc11g005330* gene in all root cell types under 35Spro, followed by phloem under AtS32pro, endodermis, and cortex under SlPEPpro (Figure 4). Additionally, the *Solyc02g089940* gene was highly expressed in all root cell types under 35Spro, then endodermis and cortex under SlPEPpro, and cortex under AtPEPpro. Moreover, the highest expression levels of the *Solyc03g031970* gene were reported for all root cell types under

35Spro, followed by epidermis and procambium under AtS18. In addition, a highly expressed *Solyc01g097450* gene was observed for all root cell types under 35Spro, followed by endodermis and cortex under SlPEPpro, epidermis, and lateral root cap under AtWER. Furthermore, the highest expression levels of the *Solyc10g011690* gene were observed for all root cell types under 35Spro, columella, and cortex under SlCO2pro, then exodermis and cortex under SlPEP. At the same time, the *Solyc04g071580* gene was highly expressed in exodermis and cortex under SlPEP, the cortex under AtPEPpro (Figure 4).

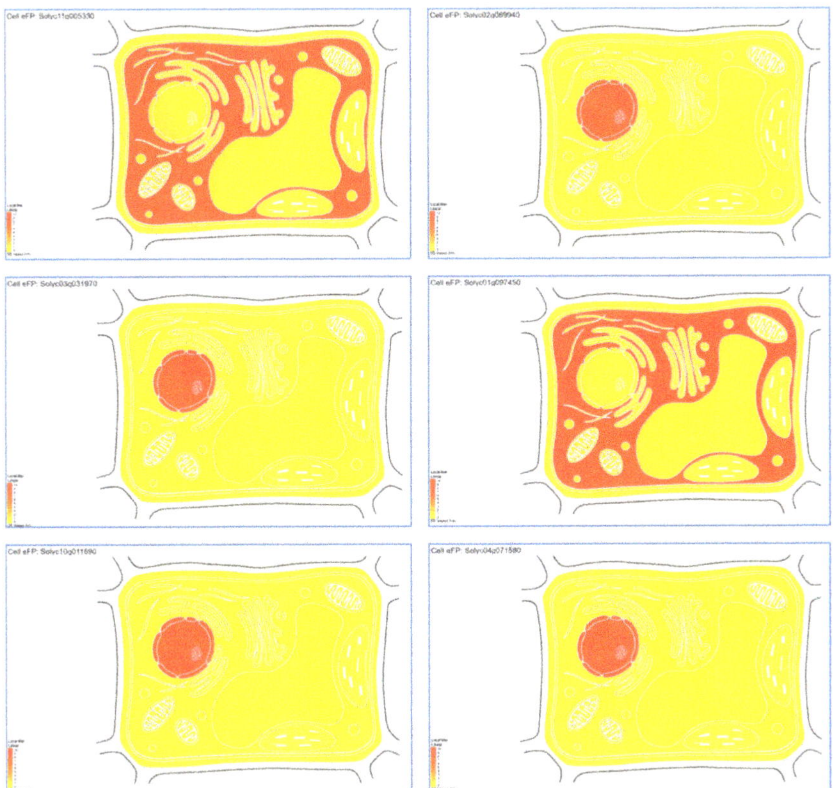

Figure 4. Putative subcellular localizations of the genes and/or' proteins for different cell organs using Cell eFP browsers in Tomato plant. The more intense the red color of the expression bar, the more gene expression detected [43].

4. Discussion

Natural variation in morphological parameters among genotypes is an essential analysis for understanding the genetic diversity that can be used to improve cultivars [47]. The current study revealed a wide range of diversity in tomatoes for most of the studied traits.

4.1. Drought Tolerance

Drought stress at different developmental stages causes various morpho-physiological changes in the plant. Water stress at the seedling stage might lead to higher dry root weights, longer roots, coleoptiles, and higher root/shoot ratios [14–16]. All of these changes are parameters of interest and have been widely used as reliable morph-physiological markers toward drought tolerance for various crop plants. Several parameters such as fresh root weight, dry root weight, root growth rate, fresh shoot weight, dry shoot weight, and shoot growth rate were measured for all tomato accessions. As expected, all morphological traits had a lower mean performance under drought stress (70% FC and 40% FC) than under normal conditions. This agreed with Vurayai et al. [48] who observed that the shoot: root ratio was significantly reduced by water stress imposed during the vegetative, flowering, and pod filling stages compared to the non-stressed control plant. Additionally, Yücel et al. [49] observed that the fresh root weight, fresh shoot weight, dry root weight, and dry shoot weight decreased by screening nine chickpea genotypes under water limited conditions. Drought stress is significantly affected by the proline concentrations in the leaf tissue for tomato accessions [8]. A highly significant increment in proline content was detected at 40%FC compared to the control treatment (100% FC). Differences between accessions and treatments showed a significantly positive relationship, which agreed with Vasquez-Robinet et al.'s study on the drought of potatoes [50]. The accumulation of osmolytes was also investigated during drought stress in durum wheat, e.g., proline was strongly upregulated by drought conditions that increased about twenty times in plants stressed at 12.5% SWC [51]. Understanding the physiological mechanisms in tomato accessions under drought stress conditions can help improve their performance and adaptation to harsh stress conditions, improving yield potential which is the ultimate target of crop breeding programs.

4.2. Morphological Characteristics

Several researchers were interested in the natural phenotypic variations which positively affect natural morphological diversity and genetic variations in a diverse population [3,27]. In this study, the analysis of variance for all the morphological characters showed a wide range of variation among tomato accessions, including fruit shape-related perimeter, fruit shape index, curved fruit shape index, and fruit shape index internal. These results are in line with Ranc et al. [52], which found a powerful link between the phenotypic variability present in the tomato germplasm and molecular polymorphisms using simple sequence repeat (SSR) markers. Additionally, Brdar-Jokanović et al. [53] studied the relationship between drought tolerance, growth type, and fruit size of different tomato accessions. They found that accessions with sizeable fruit sizes had comparatively higher water requirements.

4.3. Molecular Level

The genetic diversity analysis of crops is important for crop breeding. The selected six SSR loci that have been previously reported to be highly informative in distinguishing tomato genotypes [54,55] were used to detect the genetic diversities among the tomato accessions in our study. Additionally, Abd El-Hady et al. [56] stated that the SSR markers showed more polymorphic than RAPD markers. The average PIC value was higher than the result reported by Benor et al. [57] after testing 39 inbred lines of tomatoes using 35 polymorphic SSR loci (PIC = 0.31) and lower than the result reported by He et al. [54] after using 65 polymorphic SSR loci for testing 19 varieties of tomato (PIC = 0.37), indicating that SSR markers are of great utility for genetic diversity studies of tomatoes. This study revealed an exchange of genetic resources between farmers, particularly tomato accessions collected from Afra and Abel (southern part of Jordan). On the other hand, Rhaba accessions have special characteristics and agronomic traits such as sour taste and irregular shapes. Moreover, the results were similar to those reported by He et al. [54] _ENREF_57, but these results disagree with Smulders et al. [58], who found a positive relationship between the number of repeats and PIC in tomatoes. Taken together, we demonstrated the importance of these molecular markers and their allele detection and explained the vast genetic variation in collections from Afraa and Abeel (the southern part of Jordan).

Interestingly, our investigation detected that the plants' most frequent type of SSRs was the TA/AT. This observation was subsequently confirmed with additional studies such as those by Rajput et al. [59]. Additionally, comparing the area of tomato fruit for all accessions, the number of alleles at each locus and the heterozygosity found that accession numbers JOR111, JOR958, JOR981, JOR986, and JOR989 had the highest fruit area. In contrast, the accession numbers JOR973, JOR961, JOR950, JOR951, and JOR955 had the lowest fruit area, and these accessions had more than one allele at the most heterozygous loci (LEct018, LEta020). The highest frequency was detected for the CC genotype (at LEct018) and AA genotype (at LEta020). It has been demonstrated that the SSR markers are recommended to distinguish closely related genotypes because of their high degree of variability and, therefore, become favored in population studies [60–62]. Consequently, it assessed the genetic variability among tomato accessions using the SSR method to identify the most suitable genotype for future use in plant breeding programs.

4.4. Putative Tissue Expression Analyses of Our Target Gene-Associated SSR Markers

BAR database tools were used to generate expression pattern profiles of our candidate genes with different tissues, cell organs, and root cell types. In this context, a putative expression and recognized synthesis sites of these genes provide an excellent way to understand the epistatic relationship between our gene's synthesis site and a putative function. For example, the LEat018 SSR marker that associated with actin-related protein gene (ARP; *Solyc11g005330*), and this gene has the actin family domain (IPR004000). This domain is involved in the formation of filaments in the cytoskeletal system and plays important roles in various cellular functions in the cytoplasm and the nucleus [63]. Additionally, plants have many isoforms from actin protein which are probably involved in multiple functions such as graviperception, cell shape determination, tip growth, cytoplasmic streaming, cell wall deposition, etc. [64]. In addition, Nie et al. [65] reported that the *AtARP4* gene from *Arabidopsis thaliana* is vital for plant growth and is related to hormone response such as salicylic acid, while any mutation in this gene can cause altered transcription response in hundreds of genes that affect plant development and lead to early flowering. Moreover, from our results, we found this gene was highly expressed in different tissues, presented in the cytoplasm, and observed in all root cell types, phloem, endodermis, and cortex under 35Spro, AtS32pro, and SIPEPpro promoters which can drive expression in various root cell types and tissues throughout the root, including the elongation zone and meristematic zone (Figures 2–4). Previous studies reported that many actin-related protein (ARP) genes have distinct transcript expression patterns in different tissues (such as roots, seedlings, xylem precursor cells, pollen, flowers, leaves, and siliques) and cell organs such as cytoplasm and nucleus [66,67].

Furthermore, the LEct004 SSR marker was associated with the HOMEOBOX PROTEIN TRANSCRIPTION FACTORS gene (*Solyc02g089940*). This gene contains five domains, such as IPR009057 (homeodomain-like), PR006563 (POX domain), IPR008422 (homeobox KN domain), IPR016039 (thiolase-like), and IPR001356 (homeobox domain), that are reported to be key regulators for plant development and growth [68,69]. Additionally, many homeobox proteins were involved in transcriptional regulation and various metabolic pathways, such as *OsHOX22* and *OsHOX24* genes from rice which have a negative regulator role in abiotic stress response [70]. Moreover, this gene was found to be highly expressed in different tissues, presented in the nucleus, and observed in all root cell types, endodermis and cortex under 35Spro, SIPEPpro, and AtPEP promoters which can drive expression in various root cell types and tissues throughout the root, including the elongation zone and meristematic zone (Figures 2–4). Sakamoto et al. [71] reported that these domains could play a role in suppressing target gene expression through their function as a nuclear localization signal.

In addition, AUXIN RESPONSE FACTOR 8 gene (*Solyc03g031970*) was associated with the LEta014 SSR marker, which has four domains, such as IPR003340 (B3 DNA binding domain), IPR010525 (auxin response factor), IPR003311 (AUX/IAA protein), and IPR015300 (DNA-binding pseudo barrel domain). Recent evidence suggests that these previous domains are key regulators of auxin-modulated gene expression, such as regulating diverse cellular and developmental responses in plants, including cell expansion, division, differentiation, light responses, patterning of embryo responses, and embryonic and post-embryonic development in some plants [72,73]. Thus, we found that this gene was highly expressed in different tissues, presented in the nucleus, and observed in all root cell types and Epidermis and Procambium under 35Spro and AtS18 prompters (Figures 2–4). These results are in line with Kang et al. [74], who found that ARF was expressed and localized in the nucleus. Moreover, auxin response factors (ARFs) have played important roles in the process of plant growth and development as they increase the contents of carotenoids and enhance the tolerance to salt and drought in transgenic Arabidopsis [74]. Additionally, Bouzroud et al. [75] showed that many of ARFs genes were differentially expressed in tomato leaves and roots under salt, drought, and flooding stress conditions. Chen et al. [76] reported that the ARFs genes could play essential roles in various plant physiological processes by participating in ABA signaling pathways and regulating the expression of some genes such as SlABI5/ABF and SCL3, which influence stomatal morphology and vascular bundle development and ultimately improve tomato plant resistance to water deficit. Additionally, Salehin et al. [77] found the aliphatic Glucosinolate (GLSs) levels are regulated by the auxin-sensitive Aux/IAA repressors IAA5, IAA6, and IAA19, and any loss of these gene expressions results in reduced GLS levels and decreased drought tolerance in the Arabidopsis plant.

In addition, the LEta020 SSR marker was related to the THIOREDOXIN FAMILY TRP26 gene (Solyc01g097450), and this gene contains two domains IPR008979 (galactose-binding domain-like) and IPR010400 (PITH domain). Thioredoxin is a relatively small and very stable redox protein known to be present in many plants such as *arabidopsis thaliana*, *Brassica napus*, *Zea may*, *Oryza sativa*, *Nicotiana tabacum*, *Spinacia olerace*, and *Pisum sativum*. The higher plants have at least two types of thioredoxin, f-type, which is the only one related to activating fructose-1,6-bisphosphatase efficiently, and m-type, which can trigger the NADP-malate dehydrogenase [78]. In this context, the expression and activity of the Fructose-1,6 bisphosphatase gene are regulated in cytosolic by environmental factors such as light and drought conditions [79]. Moreover, thioredoxin plays a role in various critical biological processes, including anti-oxidative stress, cell cycle control, regulation of receptors/transcription factors, structural functions/protein folding, signal transduction (cell to cell), vacuolar inheritance, redox regulation of chloroplast enzymes, control of chloroplastic translation, the structure of the photosynthetic apparatus/folding, and other functions [80]. Additionally, from our results, we found that this gene did not show any clear expression level in any tomato tissue, while being highly expressed in the cytosol as well as observed in all root cell types, endodermis and cortex, epidermis and lateral root cap under 35Spro, SIPEPpro, and under AtWER promoters.

Additionally, the CT114 SSR marker was linked with the protein suppressor of the PHYA-105 1 (SPA1) gene (Solyc10g011690). This gene contains six domains such as IPR000719 (protein kinase domain), IPR017986 (WD40 repeat-containing domain), IPR002290 (serine/threonine/dual-specificity protein kinase, catalytic domain), IPR001680 (WD40 repeat), IPR015943 (WD40/YVTN repeat-like-containing domain) and IPR011009 (protein kinase-like domain). These domains play important roles in many cellular processes, including division, proliferation, cellular activities, apoptosis, differentiation, plant-specific developmental events, and protect cells from extreme environments [81,82]. Thus, we found that this gene was highly expressed in different tissues, presented in the nucleus, and observed in all root cell types, columella and cortex, exodermis and cortex under 35Spro, SICO2pro, and SIPEP promoters. Furthermore, the protein suppressor of PHYA-105 1 (SPA1) is involved in regulating the circadian cycle and flowering time in plants, and SPA1 has worked as a negative regulator of phytochrome A-mediated de-etiolation in seed germination and seedlings of Arabidopsis [83]. Many researchers have recently described links between anthocyanin accumulation and the CONSTITUTIVELY PHO-TOMORPHOGENIC1/SUPPRESSOR OF PHYA-105 (COP1/SPA) in plants [84]. These relationships are complex and they have a positive correlation between the increased anthocyanidin content during drought. Cirillo et al. [85] reported that in tobacco, the anthocyanin content is considered the key regulator for drought stress tolerance by playing various roles in osmotic balance, scavenging of ROS, re-assimilation of the excess of ammonium, biochemical pH-stat, and regulation of leaf gas exchange [86].

Finally, the Asr2 SSR marker was related to ABA/WDS-induced protein (ABA_WDS) gene (*Solyc10g011690*), and this gene contains Interpro domain IPR003496 (ABA/WDS induced protein). This domain is caused by water deficit stress (WDS) or abscisic acid (ABA) stress. We found that this gene was highly expressed in different tissues, presented in the nucleus, and observed in all root cell types, exodermis and cortex under SlPEPpro and AtPEP promoters. Moreover, this gene is involved in the tolerance of various abiotic stresses such as dehydration, heat, and salinity for different plant species such as durum wheat, barley, and *Pinus taeda* L. [87,88].

5. Conclusions

Conclusively, this study assessed the genetic variation among tomato accessions using the SSR markers to detect the diversity of Jordanian tomato accessions. Moreover, tomato response to increasing drought stress was apparent through a significant reduction in morphological traits in addition to physiological and biochemical alternation, e.g., increasing the proline concentration. A wide range of variations were detected among tomato accessions that are important in selection for adaptation and yield improvement. Here, we demonstrated that the SSR method effectively discovers the genetic diversity of tomato accessions, which is vital for germplasm classification, management, and further molecular and breeding utilization. The bioinformatics analysis provides excellent information for predicting the function of candidate genes in tomatoes. Furthermore, the evaluation of these accessions under different water regimes could be helpful in producing new tomato varieties coping with drought stress conditions. Further molecular and genetic validation of the candidate genes would help understand the molecular mechanisms of drought stress tolerance in tomatoes.

Supplementary Materials: The following supporting information can be downloaded at: https://www.mdpi.com/article/10.3390/horticulturae8070600/s1. Table S1: Analysis of variance (ANOVA) for all the studied parameters related to seedling tomato accessions for three treatments (control, 40%FC, 70%FC). Table S2: Natural variation for all the studied parameters of tomato accessions under drought stress at the seedling stage. Table S3: Mean value of fresh root weight, fresh shoot weight, dry root weight, dry shoot weight, root growth rate, shoot growth rate, relative fresh root weight, relative fresh shoot weight, relative shoot growth rate, and relative root growth rate. Table S4: Phenotypic variations for 20 morphological traits of tomato accessions. Figure S1: Coefficients of genetic similarity [45] for 36 local tomato accessions by using SSR markers. Figure S2: SSR patterns using primer Asr2 for 41 tomato accessions. M = molecular weight marker (100 bp). Figure S3: SSR pattern using primer LEat018 for 41 tomato accessions. M = molecular weight marker (100 bp). Figure S4: SSR patterns using primer LEct004 for 41 tomato accessions. M = molecular weight marker (100 bp). Figure S5: SSR patterns using primer LEa014 for 41 tomato accessions. M = molecular weight marker (100 bp). Figure S6: SSR patterns using primer LEta020 for 41 tomato accessions. M = molecular weight marker (100 bp). Figure S7: SSR patterns using primer CT114 for 41 tomato accessions. M = molecular weight marker (1 kp, 100 bp) using 3% Agarose gel.

Author Contributions: Conceptualization, I.M.; methodology, I.M. and S.J.; data analysis, W.A.D., S.G.T., M.A. and A.M.A.; investigation, I.M., W.A.D., S.J., S.G.T., M.A. and A.M.A.; writing—review and editing, I.M., W.A.D., S.G.T., M.A., A.A.A. and A.M.A. All authors have read and agreed to the published version of the manuscript.

Funding: The research leading to these results received funding from JUST Deanship of Research under grant agreement No188/2010.

Institutional Review Board Statement: Not applicable.

Informed Consent Statement: Not applicable.

Data Availability Statement: Not applicable.

Acknowledgments: The authors would like to thank everyone who supported this work.

Conflicts of Interest: The authors declare no conflict of interest.

References

1. Knapp, S.; Bohs, L.; Nee, M.; Spooner, D.M. Solanaceae—A model for linking genomics with biodiversity. *Comp. Funct. Genomics* **2004**, *5*, 285–291. [CrossRef] [PubMed]
2. Siddiqui, M.H.; Alamri, S.; Alsubaie, Q.D.; Ali, H.M.; Khan, M.N.; Al-Ghamdi, A.; Ibrahim, A.A.; Alsadon, A. Exogenous nitric oxide alleviates sulfur deficiency-induced oxidative damage in tomato seedlings. *Nitric Oxide* **2020**, *94*, 95–107. [CrossRef] [PubMed]
3. Gonias, E.D.; Ganopoulos, I.; Mellidou, I.; Bibi, A.C.; Kalivas, A.; Mylona, P.V.; Osanthanunkul, M.; Tsaftaris, A.; Madesis, P.; Doulis, A.G. Exploring genetic diversity of tomato (*Solanum lycopersicum* L.) germplasm of genebank collection employing SSR and SCAR markers. *Genet. Resour. Crop Evol.* **2019**, *66*, 1295–1309. [CrossRef]
4. Foolad, M.R. Genome mapping and molecular breeding of tomato. *Int. J. Plant Genom.* **2007**, *2007*, 64358. [CrossRef] [PubMed]
5. Naeem, M.; Shahzad, K.; Saqib, S.; Shahzad, A.; Nasrullah; Younas, M.; Afridi, M. The Solanum melongena COP1LIKE manipulates fruit ripening and flowering time in tomato (*Solanum lycopersicum*). *J. Plant Growth Regul.* **2022**, *96*, 14. [CrossRef]
6. Reynolds, M.; Dreccer, F.; Trethowan, R. Drought-adaptive traits derived from wheat wild relatives and landraces. *J. Exp. Bot.* **2007**, *58*, 177–186. [CrossRef]
7. Casanas, F.; Simo, J.; Casals, J.; Prohens, J. Toward an Evolved Concept of Landrace. *Front. Plant Sci.* **2017**, *8*, 145. [CrossRef]
8. Makhadmeh, I.M.; Thabet, S.G.; Ali, M.; Alabbadi, B.; Albalasmeh, A.; Alqudah, A.M. Exploring genetic variation among Jordanian *Solanum lycopersicon* L. landraces and their performance under salt stress using SSR markers. *J. Genet. Eng. Biotechnol.* **2022**, *20*, 45. [CrossRef]
9. Brake, M.H.; Al-Gharaibeh, M.A.; Hamasha, H.R.; Sakarneh, N.S.A.; Alshomali, I.A.; Migdadi, H.M.; Qaryouti, M.M.; Haddad, N.J. Assessment of genetic variability among Jordanian tomato landrace using inter-simple sequence repeats markers. *Jordan J. Biol. Sci.* **2021**, *14*, 91–95.
10. Foolad, M.R.; Zhang, L.P.; Subbiah, P. Genetics of drought tolerance during seed germination in tomato: Inheritance and QTL mapping. *Genome* **2003**, *46*, 536–545. [CrossRef]
11. Cattivelli, L.; Rizza, F.; Badeck, F.W.; Mazzucotelli, E.; Mastrangelo, A.M.; Francia, E.; Mare, C.; Tondelli, A.; Stanca, A.M. Drought tolerance improvement in crop plants: An integrated view from breeding to genomics. *Field Crops Res.* **2008**, *105*, 1–14. [CrossRef]
12. Bargali, K.; Tewari, A. Growth and water relation parameters in drought-stressed *Coriaria nepalensis* seedlings. *J. Arid. Environ.* **2004**, *58*, 505–512. [CrossRef]
13. Thabet, S.G.; Alqudah, A.M. Crops and Drought. In *eLS*; John Wiley & Sons, Ltd., Ed.; John Wiley & Sons, Ltd.: Hoboken, NJ, USA, 2019; pp. 1–8.
14. Takele, A. Seedling emergence and of growth of sorghum genotypes under variable soil moisture deficit. *Acta Agron. Hung.* **2000**, *48*, 95–102. [CrossRef]
15. Dhanda, S.S.; Sethi, G.S.; Behl, R.K. Indices of drought tolerance in wheat genotypes at early stages of plant growth. *J. Agron. Crop Sci.* **2004**, *190*, 6–12. [CrossRef]
16. Kashiwagi, J.; Krishnamurthy, L.; Upadhyaya, H.D.; Krishna, H.; Chandra, S.; Vadez, V.; Serraj, R. Genetic variability of drought-avoidance root traits in the mini-core germplasm collection of chickpea (*Cicer arietinum* L.). *Euphytica* **2005**, *146*, 213–222. [CrossRef]
17. Ali, M.A.; Jabran, K.; Awan, S.I.; Abbas, A.; Ehsanullah; Zulkiffal, M.; Acet, T.; Farooq, J.; Rehman, A. Morpho-physiological diversity and its implications for improving drought tolerance in grain sorghum at different growth stages. *Aust. J. Crop Sci.* **2011**, *5*, 308–317.
18. Wahb-Allah, M.A.; Alsadon, A.A.; Ibrahim, A.A. Drought tolerance of several tomato genotypes under greenhouse conditions. *World Appl. Sci. J.* **2011**, *15*, 933–940.
19. Ovesna, J.; Poláková, K.; Leišová, L. DNA analyses and their applications in plant breeding. *Czech J. Genet. Plant Breed.* **2002**, *38*, 29. [CrossRef]
20. Kumar, P.; Gupta, V.K.; Misra, A.K.; Modi, D.R.; Pandey, B.K. Potential of Molecular Markers in Plant Biotechnology. *Plant Omics* **2009**, *2*, 141–162.
21. Ezekiel, C.N.; Nwangburuka, C.C.; Ajibade, O.A.; Odebode, A.C. Genetic diversity in 14 tomato (*Lycopersicon esculentum* Mill.) varieties in Nigerian markets by RAPD-PCR technique. *Afr. J. Biotechnol.* **2011**, *10*, 4961–4967.
22. Tautz, D.; Renz, M. Simple sequences are ubiquitous repetitive components of eukaryotic genomes. *Nucleic Acids Res.* **1984**, *12*, 4127–4138. [CrossRef]
23. Soranzo, N.; Provan, J.; Powell, W. An example of microsatellite length variation in the mitochondrial genome of conifers. *Genome* **1999**, *42*, 158–161. [CrossRef]
24. Farooq, S.; Azam, F. Molecular markers in plant breeding-I: Concepts and characterization. *Pak. J. Biol. Sci.* **2002**, *5*, 1135–1140. [CrossRef]
25. Pearson, C.E.; Sinden, R.R. Trinucleotide repeat DNA structures: Dynamic mutations from dynamic DNA. *Curr. Opin. Struct. Biol.* **1998**, *8*, 321–330. [CrossRef]
26. Zhou, R.; Wu, Z.; Cao, X.; Jiang, F.L. Genetic diversity of cultivated and wild tomatoes revealed by morphological traits and SSR markers. *Genet. Mol. Res.* **2015**, *14*, 13868–13879. [CrossRef]

27. EL-Mansy, A.B.; Abd El-Moneim, D.; ALshamrani, S.M.; Alsafhi, F.A.; Abdein, M.A.; Ibrahim, A.A. Genetic Diversity Analysis of Tomato (*Solanum lycopersicum* L.) with Morphological, Cytological, and Molecular Markers under Heat Stress. *Horticulturae* **2021**, *7*, 65. [CrossRef]

28. Cotti, C. Molecular Markers for the Assessment of Genetic Variability in Threatened Plant Species. Ph.D Thesis, University of Bologna, Department of Experimental Evolutionary Biology, Bologna, Italy, 2008. [CrossRef]

29. Sardaro, M.L.; Marmiroli, M.; Maestri, E.; Marmiroli, N. Genetic characterization of Italian tomato varieties and their traceability in tomato food products-Sardaro-2012-Food Science & Nutrition-Wiley Online Library. *Food Sci. Nutr* **2013**, *1*, 54–62. [CrossRef]

30. Sacco, A.; Ruggieri, V.; Parisi, M.; Festa, G.; Rigano, M.M.; Picarella, M.E.; Mazzucato, A.; Barone, A. Exploring a Tomato Landraces Collection for Fruit-Related Traits by the Aid of a High-Throughput Genomic Platform. *PLoS ONE* **2015**, *10*, e0137139. [CrossRef]

31. Wang, T.; Zou, Q.D.; Qi, S.Y.; Wang, X.F.; Wu, Y.Y.; Liu, N.; Zhang, Y.M.; Zhang, Z.J.; Li, H.T. Analysis of genetic diversity and population structure in a tomato (*Solanum lycopersicum* L.) germplasm collection based on single nucleotide polymorphism markers. *Genet. Mol. Res.* **2016**, *15*, 1–12. [CrossRef]

32. Liu, S.; Cantrell, R.G.; McCarty, J.C.; Stewart, J.M. Simple sequence repeat-based assessment of genetic diversity in cotton race stock accessions. *Crop Sci.* **2000**, *40*, 1459–1469. [CrossRef]

33. Raveendren, T.J.B.; Reviews, M.B. Molecular marker technology in cotton. *Biotechnol. Mol. Biol. Rev.* **2008**, *3*, 32–45.

34. Shiri, M. Identification of informative simple sequence repeat (SSR) markers for drought tolerance in maize. *Afr. J. Biotechnol.* **2011**, *10*, 16414–16420. [CrossRef]

35. Tam, S.M.; Mhiri, C.; Vogelaar, A.; Kerkveld, M.; Pearce, S.R.; Grandbastien, M.A. Comparative analyses of genetic diversities within tomato and pepper collections detected by retrotransposon-based SSAP, AFLP and SSR. *Theor. Appl. Genet.* **2005**, *110*, 819–831. [CrossRef]

36. Bates, L.S.; Waldren, R.P.; Teare, I. Rapid determination of free proline for water-stress studies. *Plant Soil* **1973**, *39*, 205–207. [CrossRef]

37. Senthilkumar, M.; Amaresan, N.; Sankaranarayanan, A. Estimation of Proline Content in Plant Tissues. In *Plant-Microbe Interactions: Laboratory Techniques*; Senthilkumar, M., Amaresan, N., Sankaranarayanan, A., Eds.; Springer: New York, NY, USA, 2021; pp. 95–98.

38. He, C.; Poysa, V.; Yu, K.J.T. Development and characterization of simple sequence repeat (SSR) markers and their use in determining relationships among *Lycopersicon esculentum* cultivars. *Theor. Appl. Genet.* **2003**, *106*, 363–373. [CrossRef]

39. Asif, M.; Rahman, M.; Mirza, J.; Zafar, Y. High resolution metaphor agarose gel electrophoresis for genotyping with microsatellite markers. *Pak. J. Agric. Sci.* **2008**, *45*, 75–79.

40. Anderson, J.A.; Churchill, G.A.; Autrique, J.E.; Tanksley, S.D.; Sorrells, M.E. Optimizing parental selection for genetic linkage maps. *Genome* **1993**, *36*, 181–186. [CrossRef]

41. Dice, L.R. Measures of the Amount of Ecologic Association between Species. *Ecology* **1945**, *26*, 297–302. [CrossRef]

42. Rohlf, F.J. *NTSYSpc Numerical Taxonomy and Multivariate Analysis System Version 2.0 User Guide*; Exeter Software Publishers Ltd.: Setauket, NY, USA, 1998.

43. Fucile, G.; Di Biase, D.; Nahal, H.; La, G.; Khodabandeh, S.; Chen, Y.; Easley, K.; Christendat, D.; Kelley, L.; Provart, N.J. ePlant and the 3D data display initiative: Integrative systems biology on the world wide web. *PLoS ONE* **2011**, *6*, e15237. [CrossRef]

44. Kimura, M.; Crow, J.F. The number of alleles that can be maintained in a finite population. *Genetics* **1964**, *49*, 725. [CrossRef]

45. Nei, M. Analysis of gene diversity in subdivided populations. *Proc. Natl. Acad. Sci. USA* **1973**, *70*, 3321–3323. [CrossRef]

46. Lewontin, R.C. Testing the theory of natural selection. *Nature* **1972**, *236*, 181–182. [CrossRef]

47. Luo, C.; He, X.H.; Chen, H.; Ou, S.J.; Gao, M.P. Analysis of diversity and relationships among mango cultivars using Start Codon Targeted (SCoT) markers. *Biochem. Syst. Ecol.* **2010**, *38*, 1176–1184. [CrossRef]

48. Vurayai, R.; Emongor, V.; Moseki, B. Effect of water stress imposed at different growth and development stages on morphological traits and yield of bambara groundnuts (*Vigna subterranea* L. Verdc). *Am. J. Plant Physiol.* **2011**, *6*, 17–27. [CrossRef]

49. Yücel, D.; Anlarsal, A.; Mart, D.; Yücel, C. Effects of drought stress on early seedling growth of chickpea (*Cicer arietinum* L.) genotypes. *World Appl. Sci. J.* **2010**, *11*, 478–485.

50. Vasquez-Robinet, C.; Mane, S.P.; Ulanov, A.V.; Watkinson, J.I.; Stromberg, V.K.; De Koeyer, D.; Schafleitner, R.; Willmot, D.B.; Bonierbale, M.; Bohnert, H.J.; et al. Physiological and molecular adaptations to drought in Andean potato genotypes. *J. Exp. Bot.* **2008**, *59*, 2109–2123. [CrossRef]

51. Mastrangelo, A.; Rascio, A.; Mazzucco, L.; Russo, M.; Cattivelli, L.; Di Fonzo, N. Series A. Molecular aspects of abiotic stress resistance in durum wheat. *Options Méditerranéennes Ser. A* **2000**, *40*, 207–213.

52. Ranc, N.; Munos, S.; Santoni, S.; Causse, M. A clarified position for *Solanum lycopersicum* var. cerasiforme in the evolutionary history of tomatoes (solanaceae). *BMC Plant Biol.* **2008**, *8*, 130. [CrossRef]

53. Brdar-Jokanović, M.; Girek, Z.; Pavlović, S.; Ugrinović, M.; Zdravković, J. Traits related to drought tolerance in tomato accessions of different growth type and fruit size. *J. Anim. Plant Sci.* **2017**, *27*, 869–876.

54. Korir, N.K.; Diao, W.; Tao, R.; Li, X.; Kayesh, E.; Li, A.; Zhen, W.; Wang, S. Genetic diversity and relationships among different tomato varieties revealed by EST-SSR markers. *Genet. Mol. Res.* **2014**, *13*, 43–53. [CrossRef]

55. Abd El-Hady, E.A.; Haiba, A.A.; Abd El-Hamid, N.R.; Rizkalla, A.A. Phylogenetic diversity and relationships of some tomato varieties by electrophoretic protein and RAPD analysis. *J. Am. Sci.* **2010**, *6*, 434–441.

56. Benor, S.; Zhang, M.Y.; Wang, Z.F.; Zhang, H.S. Assessment of genetic variation in tomato (*Solanum lycopersicum* L.) inbred lines using SSR molecular markers. *J. Genet. Genom.* **2008**, *35*, 373–379. [CrossRef]

57. Smulders, M.J.M.; Bredemeijer, G.; RusKortekaas, W.; Arens, P.; Vosman, B. Use of short microsatellites from database sequences to generate polymorphisms among *Lycopersicon esculentum* cultivars and accessions of other Lycopersicon species. *Theor. Appl. Genet.* **1997**, *94*, 264–272. [CrossRef]

58. Rajput, S.G.; Plyler-Harveson, T.; Santra, D.K.J.A.J.o.P.S. Development and characterization of SSR markers in proso millet based on switchgrass genomics. *Am. J. Plant Sci.* **2014**, *5*, 175–186. [CrossRef]

59. Smith, D.; Devey, M.E. Occurrence and inheritance of microsatellites in *Pinus radiata*. *Genome* **1994**, *37*, 977–983. [CrossRef]

60. Vargas, J.E.E.; Aguirre, N.C.; Coronado, Y.M. Study of the genetic diversity of tomato (*Solanum* spp.) with ISSR markers. *Revista Ceres* **2020**, *67*, 199–206. [CrossRef]

61. Alzahib, R.H.; Migdadi, H.M.; Al Ghamdi, A.A.; Alwahibi, M.S.; Afzal, M.; Elharty, E.H.; Alghamdi, S.S. Exploring Genetic Variability among and within Hail Tomato Landraces Based on Sequence-Related Amplified Polymorphism Markers. *Diversity* **2021**, *13*, 135. [CrossRef]

62. Mittler, R.; Vanderauwera, S.; Gollery, M.; Van Breusegem, F. Reactive oxygen gene network of plants. *Trends Plant Sci.* **2004**, *9*, 490–498. [CrossRef]

63. Dominguez, R.; Holmes, K.C. Actin structure and function. *Annu. Rev. Biophys.* **2011**, *40*, 169–186. [CrossRef]

64. Nie, W.F.; Wang, J. Actin-Related Protein 4 Interacts with PIE1 and Regulates Gene Expression in Arabidopsis. *Genes* **2021**, *12*, 520. [CrossRef]

65. McKinney, E.C.; Kandasamy, M.K.; Meagher, R.B. Arabidopsis contains ancient classes of differentially expressed actin-related protein genes. *Plant Physiol.* **2002**, *128*, 997–1007. [CrossRef] [PubMed]

66. Mittler, R.; Vanderauwera, S.; Gollery, M.; Van Breusegem, F. Questions and future challenges. *Trends Plant Sci.* **2004**, *10*, 490–498. [CrossRef] [PubMed]

67. Nagasaki, H.; Sakamoto, T.; Sato, Y.; Matsuoka, M. Functional analysis of the conserved domains of a rice KNOX homeodomain protein, OSH15. *Plant Cell* **2001**, *13*, 2085–2098. [CrossRef] [PubMed]

68. Ito, Y.; Hirochika, H.; Kurata, N. Organ-specific alternative transcripts of KNOX family class 2 homeobox genes of rice. *Gene* **2002**, *288*, 41–47. [CrossRef]

69. Clark, S.E.; Jacobsen, S.E.; Levin, J.Z.; Meyerowitz, E.M. The CLAVATA and SHOOT MERISTEMLESS loci competitively regulate meristem activity in Arabidopsis. *Development* **1996**, *122*, 1567–1575. [CrossRef]

70. Bhattacharjee, A.; Khurana, J.P.; Jain, M. Characterization of Rice Homeobox Genes, OsHOX22 and OsHOX24, and Overexpression of OsHOX24 in Transgenic Arabidopsis Suggest Their Role in Abiotic Stress Response. *Front Plant Sci.* **2016**, *7*, 627. [CrossRef]

71. Sakamoto, T.; Nishimura, A.; Tamaoki, M.; Kuba, M.; Tanaka, H.; Iwahori, S.; Matsuoka, M. The conserved KNOX domain mediates specificity of tobacco KNOTTED1-type homeodomain proteins. *Plant Cell* **1999**, *11*, 1419–1432. [CrossRef]

72. Liscum, E.; Reed, J.W. Genetics of Aux/IAA and ARF action in plant growth and development. *Plant Mol. Biol.* **2002**, *49*, 387–400. [CrossRef]

73. Kulaeva, O.N.; Prokoptseva, O.S. Recent advances in the study of mechanisms of action of phytohormones. *Biochemistry* **2004**, *69*, 233–247. [CrossRef]

74. Kang, C.; He, S.; Zhai, H.; Li, R.; Zhao, N.; Liu, Q. A Sweetpotato Auxin Response Factor Gene (IbARF5) Is Involved in Carotenoid Biosynthesis and Salt and Drought Tolerance in Transgenic Arabidopsis. *Front Plant Sci.* **2018**, *9*, 1307. [CrossRef]

75. Bouzroud, S.; Gouiaa, S.; Hu, N.; Bernadac, A.; Mila, I.; Bendaou, N.; Smouni, A.; Bouzayen, M.; Zouine, M. Auxin Response Factors (ARFs) are potential mediators of auxin action in tomato response to biotic and abiotic stress (*Solanum lycopersicum*). *PLoS ONE* **2018**, *13*, e0193517. [CrossRef] [PubMed]

76. Chen, M.; Zhu, X.; Liu, X.; Wu, C.; Yu, C.; Hu, G.; Chen, L.; Chen, R.; Bouzayen, M.; Zouine, M.; et al. Knockout of Auxin Response Factor SlARF4 Improves Tomato Resistance to Water Deficit. *Int. J. Mol. Sci.* **2021**, *22*, 3347. [CrossRef] [PubMed]

77. Salehin, M.; Li, B.; Tang, M.; Katz, E.; Song, L.; Ecker, J.R.; Kliebenstein, D.J.; Estelle, M. Auxin-sensitive Aux/IAA proteins mediate drought tolerance in Arabidopsis by regulating glucosinolate levels. *Nat. Commun.* **2019**, *10*, 4021. [CrossRef] [PubMed]

78. Maeda, K.; Tsugita, A.; Dalzoppo, D.; Vilbois, F.; Schurmann, P. Further characterization and amino acid sequence of m-type thioredoxins from spinach chloroplasts. *Eur. J. Biochem.* **1986**, *154*, 197–203. [CrossRef]

79. Daie, J. Cytosolic fructose-1,6-bisphosphatase: A key enzyme in the sucrose biosynthetic pathway. *Photosynth Res.* **1993**, *38*, 5–14. [CrossRef] [PubMed]

80. Jacquot, J.J.; Lancelin, J.M.; Meyer, Y. Thioredoxins: Structure and function in plant cells. *New Phytol.* **1997**, *136*, 543–570. [CrossRef]

81. Li, D.; Roberts, R.J.C. Human Genome and Diseases: WD-repeat proteins: Structure characteristics, biological function, and their involvement in human diseases. *Cell. Mol. Life Sci. CMLS* **2001**, *58*, 2085–2097. [CrossRef]

82. Datta, S.; Ikeda, T.; Kano, K.; Mathews, F.S. Structure of the phenylhydrazine adduct of the quinohemoprotein amine dehydrogenase from *Paracoccus denitrificans* at 1.7 Å resolution. *Acta Crystallogr. Sect. D Biol. Crystallogr.* **2003**, *59*, 1551–1556. [CrossRef]

83. Ishikawa, M.; Kiba, T.; Chua, N.H. The Arabidopsis SPA1 gene is required for circadian clock function and photoperiodic flowering. *Plant J.* **2006**, *46*, 736–746. [CrossRef]

84. Maier, A.; Schrader, A.; Kokkelink, L.; Falke, C.; Welter, B.; Iniesto, E.; Rubio, V.; Uhrig, J.F.; Hulskamp, M.; Hoecker, U. Light and the E3 ubiquitin ligase COP1/SPA control the protein stability of the MYB transcription factors PAP1 and PAP2 involved in anthocyanin accumulation in Arabidopsis. *Plant J.* **2013**, *74*, 638–651. [CrossRef]
85. Cirillo, V.; D'Amelia, V.; Esposito, M.; Amitrano, C.; Carillo, P.; Carputo, D.; Maggio, A. Anthocyanins are Key Regulators of Drought Stress Tolerance in Tobacco. *Biology* **2021**, *10*, 139. [CrossRef] [PubMed]
86. Hinojosa-Gomez, J.; San Martin-Hernandez, C.; Heredia, J.B.; Leon-Felix, J.; Osuna-Enciso, T.; Muy-Rangel, M.D. Anthocyanin Induction by Drought Stress in the Calyx of Roselle Cultivars. *Molecules* **2020**, *25*, 1555. [CrossRef] [PubMed]
87. Yacoubi, I.; Hamdi, K.; Fourquet, P.; Bignon, C.; Longhi, S. Structural and Functional Characterization of the ABA-Water Deficit Stress Domain from Wheat and Barley: An Intrinsically Disordered Domain behind the Versatile Functions of the Plant Abscissic Acid, Stress and Ripening Protein Family. *Int. J. Mol. Sci.* **2021**, *22*, 2314. [CrossRef]
88. Hamdi, K.; Brini, F.; Kharrat, N.; Masmoudi, K.; Yakoubi, I. Abscisic Acid, Stress, and Ripening (TtASR1) Gene as a Functional Marker for Salt Tolerance in Durum Wheat. *Biomed. Res. Int.* **2020**, *2020*, 7876357. [CrossRef] [PubMed]

Article

Distinctive Physio-Biochemical Properties and Transcriptional Changes Unfold the Mungbean Cultivars Differing by Their Response to Drought Stress at Flowering Stage

Gunasekaran Ariharasutharsan [1,2,†], Adhimoolam Karthikeyan [3,†], Vellaichamy Gandhimeyyan Renganathan [2,3], Vishvanathan Marthandan [3,4], Manickam Dhasarathan [5], Ayyavoo Ambigapathi [2], Manoharan Akilan [2,6], Subramani Palaniyappan [1,2], Irulappan Mariyammal [2], Muthaiyan Pandiyan [7,*] and Natesan Senthil [8,*]

[1] Department of Genetics and Plant Breeding, Centre for Plant Breeding and Genetics, Tamil Nadu Agricultural University, Coimbatore 641003, India; sutharsang7@gmail.com (G.A.); palanimalliga3@gmail.com (S.P.)

[2] Department of Plant Breeding and Genetics, Agricultural College and Research Institute, Tamil Nadu Agricultural University, Madurai 625105, India; vgrenga@gmail.com (V.G.R.); ambikabathya@gmail.com (A.A.); akilanmkarur@gmail.com (M.A.); mari.tnau@gmail.com (I.M.)

[3] Department of Biotechnology, Centre of Innovation, Agricultural College and Research Institute, Tamil Nadu Agricultural University, Madurai 625105, India; karthik2373@gmail.com (A.K.); v_marthandan@cb.amrita.edu (V.M.)

[4] Department of Seed Science and Technology, Amrita School of Agricultural Sciences, Amrita Vishwa Vidyapeetham, Coimbatore 641032, India

[5] Agro Climate Research Centre, Directorate of Crop Management, Tamil Nadu Agricultural University, Coimbatore 641003, India; plantdr.dhasarathan@gmail.com

[6] Department of Plant Breeding and Genetics, Anbil Dharmalingam Agricultural College and Research Institute, Tamil Nadu Agricultural University, Trichy 620009, India

[7] Regional Research Station, Tamil Nadu Agricultural University, Virudhachalam 606001, India

[8] Department of Plant Molecular Biology and Bioinformatics, Centre for Plant Molecular Biology and Biotechnology, Tamil Nadu Agricultural University, Coimbatore 641003, India

[*] Correspondence: mpandiyan8@yahoo.co.in (M.P.); senthil_natesan@tnau.ac.in (N.S.)

[†] These authors contributed equally to this work.

Citation: Ariharasutharsan, G.; Karthikeyan, A.; Renganathan, V.G.; Marthandan, V.; Dhasarathan, M.; Ambigapathi, A.; Akilan, M.; Palaniyappan, S.; Mariyammal, I.; Pandiyan, M.; et al. Distinctive Physio-Biochemical Properties and Transcriptional Changes Unfold the Mungbean Cultivars Differing by Their Response to Drought Stress at Flowering Stage. *Horticulturae* **2022**, *8*, 424. https://doi.org/10.3390/horticulturae8050424

Academic Editors: Stefania Toscano, Giulia Franzoni and Sara Álvarez

Received: 13 February 2022
Accepted: 26 April 2022
Published: 10 May 2022

Publisher's Note: MDPI stays neutral with regard to jurisdictional claims in published maps and institutional affiliations.

Abstract: Mungbean is a nutritionally and economically important pulse crop cultivated around Asia, mainly in India. The crop is sensitive to drought at various developmental stages of its growing period. However, there is limited or almost no research on a comparative evaluation of mung-bean plants at the flowering stage under drought conditions. Hence, the aim of this research was to impose the drought stress on two mungbean cultivars VRM (Gg) 1 and CO6 at the flowering stage and assess the physio-biochemical and transcriptional changes. After imposing the drought stress, we found that VRM (Gg) 1 exhibited a low reduction in physiological traits (Chlorophyll, relative water content, and plant dry mass) and high proline content than CO6. Additionally, VRM (Gg) 1 has a low level of H_2O_2 and MDA contents and higher antioxidant enzymes (SOD, POD, and CAT) activity than CO6 during drought stress. The transcriptional analysis of photosynthesis (*PS II-PsbP, PS II-LHC, PS I-PsaG/PsaK,* and *PEPC 3*), antioxidant (*SOD 2, POD, CAT 2*), and drought-responsive genes (*HSP-90, DREB2C, NAC 3* and *AREB 2*) show that VRM (Gg) 1 had increased transcripts more than CO6 under drought stress. Taken together, VRM (Gg) 1 had a better photosynthetic performance which resulted in fewer reductions in chlorophyll, relative water content, and plant dry mass during drought stress. In addition, higher antioxidative enzyme activities led to lower H_2O_2 and MDA levels, limiting oxidative damage in VRM (Gg) 1. This was positively correlated with increased transcripts of photosynthesis and antioxidant-related genes in VRM (Gg) 1. Further, the increased transcripts of drought-responsive genes indicate that VRM (Gg) 1 has a better genetic basis against drought stress than CO6. These findings help to understand the mungbean response to drought stress and will aid in the development of genotypes with greater drought tolerance by utilizing natural genetic variants.

Keywords: abiotic stresses; biochemical response; drought; legumes; mungbean

1. Introduction

With the effects of global warming and drastic climate changes, drought is the major abiotic stress that affects crop production in arid and semi-arid regions of the world. It is brought about by a scarcity of rain or a vast difference in rainfall quantity [1,2]. Drought impairs plant growth and development and accounts for over 70% of agriculture yield losses worldwide. However, it relies on the drought intensity, duration, phenophases of the crop, and environmental stress factors. An increasingly warming climate and decreased water availability are likely to upsurge the occurrence and severity of drought in the near future. Therefore, boosting the tolerance to drought is a major aim of crop improvement programs. Much progress has been made in understanding the effect of drought stress on plants. Decoding the molecular mechanisms underpinning plant response during drought is not easy due to the intricacy of drought behavior, environmental factors, and their interactions [3]. When subject to drought, plants undergo a series of morphological, physiological, and biochemical changes that seriously reduce plant growth and development [4,5] Physiological responses include (i) a reduction in the content of chlorophyll, rate of photosynthesis, and transpiration, (ii) stomatal closure, (iii) dehydration of cells [6–8]. Drought stress causes increased peroxidation of lipid membranes and mass accumulation of reactive oxygen species (ROS) [9–11]. The augmented ROS accumulation causes damage to proteins, lipids, cell membranes, carbohydrates, and nucleic acids and leads to disruption of cellular homeostasis and subsequently cellular death [12]. Both enzymatic and non-enzymatic antioxidant systems are fundamental to protecting the cells against toxic ROS and minimizing the oxidative stress effects [13–17]. Previously, it was shown that ROS production under drought stress can be minimized by increasing the antioxidant enzymatic activities in mungbean [18]. Masoumi et al. [19] reported that tolerant soybean plants enhanced their antioxidant enzyme activities and antioxidant contents in response to drought stress, whereas drought-sensitive plants were unable to do so. The lower level of MDA along with enhanced activities of SOD and CAT in black gram plants can be linked to its ability to cope up with water scarcity by limiting the damaging effects of drought through up-regulation of antioxidant enzymes [20].

Physiological and transcriptome responses of soybean to drought stress were investigated by Xu et al. [21]. Physiological traits such as photosynthetic rate, stomatal conductance, transpiration, and water potential were reduced, while SOD and CAT activities were enhanced, and POD activity remained unchanged. Furthermore, in drought-stressed plants, a total of 2771 differentially expressed genes were identified, and they were involved in different biochemical and molecular pathways, including ABA biogenesis, compatible compound accumulation, secondary metabolite synthesis, fatty acid desaturation, and transcription factors. In another study, Mahdavi Mashaki, et al. [22] employed RNA-Seq to investigate transcriptome profiles in drought-responsive contrasting genotypes of Iranian kabuli chickpea under drought stress in root and shoot tissues at the early flowering stage. Of these, 261 and 169 drought stress-responsive genes were identified in the shoots and the roots, respectively, and 17 genes were common in the shoots and the roots. Several molecular mechanisms are involved in the stress response and their corresponding drought-related pathway, (i.e., ABA, proline, and flavonoid biosynthesis). Lopez et al. [23] showed the importance of phosphorous homeostasis, as well as several other key factors, in response to drought stress in the common bean. Upregulation of several key transcription factors, remodeling of cell walls, synthesis of osmoprotectant oligosaccharides, protection of the photosynthetic apparatus, and downregulation of genes involved in cell expansion were all revealed by RNA-seq analysis of the drought-tolerant landrace PHA-683 in response to drought, but there was a significant proportion of DEGs related to phosphate starvation response.

Mungbean (*Vigna radiata*), native to India, is a short-duration grain legume and is extensively cultivated in Asia. This crop mainly features high protein (25%) and nutrient (carbohydrates, lipids, minerals, and vitamins) contents. It also improves soil fertility by fixing atmospheric nitrogen [24]. India is the world's leading mungbean grower, with

2.17 million tonnes of grains from a 4.32 m ha area. Mungbean yield in India is still low (502 kg/ha), considerably lower than the productivity of other main pulse crops [25]. In India, mungbean is mainly grown under rainfed conditions at high temperatures, with low humidity and rainfall. Thus, the mungbean is exposed to drought at various developmental stages of its growing period [26,27]. This crop is comparatively surviving under drought conditions. However, in comparison to other developmental stages, mungbean growth is sensitive to drought during flowering and post-flowering. Drought stress during these stages can decrease the grain yield ranging from 30% to 80%. Regardless of their importance, the studies that investigated the impacts of drought-influenced mungbean are limited [28,29]. Under drought, comparative evaluation of physio-biochemical and transcriptional changes between mungbean cultivars at the flowering stage lacks existing information in the literature. Taking into account the above, in this research, we aimed at revealing the physio-biochemical and transcriptional alterations in two mungbean cultivars (VRM (Gg) 1 and CO6) at the flowering stage under drought conditions.

2. Materials and Methods

2.1. Plant Materials and Drought Treatment

Seeds of mungbean varieties CO6 and VRM (Gg) 1 were provided by Agricultural Research Station, Tamil Nadu Agricultural University, Virinjipuram, Tamil Nadu, India. The seeds for two varieties were sown in pots (15 L, 30 cm height, 33.0 cm diameter at top and 25.5 cm bottom diameter) containing 3:1 ratio of soil and compost in a greenhouse at Agricultural College and Research Station, Madurai, Tamil Nadu, India. The experimental design was a completely randomized design (CRD) with three replications of 15 plants (5 plants per pot). During the experiment period, the temperature was maintained at 28 °C, and relative humidity was 65%. Three different sets of plants were maintained. Watering was done regularly until the flowering started. When the flowering was observed, drought stress for 6 days (soil moisture, 50%) and 12 days (soil moisture, 25%) were imposed in two sets; the third set was kept as the control 0 days (soil moisture, >80%). Soil moisture was measured using Lutron PMS-714 soil moisture meter.

2.2. Plant Sampling

The fully expanded leaves from three to five plants were sampled following the 0, 6, and 12 days of drought stress and immediately frozen into liquid nitrogen and then stored at −80 °C. Leaf relative water content (RWC), chlorophyll content, leaf gas exchange parameters, and plant dry mass were used to estimate at 0 and 12 days of drought stress plants. Proline, protein, hydrogen peroxide (H_2O_2), malondialdehyde (MDA), enzymatic (SOD, POD, and CAT), and non-enzymatic (Ascorbic acid) antioxidants assays were conducted at 0, 6, and 12 days of drought stress plants. Quantitative real-time polymerase chain reaction (qRT-PCR) analysis was conducted at 0 and 12 days.

2.3. Physiological Traits

Total chlorophyll content was measured according to Arnon [30]. Plant dry mass was measured after drying the plant at 80 °C to a constant weight. RWC was measured as the standard method described by Barrs and Weatherly [31]. RWC = (FW − DW)/(TW − DW) × 100, where FW is the fresh leaf weight, DW is the dry leaf weight, and TW is the turgid weight of the leaves. Photosynthetic gas exchange parameters were measured by portable photosynthesis system Li-6400 (LiCor, Lincoln, NE, USA).

2.4. Proline Estimation

Proline content was estimated according to the methodology described by Bates et al. [32]. Leaf samples (1 g) were taken and homogenized in 3% aqueous sulphosalicylic acid and centrifuged at 10,000 rpm for 15 min at 4 °C. The supernatant was carefully taken, and then the acid-ninhydrin solution (1.25 g of ninhydrin in 30 mL glacial acetic acid) was added. The mixture was then incubated for an hour at 100 °C and cooled in ice to stop the reaction. For extracting the reaction mixture 4 mL of toluene was added and vortexed thoroughly for 2 min. The solute is then measured for absorbance at 520 nm. Toluene was considered as a blank, and the content of proline was calculated giving to the formula: [(μg proline/mL × mL toluene)/115.5 μg/μmole]/[(g sample)/5] = μg proline/g FW.

2.5. Damage Index; Determination of MDA and H_2O_2 Contents

MDA content estimation was done following the Stewart and Bewly method [33] to determine lipid peroxidation. Leaf samples (1 g) were homogenized in 0.1% of trichloroacetic acid (TCA) and then centrifuged at 12,000 rpm for 15 min at 4 °C. After centrifugation, 0.5 mL of the supernatant was collected and mixed with a 1 mL volume of 20% TCA containing a 0.5% thiobarbituric acid (TBA) solution. The sample was incubated for another 30 min at 95 °C and placed in ice bath to stop the reaction followed by centrifuging at 12,000 rpm for 10 min. The resulting solute was measured for absorbance at 532 nm. MDA content was estimated by the 155 mM^{-1} cm^{-1} extinction coefficient. Results were stated as μmol/g fresh weight. The H_2O_2 content was determined following the method previously elucidated by Loreto and Velikova [34]. Leaf samples (1 g) were homogenized in ice-cold 0.1% TCA and centrifuged at 12,000 rpm for 15 min at 4 °C. After that, 0.5 mL of 10 mM potassium phosphate buffer and 0.75 mL of 1 M KI were added to 0.5 mL of the supernatant. The absorbance was measured at 390 nm against a blank, and the H_2O_2 content was inferred by a standard calibration curve, previously made solutions with known H_2O_2 concentration. H_2O_2 concentration was expressed as μmol/g fresh weight.

2.6. Assay of Antioxidant Enzymes

Approximately 1 g of sampled leaves were weighed and finely chopped to powder with liquid nitrogen. About 10 mL of ice-cold 50 mM potassium phosphate buffer with 0.1-mM Na_2 EDTA and 1% (w/v) polyvinylpyrrolidone (PVP) was used to homogenize the powder. The pH of the buffer should be 7.8 for SOD and POD assays whereas it is 7.0 for CAT assay. After filtering the homogenate with a 4-layered muslin cloth, it was centrifuged at 12,000 rpm at 4 °C for 15 min. The aliquot part in the supernatant was gathered for enzyme activity assays. For recording the SOD activity, the protocol of Madamanchi and Alscher [35] was followed. The amount of enzyme that reduces nitroblue tetrazolium (NBT) to half was referred to be one unit of SOD. The solute was read at an absorbance of 560 nm. As guaiacol was the electron donor, the POD activity was recorded at 470 nm as proposed by Chance and Maehly [36] in which one unit POD function was referred to as one unit change in absorbance of 0.01 unit in a minute. As described by Aebi [37], the activity of CAT was found. In an interval of 2 min, a reduction in the absorbance at 240 nm was recorded after the digestion of H_2O_2. It is found that one unit of CAT produces 1 mM of H_2O_2 in a minute for which the results are expressed in units/mg of protein.

2.7. Ascorbic Acid Content

Ascorbic acid (referred to as vitamin C) was measured following the method previously reported by Arakawa et al. [38] with minor modifications. Leaf samples and 5 mL of 5% trichloroacetic acid (TCA) were homogenized in the mortar and centrifuged at 10,000 rpm for 10 min at 4 °C. Then, 1 mL of clear supernatant, 1 mL of 5% TC, 1 mL alcohol, 0.5 mL 0.4% phosphoric acid (H_3PO_4)-alcohol, 1 mL of 0.5% 4,7-diphenyl-1,10-phenanthroline (BP)-alcohol, and 0.5 mL 0.03% ferric trichloride ($FeCl_3$)-alcohol were added into to a tube and incubated at 40 °C for 1 h. The reaction was ended at room temperature, and absorbance was measured at 534 nm with a spectrophotometer (Unicam UV-330, Cambridge, UK). Results were expressed as the unit's μmol/g fresh weight.

2.8. RNA Isolation, cDNA Synthesis, and qRT-PCR Analysis

According to the manufacturer's guidelines, total RNA was isolated using an RNeasy plant mini kit (Qiagen, Hilden, Germany) and treated with RNase-free DNAseI (Promega, Madison, WI, USA). The RNA quantity was assessed using the bio spectrometer (Eppendorf, Hamburg, Germany) based on the absorbance ratio at 280 nm. Further, the quality of RNA was tested on 1% agarose gel via electrophoresis. The first-strand cDNA was done by transcriptor First Strand cDNA Synthesis Kit (Roche Applied Science, Penzberg, Germany) following the manufacturer's guidelines. Sequence information of the following genes, photosystem II oxygen-evolving complex protein (*PS II-PsbP*), photosystem II chlorophyll A/B binding protein (*PS II-LHC*), photosystem I PsaG/PsaK (*PS I-PsaG/PsaK*), phosphoenolpyruvate carboxylase 3 (*PEPC 3*), superoxide dismutase 2 (*SOD 2*), peroxidase (*POD*), catalase-2 (*CAT 2*), heat shock protein-90 (HSP-90), dehydration responsive element-binding transcription factor (*DREB2C*), NAC transcription factor 3 (*NAC 3*), and abscisic acid-responsive elements-binding factor 2 (*ABF 2*) were obtained from NCBI database [https://www.ncbi.nlm.nih.gov/ (accessed on 29 May 2019); *Vigna radiata* var. radiata (Mungbean)]. Primer 5.0 software was used to design the corresponding primer pairs of the concerned gene sequences for qRT-PCR reaction (Table S1) and were verified to produce a single peak in the melting curve by using a Light Cycler 480® Real-Time PCR System (Roche Applied Science, Penzberg, Germany). Aliquots for qRT-PCR reactions included 10 µL of final volumes containing 1 µL of cDNA, 0.5 µL each primer (10 µM), and 5 µL (2×) of FastStart Essential DNA Green Master mix (Roche Applied Science, Penzberg, Germany) and 3 µL of ddH$_2$O. All reactions were performed in 96-well plates using a Light Cycler 480® Real-Time PCR System with three technical replicates. The thermal conditions are as follows: 95 °C for 5 min, followed by 40 cycles of 95 °C for 10 s, 60 °C for 30 s, and then 72 °C for 30 s. Actin gene (internal control) from mungbean was used to normalize, and transcripts change was calculated using the $2^{-\Delta\Delta CT}$ method.

2.9. Statistical Analysis

The experimental data are presented as the mean and standard error of the mean. All statistical analyses were performed using SPSS statistical package (SPSS Inc., Chicago, IL, USA). In order to find out the differences among the groups, Duncan's multiple range test for one-way ANOVA was performed at a *p*-value < 0.05 statistical significance.

3. Results

3.1. The Effect of Drought Stress on Physiological Traits

Chlorophyll content, plant dry mass, and RWC were observed in two mungbean cultivars following the 12 days of drought stress and compared with the control (Figure 1). After 12 days under drought stress, CO6 plants showed severe wilting, whereas a few leaves of the VRM (Gg) 1 plants had slowly begun to curl. Notably, considerable reductions in chlorophyll content and plant dry mass were observed in the CO6 compared to the respective control. Next, we determined the RWC at the control and drought stress conditions. Mungbean cultivar VRM (Gg) 1 did not show any considerable changes in RWC when subjected to drought stress. In contrast, RWC had a considerable decrease in the CO6 after 12 days of drought stress.

During the 12 days of drought stress, photosynthetic gas exchange parameters (leaf net photosynthesis rate, stomatal conductance, and intercellular CO_2 concentration) were determined in both cultivars (Figure 1). After 6 days of the drought stress, no major difference in the leaf net photosynthetic rate among VRM (Gg) 1 and CO6 was observed compared to their respective control. However, after 12 days of drought stress, the reduction in CO6 was higher than observed in VRM (Gg) 1. The differences observed in stomatal conductance after drought stress were also similar to the differences seen in leaf net photosynthetic rate, with the same trend as in leaf net photosynthetic rate but to a smaller extent. However, after drought stress, CO6 plants showed an increased intercellular CO2 concentration than in VRM (Gg) 1 revealing the opposite relationship between stomatal conductance and leaf net photosynthetic rate.

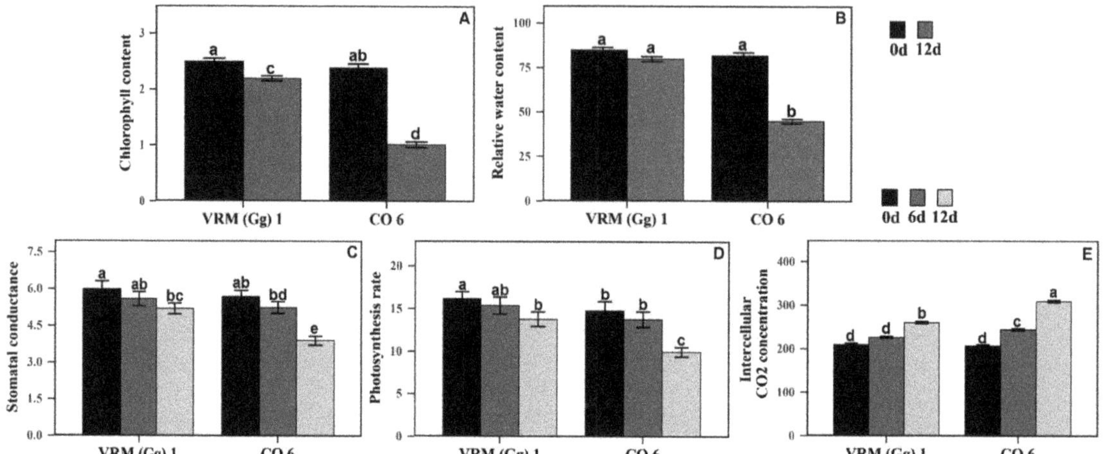

Figure 1. (**A**) Chlorophyll content (mg g^{-1} FW), (**B**) relative water content (%), (**C**) stomatal conductance (mol H$_2$O m^{-2} s^{-1}), (**D**) photosynthesis rate (μmol CO$_2$ m^{-2} s^{-1}), and (**E**) intercellular CO$_2$ concentration (ppm) of two mungbean cultivars (VRM (Gg) 1 and CO6) grown under control and drought-stressed conditions. Values followed by the same letter are not significantly different ($p \leq 0.05$) according to duncan's multiple range test. Bars present means $\pm$ SE (n = 3).

3.2. Proline Content

Proline accumulation is an eminent metabolic response of plants to drought and is also used as an indicator to determine drought tolerance. The difference in proline content during the 12 days of drought stress was estimated in both cultivars. Under the control conditions, a slight difference was seen in the proline content of both cultivars. VRM (Gg) 1 showed a considerably high amount of proline than CO6 after 6 and 12 days of drought stress. The proline content of VRM (Gg) 1 and CO6 under drought stress are presented in Figure 2.

3.3. Changes in MDA and H$_2$O$_2$ Accumulation

During the 12 days of drought stress, MDA, a product of lipid peroxidation was detected among two cultivars. Compared to their respective controls, the amount of MDA content was increased in both cultivars under stress. However, the amounts of upsurge were the degree of difference. After 12 days of stress, the content of MDA was considerably increased in CO6 (119%) than in VRM (Gg) 1 (49%), further revealing that the VRM (Gg) 1 plants cope with a smaller amount of membrane damage compared to CO6 (Figure 2). Likewise, the accumulation level of H$_2$O$_2$ in plants as a response to drought stress imposed on VRM (Gg) 1 and CO6 was also estimated (Figure 2). Six days after drought stress, no major changes in H$_2$O$_2$ content were obtained in both cultivars under stress and control conditions. However, after 12 days, H$_2$O$_2$ levels exhibited a significant increase between both types of plants compared with respective controls. Notably, the magnitudes of increase were different in CO6 (139%) and VRM (Gg) 1 (51.75%). Under control conditions, both cultivars showed no major difference in MDA and H$_2$O$_2$ (Figure 2).

3.4. Activity of Enzymatic and Non-Enzymatic Antioxidants

The effect of drought on enzymatic antioxidants viz., SOD, POD, and CAT were evaluated on the mungbean cultivars (Figure 3). After 12 days of drought stress, the three antioxidant enzyme activities were higher in both cultivars. No major changes in SOD activity were found 6 days after drought stress in the VRM (Gg) 1 and CO6 compared to their respective control. However, a significant increase was found in 12 days after drought stress in both cultivars. The increase in VRM (Gg) 1 was high compared to CO6. Unlike

SOD, POD and CAT activities in the VRM (Gg) 1 and CO6 significantly increased 6 days after drought stress and maintained a higher level after 12 days of drought stress compared to their respective controls. However, the increase in VRM (Gg) 1 was high compared to CO6. Ascorbic acid is one of the most abundant water-soluble antioxidant compounds in plants. The response of ascorbic acid content for drought stress in VRM (Gg) 1 and CO6 was estimated and is presented in Figure 3. The ascorbic acid content was higher in the VRM (Gg) 1 than the CO6. After 12 days of drought stress, the ascorbic acid content was slightly increased in the VRM (Gg) 1. No significant changes were seen in the CO6 than their respective controls.

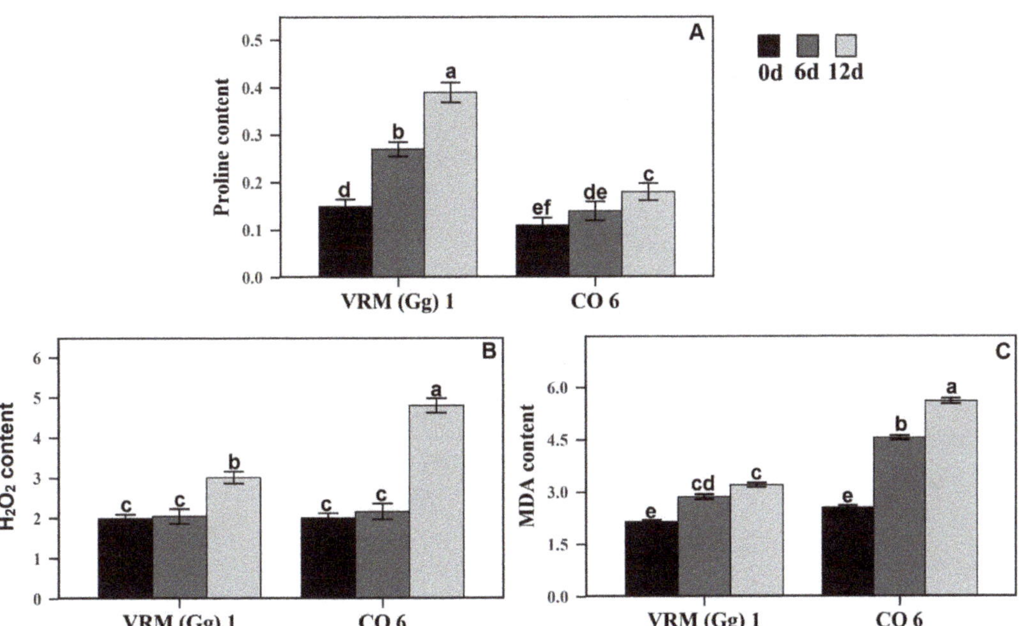

Figure 2. (**A**) Effect of drought stress on proline (μg g^{-1} FW), (**B**) H_2O_2 (μmol g^{-1} FW), and (**C**) MDA contents (μmol g^{-1} FW) in two mungbean cultivars VRM (Gg) 1 and CO6. Values followed by the same letter are not significantly different ($p \leq 0.05$) according to duncan's multiple range test. Bars present means $\pm$ SE ($n = 3$).

3.5. Transcriptional Profiling of Photosynthesis, Antioxidant, and Stress-Responsive Genes

After 12 days of drought stress, four photosynthesis-related genes (*PS II-PsbP, PS II-LHCB, PS I-PsaG/PsaK*, and *PEPC 3*) transcripts levels were analyzed between VRM (Gg) 1 and CO6 by qRT-PCR analysis (Figure 4). Under control conditions, the transcripts of photosynthesis-related genes in VRM (Gg) 1 and CO6 showed no major differences. However, after 12 days of drought stress, transcripts level increased in both cultivars compared with respective controls. Noticeably, the transcripts level was higher in VRM (Gg) 1 than CO6. Further, we analyzed the three antioxidants (*SOD 2, POD*, and *CAT 2*) and four drought stress-responsive genes (*HSP-90, DREB2C, NAC 3*, and *AREB 2*) in both cultivars. Like photosynthesis genes, transcripts of antioxidant and drought stress-responsive genes also did not exhibit major differences in VRM (Gg) 1 and CO6 under controlled conditions. However, the transcripts levels were increased in VRM (Gg) 1 than CO6 after 12 days of drought stress. Together, these results suggest that VRM (Gg) 1 exhibited better performance under drought stress compared to CO6, owing to the transcripts differences of these genes under drought stress.

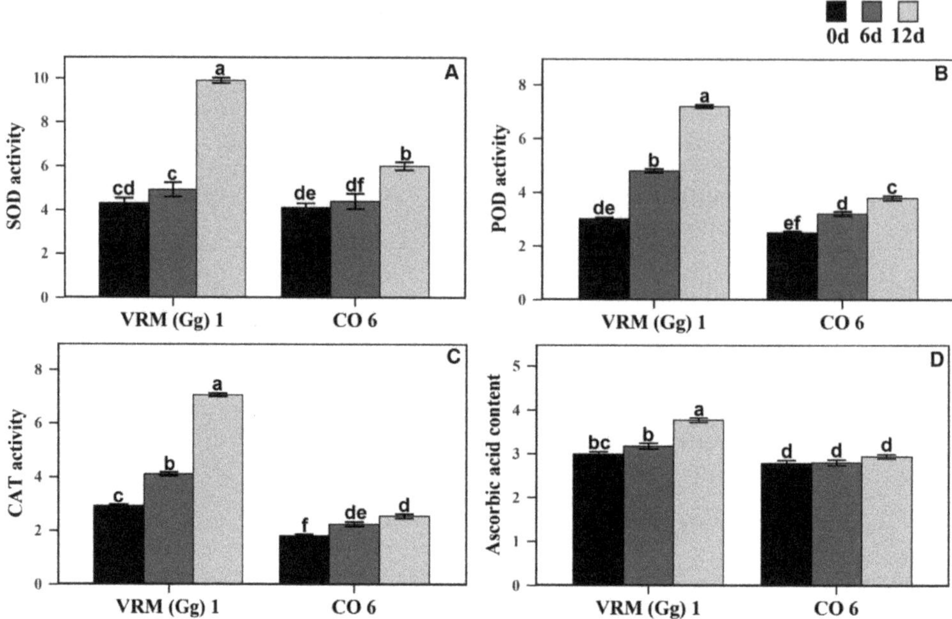

Figure 3. Activities of enzymatic and non-enzymatic antioxidants in two mungbean cultivars VRM (Gg) 1 and CO6. (**A**) SOD (Umg^{-1} Protein), (**B**) POD (Umg^{-1} Protein), (**C**) CAT (Umg^{-1} Protein), and (**D**) ascorbic acid (μmolg^{-1} FW). Values followed by the same letter are not significantly different ($p \leq 0.05$) according to duncan's multiple range test. Bars present means $\pm$ SE ($n = 3$).

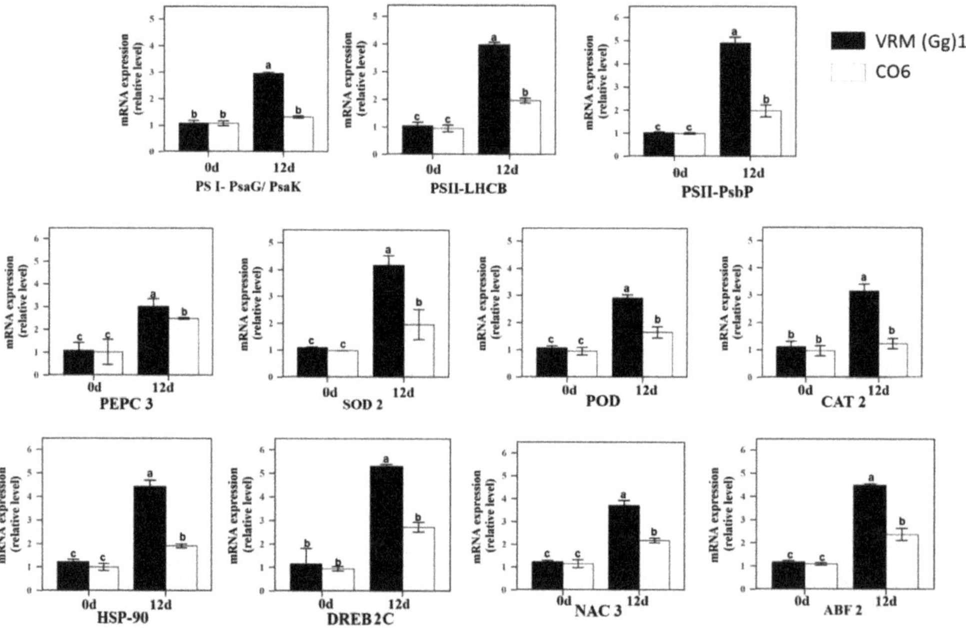

Figure 4. Effect of drought stress on the relative expression level of photosynthesis (*PS II-PsbP, PS II-LHCB, PS I-PsaG/PsaK,* and *PEPC 3*), antioxidants (*SOD 2, POD,* and *CAT 2*), and stress (*HSP-90,*

DREB2C, NAC 3 and *AREB 2*) related genes in two mungbean cultivars VRM (Gg) 1 and CO6. Values followed by the same letter are not significantly different ($p \leq 0.05$) according to duncan's multiple range test. Bars present means $\pm$ SE ($n = 3$).

4. Discussion

In the present study, the response of mungbean cultivars to drought stress was investigated in terms of analyzing the physio-biochemical and transcriptional changes. We found that chlorophyll content and plant dry mass were decreased during drought stress, and the cultivar VRM (Gg) 1 showed a lower decrease compared to CO6, indicating improved photosynthesis and plant growth development. Additionally, VRM (Gg) 1 subjected to drought stress did not show any major changes in RWC. However, the RWC had a significant decrease in CO6, suggesting better water maintaining capacity in VRM (Gg) 1. We also found that when VRM (Gg) 1 and CO6 were exposed to drought stress, the photosynthetic gas exchange parameters (leaf net photosynthetic rate and stomatal conductance) were decreased. Notably, the stomatal conductance in VRM (Gg) 1 plants showed lower decreases than that in CO6 under drought stress. This changing pattern in stomatal conductance is comparable to that in leaf net photosynthetic rate among the plants of VRM (Gg) 1 and CO6, showing that the better leaf net photosynthetic rate in VRM (Gg) 1 was related to the regulation of stomatal conductance. In contrast, the intercellular CO2 concentration of VRM (Gg) 1 was lower than CO6 plants. This fact was because of the varied reduction of leaf net photosynthetic rate in VRM (Gg) 1 and CO6. It might be the reason for increased CO_2 assimilation and decreased intercellular CO_2 concentration in the VRM (Gg) 1 plants compared to CO6. Therefore, photosynthesis and growth in the VRM (Gg) 1 were better when imposed the drought stress. Moreover, we investigated the transcripts level differences of photosynthesis-related genes under drought stress. *PS II-PsbP, PS II-LHCB, PS I-PsaG/PsaK,* and *PEPC 3* are major genes related to photosynthesis. In our study, following 12 days of drought stress, the transcripts levels of all the genes excluding *PS I-PsaG/PsaK* considerably increased in both cultivars compared to their control. Notably, the transcripts level in VRM (Gg) 1 was high compared to CO6, suggesting that VRM (Gg) 1 had better photosynthetic capacity than CO6 under drought stress. Proline accumulation is an important metabolic response to drought in plants and it is also employed as an indicator to regulate the drought tolerance. After 6 and 12 days of drought stress, VRM (Gg) 1 had a much higher level of proline than CO6. Collectively, these results are in line with the reports of Li et al. [39]. Ansari et al. [40] Favero Peixoto-Junior et al. [41], who described that the genotype is referred to as tolerant to drought stress if it keeps better photosynthetic performance, chlorophyll content, RWC, plant dry mass, and proline under stress conditions.

Many studies showed that drought stress causes oxidative damage, characterized as an accumulation of H_2O_2 and MDA [42,43]. Our results showed that, after the drought stress, the accumulation of H_2O_2 and MDA was low in VRM (Gg) 1 whereas high in CO6. A lower H_2O_2 and MDA content in VRM (Gg) 1 specified that it has stable ROS scavenging and better protective mechanism. In several crops, including mungbean, wheat, and muskmelon [28,40,44] under drought stress, the genotypes with contrasting drought tolerance showed differences in H_2O_2 and MDA content. Additionally, a higher accumulation of proline in VRM (Gg) 1 was vital, and it acts as a compatible solute that prevents the protein and membrane structure while also scavenging ROS to maintain the cellular redox level under drought stress and agrees with the statement of Yamada et al. [45]. Plants with tolerance to abiotic stress possess a robust antioxidant system to defend them from oxidative stress by keeping increased antioxidant enzymes and antioxidant molecule activity and contents under stress conditions [46]. SOD, POD, and CAT are major enzymes protecting the plants against ROS-induced oxidative damage [14,47]. Many research reports detailed that the up-regulated expression of SOD, POD, and CAT leads to decreased ROS production under stress conditions [44,48]. In our study, SOD, POD, and CAT activities were heightened over time in VRM (Gg) 1 compared with the CO6 under drought stress and corroborate with the low ROS production observed in VRM (Gg) 1. Abid et al. [44] and Ali et al. [28] showed SOD, POD, and CAT activities were higher in VRM (Gg) 1 than CO6 under drought stress which was in concurrence with our findings. Likewise, the accumulation of non-enzymatic antioxidant ascorbic acid was higher in VRM (Gg) 1 than in CO6. However, the increase was non-significant, and only a marginal increase was observed. Taking together, we conclude that VRM (Gg) 1 has a stronger antioxidant system than CO6.

Drought stress regulates the expression of genes in plants at both transcriptional and post-transcriptional levels. Drought tolerance in plants is thought to be mediated by many genes and biological pathways. Heat-shock proteins serve as molecular chaperones for various client proteins in abiotic stress response and play a significant role in preventing the plants against abiotic stresses. The plant's *HSP90* genes had a major role in response to abiotic stresses, including drought [49,50]. Song et al. [51] reported that the overexpression of *Hsp90* in *Arabidopsis thaliana* improved the plant's sensitivity to drought stresses. VRM (Gg) 1 exhibited higher expression of *Hsp90* in leaves than CO6 during drought stress, suggesting the possible role of preventing the cells from oxidative damage in VRM (Gg) 1 plants. *DREB* is the key transcription factor playing a pivotal role in drought stress response and tolerance to drought [52–55]. In the present study, the *DREB2C* transcription factor was examined in VRM (Gg) 1 and CO6 under drought stress. After ten days of drought stress, the expression level of *DREB2C* in leaves was expressed considerably higher in VRM (Gg) 1 than in CO6. Similarly, *NAC 3* and *AREB 2*, which play critical roles during various abiotic stresses, also showed higher expression in VRM (Gg) 1 compared to CO6. Previously, several studies demonstrated the possible involvement of *NAC 3* and *AREB 2* in drought tolerance [56,57]. From these outcomes, we inferred that it might be possible that the higher expression of *Hsp90, DREB1, NAC 3*, and *AREB 2* is likely to contribute to the better performance of VRM (Gg) 1 during drought stress.

VRM (Gg) 1 had better photosynthetic activity during drought stress, resulting in fewer losses in chlorophyll, relative water content, and plant dry mass. Furthermore, enhanced antioxidative enzyme activities resulted in decreased H_2O_2 and MDA levels in VRM (Gg) 1, limiting oxidative damage. These physio-biochemical alterations positively correlated with increased transcripts of photosynthesis and antioxidant-related genes in VRM (Gg) 1 and were consistent with the earlier studies on, mungbean, faba bean, and alfalfa responses to drought stress [58–60]. The increased transcripts of drought-responsive genes suggest that VRM (Gg) 1 has a stronger genetic base for drought tolerance than CO6. However, supplement research is needed to understand the exact genetic and molecular mechanism underlying drought tolerance.

5. Conclusions

In this study, we found that the mungbean cultivar VRM (Gg) 1 performed well and exhibited tolerance to drought stress compared to CO6, as supported by the physio-biochemical and gene transcriptional changes. In the future, VRM (Gg) 1 will need to be tested under field conditions before being employed in mungbean drought-tolerant breeding programs. VRM (Gg) 1 is a potential source to detect the quantitative trait locus (QTL)/gene (s) associated with drought tolerance. Collectively, the obtained results from our study could be used in the future search for drought-tolerant genotypes or in breeding programs with an aim to obtain tolerant mungbean genotypes.

Supplementary Materials: The following supporting information can be downloaded at: https://www.mdpi.com/article/10.3390/horticulturae8050424/s1. Table S1: Details of primers used for quantitative real-time PCR (qRT-PCR) analysis. Reference [61] is cited in the supplementary materials.

Author Contributions: Conceptualization, A.K., N.S. and M.P.; methodology, A.K., V.G.R. and G.A.; formal analysis, S.P., V.M. and M.D.; investigation, G.A., A.K., A.A. and I.M.; resources, M.P. and N.S.; data curation, G.A., M.A. and A.K.; writing—review and editing, G.A. and A.K.; supervision, M.P. and N.S.; funding acquisition, A.K. and N.S. All authors have read and agreed to the published version of the manuscript.

Funding: A.K. and N.S. acknowledged the support of the Science and Engineering Research Board (SERB), Department of Science and Technology (DST), Government of India for the DST-SERB NPDF fellowship program (PDF/2016/003676).

Acknowledgments: All the authors wish to acknowledge NationalAgricultural Development Programme (NADP)/Rashtriya KrishiVikas Yojana (RKVY)—Government of Tamil Nadu, and Centre of Innovation (CI), Agricultural Collegeand Research Institute, Tamil Nadu Agricultural University, Madurai, for providing instrumentation facilities.

Conflicts of Interest: The authors declare that the research was conducted in the absence of any commercial or financial relationships that could be construed as a potential conflicts of interest.

References

1. Saini, H.S.; Westgate, M.E. Reproductive Development in Grain Crops during Drought. *Adv. Agron.* **1999**, *68*, 59–96. [CrossRef]
2. Zarei, A.R.; Moghimi, M.M.; Mahmoudi, M.R. Parametric and non-parametric trend of drought in arid and semi-arid regions using RDI index. *Water Resour. Manag.* **2016**, *30*, 5479–5500. [CrossRef]
3. Varshney, R.K.; Tuberosa, R.; Tardieu, F. Progress in understanding drought tolerance: From alleles to cropping systems. *J. Exp. Bot.* **2018**, *69*, 3175–3179. [CrossRef]
4. Shinozaki, K.; Yamaguchi-Shinozaki, K. Gene networks involved in drought stress response and tolerance. *J. Exp. Bot.* **2007**, *58*, 221–227. [CrossRef] [PubMed]
5. Toscano, S.; Farieri, E.; Ferrante, A.; Romano, D. Physiological and biochemical responses in two ornamental shrubs to drought stress. *Front. Plant Sci.* **2016**, *7*, 645. [CrossRef] [PubMed]
6. Hoshika, Y.; Omasa, K.; Paoletti, E. Both ozone exposure and soil water stress are able to induce stomatal sluggishness. *Environ. Exp. Bot.* **2013**, *88*, 19–23. [CrossRef]
7. Hu, L.; Wang, Z.; Huang, B. Diffusion limitations and metabolic factors associated with inhibition and recovery of photosynthesis from drought stress in a C3 perennial grass species. *Physiol. Plant.* **2010**, *139*, 93–106. [CrossRef]
8. Manes, F.; Vitale, M.; Donato, E.; Giannini, M.; Puppi, G. Different ability of three Mediterranean oak species to tolerate progressive water stress. *Photosynthetica* **2006**, *44*, 387–393. [CrossRef]
9. Carvalho, M.; Castro, I.; Moutinho-Pereira, J.; Correia, C.; Egea-Cortines, M.; Matos, M.; Rosa, E.; Carnide, V.; Lino-Neto, T. Evaluating stress responses in cowpea under drought stress. *J. Plant Physiol.* **2019**, *241*, 153001. [CrossRef]
10. Dhanda, S.; Sethi, G.; Behl, R. Indices of drought tolerance in wheat genotypes at early stages of plant growth. *J. Agron. Crop Sci.* **2004**, *190*, 6–12. [CrossRef]
11. Miller, G.; Suzuki, N.; Ciftci-Yilmaz, S.; Mittler, R. Reactive oxygen species homeostasis and signalling during drought and salinity stresses. *Plant Cell Environ.* **2010**, *33*, 453–467. [CrossRef] [PubMed]
12. Ibrahim, W.; Zhu, Y.M.; Chen, Y.; Qiu, C.W.; Zhu, S.; Wu, F. Genotypic differences in leaf secondary metabolism, plant hormones and yield under alone and combined stress of drought and salinity in cotton genotypes. *Physiol. Plant.* **2019**, *165*, 343–355. [CrossRef] [PubMed]
13. Al Hassan, M.; Chaura, J.; Donat-Torres, M.P.; Boscaiu, M.; Vicente, O. Antioxidant responses under salinity and drought in three closely related wild monocots with different ecological optima. *AoB Plants* **2017**, *9*, plx009. [CrossRef] [PubMed]
14. Apel, K.; Hirt, H. Reactive oxygen species: Metabolism, oxidative stress, and signal transduction. *Annu. Rev. Plant Biol.* **2004**, *55*, 373–399. [CrossRef]
15. Dudziak, K.; Zapalska, M.; Börner, A.; Szczerba, H.; Kowalczyk, K.; Nowak, M. Analysis of wheat gene expression related to the oxidative stress response and signal transduction under short-term osmotic stress. *Sci. Rep.* **2019**, *9*, 2743. [CrossRef]
16. Manavalan, L.P.; Nguyen, H.T. Drought tolerance in crops: Physiology to genomics. In *Plant Stress Physiology*; Shabala, S., Ed.; CABI: Wallingford, UK, 2017; pp. 1–23.
17. Wu, Q.S.; Zou, Y.N.; Xia, R.X. Effects of water stress and arbuscular mycorrhizal fungi on reactive oxygen metabolism and antioxidant production by citrus (*Citrus tangerine*) roots. *Eur. J. Soil Biol.* **2006**, *42*, 166–172. [CrossRef]
18. Alderfasi, A.A.; Alzarqaa, A.A.; AL-Yahya, F.A.; Roushdy, S.S.; Dawabah, A.A.; Alhammad, B.A. Effect of combined biotic and abiotic stress on some physiological aspects and antioxidant enzymatic activity in mungbean (*Vigna radiate* L.). *Afr. J. Agric. Res.* **2017**, *12*, 700–705.
19. Masoumi, H.; Darvish, F.; Daneshian, J.; Normohammadi, G.; Habibi, D. Effects of water deficit stress on seed yield and antioxidants content in soybean (*Glycine max* L.) cultivars. *Afr. J. Agric. Res.* **2011**, *6*, 1209–1218.
20. Baroowa, B.; Gogoi, N. The effect of osmotic stress on anti-oxidative capacity of black gram (*Vigna mungo* L.). *Exp. Agric.* **2017**, *53*, 84–99. [CrossRef]

21. Xu, C.; Xia, C.; Xia, Z.; Zhou, X.; Huang, J.; Huang, Z.; Liu, Y.; Jiang, Y.; Casteel, S.; Zhang, C. Physiological and transcriptomic responses of reproductive stage soybean to drought stress. *Plant Cell Rep.* **2018**, *37*, 1611–1624. [CrossRef]

22. Mahdavi Mashaki, K.; Garg, V.; Nasrollahnezhad Ghomi, A.A.; Kudapa, H.; Chitikineni, A.; Zaynali Nezhad, K.; Yamchi, A.; Soltanloo, H.; Varshney, R.K.; Thudi, M. RNA-Seq analysis revealed genes associated with drought stress response in kabuli chickpea (*Cicer arietinum* L.). *PLoS ONE* **2018**, *13*, e0199774. [CrossRef] [PubMed]

23. López, C.M.; Pineda, M.; Alamillo, J.M. Transcriptomic response to water deficit reveals a crucial role of phosphate acquisition in a drought-tolerant common bean landrace. *Plants* **2020**, *9*, 445. [CrossRef] [PubMed]

24. Karthikeyan, A.; Shobhana, V.; Sudha, M.; Raveendran, M.; Senthil, N.; Pandiyan, M.; Nagarajan, P. Mungbean yellow mosaic virus (MYMV): A threat to green gram (*Vigna radiata*) production in Asia. *Int. J. Pest Manag.* **2014**, *60*, 314–324. [CrossRef]

25. Project Coordinators Report. *All India Coordinated Research Project on MULLaRP (Mungbean, Urdbean, Lentil, Lathyrus, Rajmash, Fieldpea)*; ICAR-Indian Institute of Pulses Research: Kanpur, India, 2018; pp. 1–46.

26. Ahmad, A.; Selim, M.M.; Alderfasi, A.; Afzal, M. Effect of drought stress on mung bean (*Vigna radiata* L.) under arid climatic conditions of Saudi Arabia. In *Ecosystems and Sustainable Development*; Garcia, M., Brebbia, C.A., Eds.; WIT Press: Southampton, UK, 2015; Volume 192, pp. 185–193.

27. Dutta, P.; Bera, A. Screening of mungbean genotypes for drought tolerance. *Legume Res.* **2008**, *31*, 145–148.

28. Ali, Q.; Javed, M.T.; Noman, A.; Haider, M.Z.; Waseem, M.; Iqbal, N.; Waseem, M.; Shah, M.S.; Shahzad, F.; Perveen, R. Assessment of drought tolerance in mung bean cultivars/lines as depicted by the activities of germination enzymes, seedling's antioxidative potential and nutrient acquisition. *Arch. Agron. Soil Sci.* **2018**, *64*, 84–102. [CrossRef]

29. Bangar, P.; Chaudhury, A.; Tiwari, B.; Kumar, S.; Kumari, R.; Bhat, K.V. Morphophysiological and biochemical response of mungbean [*Vigna radiata* (L.) Wilczek] varieties at different developmental stages under drought stress. *Turk. J. Biol.* **2019**, *43*, 58–69. [CrossRef]

30. Arnon, D.I. Copper enzymes in isolated chloroplasts. Polyphenoloxidase in Beta vulgaris. *Plant Physiol.* **1949**, *24*, 1–15. [CrossRef]

31. Barrs, H.; Weatherley, P. A re-examination of the relative turgidity technique for estimating water deficits in leaves. *Aust. J. Biol. Sci.* **1962**, *15*, 413–428. [CrossRef]

32. Bates, L.S.; Waldren, R.P.; Teare, I. Rapid determination of free proline for water-stress studies. *Plant Soil* **1973**, *39*, 205–207. [CrossRef]

33. Stewart, R.R.; Bewley, J.D. Lipid peroxidation associated with accelerated aging of soybean axes. *Plant Physiol.* **1980**, *65*, 245–248. [CrossRef]

34. Loreto, F.; Velikova, V. Isoprene produced by leaves protects the photosynthetic apparatus against ozone damage, quenches ozone products, and reduces lipid peroxidation of cellular membranes. *Plant Physiol.* **2001**, *127*, 1781–1787. [CrossRef] [PubMed]

35. Madamanchi, N.R.; Alscher, R.G. Metabolic bases for differences in sensitivity of two pea cultivars to sulfur dioxide. *Plant Physiol.* **1991**, *97*, 88–93. [CrossRef] [PubMed]

36. Chance, B.; Maehly, A. Assay of catalases and peroxidases. *Methods Enzymol.* **1955**, *2*, 764–775.

37. Aebi, H. Catalase in vitro. *Methods Enzymol.* **1984**, *105*, 121–126.

38. Arakawa, N.; Otsuka, M.; Kurata, T.; Inagaki, C. Separative determination of ascorbic acid and erythorbic acid by high-performance liquid chromatography. *J. Nutr. Sci. Vitaminol.* **1981**, *27*, 1–7. [CrossRef]

39. Li, R.-H.; Guo, P.-G.; Michael, B.; Stefania, G.; Salvatore, C. Evaluation of chlorophyll content and fluorescence parameters as indicators of drought tolerance in barley. *Agric. Sci. Chin.* **2006**, *5*, 751–757. [CrossRef]

40. Ansari, W.; Atri, N.; Singh, B.; Pandey, S. Changes in antioxidant enzyme activities and gene expression in two muskmelon genotypes under progressive water stress. *Biol. Plant.* **2017**, *61*, 333–341. [CrossRef]

41. Fávero Peixoto-Junior, R.; Mara de Andrade, L.; dos Santos Brito, M.; Macedo Nobile, P.; Palma Boer Martins, A.; Domingues Carlin, S.; Vasconcelos Ribeiro, R.; de Souza Goldman, M.H.; Nebo Carlos de Oliveira, J.F.; Vargas de Oliveira Figueira, A.; et al. Overexpression of ScMYBAS1 alternative splicing transcripts differentially impacts biomass accumulation and drought tolerance in rice transgenic plants. *PLoS ONE* **2018**, *13*, e0207534. [CrossRef]

42. Amoah, J.N.; Ko, C.S.; Yoon, J.S.; Weon, S.Y. Effect of drought acclimation on oxidative stress and transcript expression in wheat (*Triticum aestivum* L.). *J. Plant Interact.* **2019**, *14*, 492–505. [CrossRef]

43. Nxele, X.; Klein, A.; Ndimba, B. Drought and salinity stress alters ROS accumulation, water retention, and osmolyte content in sorghum plants. *S. Afr. J. Bot.* **2017**, *108*, 261–266. [CrossRef]

44. Abid, M.; Ali, S.; Qi, L.K.; Zahoor, R.; Tian, Z.; Jiang, D.; Snider, J.L.; Dai, T. Physiological and biochemical changes during drought and recovery periods at tillering and jointing stages in wheat (*Triticum aestivum* L.). *Sci. Rep.* **2018**, *8*, 1–15. [CrossRef] [PubMed]

45. Yamada, M.; Morishita, H.; Urano, K.; Shiozaki, N.; Yamaguchi-Shinozaki, K.; Shinozaki, K.; Yoshiba, Y. Effects of free proline accumulation in petunias under drought stress. *J. Exp. Bot.* **2005**, *56*, 1975–1981. [CrossRef] [PubMed]

46. Chang-Quan, W.; Rui-Chang, L. Enhancement of superoxide dismutase activity in the leaves of white clover (*Trifolium repens* L.) in response to polyethylene glycol-induced water stress. *Acta Physiol. Plant.* **2008**, *30*, 841–847. [CrossRef]

47. Wang, S.Y.; Chen, H.; Ehlenfeldt, M.K. Variation in antioxidant enzyme activities and nonenzyme components among cultivars of rabbiteye blueberries (*Vaccinium ashei Reade*) and *V. ashei* derivatives. *Food Chem.* **2011**, *129*, 13–20. [CrossRef]

48. Huo, Y.; Wang, M.; Wei, Y.; Xia, Z. Overexpression of the maize psbA gene enhances drought tolerance through regulating antioxidant system, photosynthetic capability, and stress defense gene expression in tobacco. *Front. Plant Sci.* **2016**, *6*, 1223. [CrossRef]

49. Song, H.; Wang, H.; Xu, X. Overexpression of AtHsp90. 3 in Arabidopsis thaliana impairs plant tolerance to heavy metal stress. *Biol. Plant.* **2012**, *56*, 197–199. [CrossRef]
50. Xu, J.; Xue, C.; Xue, D.; Zhao, J.; Gai, J.; Guo, N.; Xing, H. Overexpression of GmHsp90s, a heat shock protein 90 (Hsp90) gene family cloning from soybean, decrease damage of abiotic stresses in Arabidopsis thaliana. *PLoS ONE* **2013**, *8*, e69810. [CrossRef]
51. Song, H.; Zhao, R.; Fan, P.; Wang, X.; Chen, X.; Li, Y. Overexpression of AtHsp90. 2, AtHsp90. 5 and AtHsp90. 7 in Arabidopsis thaliana enhances plant sensitivity to salt and drought stresses. *Planta* **2009**, *229*, 955–964. [CrossRef]
52. Chen, Y.; Huang, L.; Yan, H.; Zhang, X.; Xu, B.; Ma, X. Cloning and characterization of an ABA-independent DREB transcription factor gene, HcDREB2, in Hemarthria compressa. *Hereditas* **2016**, *153*, 1–7. [CrossRef]
53. Nakashima, K.; Shinwari, Z.K.; Sakuma, Y.; Seki, M.; Miura, S.; Shinozaki, K.; Yamaguchi-Shinozaki, K. Organization and expression of two Arabidopsis DREB2 genes encoding DRE-binding proteins involved in dehydration-and high-salinity-responsive gene expression. *Plant Mol. Biol.* **2000**, *42*, 657–665. [CrossRef]
54. Qin, Q.-L.; Liu, J.-G.; Zhang, Z.; Peng, R.-H.; Xiong, A.-S.; Yao, Q.-H.; Chen, J.-M. Isolation, optimization, and functional analysis of the cDNA encoding transcription factor OsDREB1B in *Oryza Sativa* L. *Mol. Breed.* **2007**, *19*, 329–340. [CrossRef]
55. Sakuma, Y.; Maruyama, K.; Osakabe, Y.; Qin, F.; Seki, M.; Shinozaki, K.; Yamaguchi-Shinozaki, K. Functional analysis of an Arabidopsis transcription factor, DREB2A, involved in drought-responsive gene expression. *Plant Cell* **2006**, *18*, 1292–1309. [CrossRef] [PubMed]
56. Fang, Y.; Liao, K.; Du, H.; Xu, Y.; Song, H.; Li, X.; Xiong, L. A stress-responsive NAC transcription factor SNAC3 confers heat and drought tolerance through modulation of reactive oxygen species in rice. *J. Exp. Bot.* **2015**, *66*, 6803–6817. [CrossRef] [PubMed]
57. Yoshida, T.; Fujita, Y.; Sayama, H.; Kidokoro, S.; Maruyama, K.; Mizoi, J.; Shinozaki, K.; Yamaguchi-Shinozaki, K. AREB1, AREB2, and ABF3 are master transcription factors that cooperatively regulate ABRE-dependent ABA signaling involved in drought stress tolerance and require ABA for full activation. *Plant J.* **2010**, *61*, 672–685. [CrossRef]
58. Abid, G.; M'hamdi, M.; Mingeot, D.; Aouida, M.; Aroua, I.; Muhovski, Y.; Sassi, K.; Souissi, F.; Mannai, K.; Jebara, M. Effect of drought stress on chlorophyll fluorescence, antioxidant enzyme activities and gene expression patterns in faba bean (*Vicia faba* L.). *Arch. Agron. Soil Sci.* **2017**, *63*, 536–552. [CrossRef]
59. Filippou, P.; Antoniou, C.; Fotopoulos, V. Effect of drought and rewatering on the cellular status and antioxidant response of Medicago truncatula plants. *Plant Signal. Behav.* **2011**, *6*, 270–277. [CrossRef]
60. Kumar, S.; Ayachit, G.; Sahoo, L. Screening of mungbean for drought tolerance and transcriptome profiling between drought-tolerant and susceptible genotype in response to drought stress. *Plant Physiol. Biochem.* **2020**, *157*, 229–238. [CrossRef]
61. Li, S.; Wang, R.; Jin, H.; Ding, Y.; Cai, C. Molecular Characterization and Expression Profile Analysis of Heat Shock Transcription Factors in Mungbean. *Front. Genet.* **2019**, *9*, 736. [CrossRef]

 horticulturae

Article

Graded Moisture Deficit Effect on Secondary Metabolites, Antioxidant, and Inhibitory Enzyme Activities in Leaf Extracts of *Rosa damascena* Mill. var. *trigentipetala*

Kamel Hessini [1,*], Hanen Wasli [2], Hatim M. Al-Yasi [1], Esmat F. Ali [1], Ahmed A. Issa [1], Fahmy A. S. Hassan [3] and Kadambot H. M. Siddique [4]

[1] Department of Biology, College of Sciences, Taïf University, P.O. Box 888, Taif 21974, Saudi Arabia; h.alyasi@tu.edu.sa (H.M.A.-Y.); a.esmat@tu.edu.sa (E.F.A.); a.hissa@tu.edu.sa (A.A.I.)
[2] Laboratoire des Plantes Aromatiques et Médicinales, Centre de Biotechnologie de Borj-Cédria (LR15CBBC06), BP 901, Hammam Lif 2050, Tunisia; hanenwasli@gmail.com
[3] Horticulture Department, Faculty of Agriculture, Tanta University, Tanta 31527, Egypt; fahmy_hssn@yahoo.com
[4] The UWA Institute of Agriculture, The University of Western Australia, Perth, WA 6001, Australia; kadambot.siddique@uwa.edu.au
* Correspondence: k.youssef@tu.edu.sa

Citation: Hessini, K.; Wasli, H.; Al-Yasi, H.M.; Ali, E.F.; Issa, A.A.; Hassan, F.A.S.; Siddique, K.H.M. Graded Moisture Deficit Effect on Secondary Metabolites, Antioxidant, and Inhibitory Enzyme Activities in Leaf Extracts of *Rosa damascena* Mill. var. *trigentipetala*. *Horticulturae* **2022**, *8*, 177. https://doi.org/10.3390/horticulturae8020177

Academic Editors: Stefania Toscano, Giulia Franzoni and Sara Álvarez

Received: 22 January 2022
Accepted: 17 February 2022
Published: 21 February 2022

Abstract: Drought affects plant growth and yield in many agricultural areas worldwide by producing negative water potentials in the root zone that reduce water availability, affecting plant development and metabolism. This study investigated the effect of varying moisture regimes (100% field capacity (FC), well-watered plants, 50% FC (moderate water stress), and 25% FC (severe water stress)) on growth parameters, chlorophyll content, and bioactive molecule patterns, and the impact on antioxidant, lipoxygenase (LOX), and acetylcholinesterase (AChE) activities in *Rosa damascena*. The water deficit treatments reduced biomass production for both treatments (-29 and -33%, respectively, for MWS and SWS) and total chlorophyll (-18 and -38% respectively for MWS and SWS), relative to the control. The 50% FC treatment had the greatest effect on the phenolic profiles and their respective functionalities, with significant increases in the levels of total phenolic, benzoic (gallic, *p*-coumaric, and syringic acids) ($+32\%$), and cinnamic (caffeic and *trans*-cinnamic acid) acids ($+19\%$) and flavonoids (epicatechin-3-*O*-gallate) ($+15\%$) compared to well-watered leaves (control leaves). The 50% FC treatment also exhibited the highest potential antioxidant activities (apart from NO-quenching activity), evidenced by the lowest IC_{50} and EC_{50} values. The inhibitory LOX and AChE capacities varied depending on the severity of stress, with superior activity in the 50% FC treatment. Overall, the drought tolerance in rose was associated mainly with its suitable manipulation of antioxidant production and orderly regulation of LOX and AChE activities.

Keywords: AChE activity; antioxidant; drought stress; LOX activity; phenolics; damask rose

1. Introduction

The severity and incidence of drought are expected to increase with the predicted change in typical precipitation patterns associated with climate change [1]. Water deficits are anticipated to reduce world crop production by up to 30% by 2025 compared to current yields [2]. In arid and semi-arid zones, the potential of water resources to expand landscapes and grow ornamental plants is threatened. Water distribution to the floral industry is in strong competition with other demands, such as agriculture, urban management, and human consumption [3], and should be used optimally and with high efficiency [4].

Limited water supply to plants incites a chemical signal in the aerial system through xylem sap, eliciting partial stomatal closure to avert water loss by evaporation. As a result, plants shift to a water-saving strategy that decreases intracellular CO_2, reducing the amount of $NADPH^+$, H^+, and ATP available for CO_2 fixation within the Calvin cycle, thus decreasing $NADP^+$ regeneration and affecting the photosynthetic electron transport chain [5,6]. Such effects promote the production of free radicals and reactive oxygen species (ROS), generating oxidative stress [7].

Alternatively, to manage water status, a non-antioxidant system (phenolic compounds) can be synthesized in different plant parts to keep ROS production below toxic levels during abiotic/biotic stresses. Depending on the stress intensity and plant efficiency to trigger these mechanisms, the production of such metabolites can increase or decrease under drought conditions [8].

To confront drought constraints, dehydrated plants could normalize a latent source of phenols valuable for economic exploitation. Nonetheless, abiotic constraints have the opposite effect on polyphenol yields, i.e., the increment of polyphenol quantity in tissues is negatively correlated with plant biomass production [9]. Furthermore, water scarcity may be related to increases in phenolic pools by reallocating the incorporated C as plant growth progressively declines [10]; thus, optimal polyphenol yields would be required with respect to stress-tolerant species [11].

Saudi Arabia is rich in flora, including a multiplicity of aromatic and ornamental species such as Damask rose (*Rosa damascena* Mill. var. *trigentipetala*) that belongs to the Rosaceae family, a perennial bushy shrub renowned in the perfumery, cosmetic trade, and food industries [12].

The major bioactive molecules isolated from different organs of *Rosa damascena* are flavonoids, glycosides (kaempferol, cyanidin 3,5, D-glycoside, and quercetin), gallic acid, terpenes, and anthocyanins. The leaves are noteworthy sources of vitamins C, A, B, and K, pectin, tannins, and carotenoids [13].

The flowers of *R. damascena* have analgesic, anti-inflammatory, astringent, antibacterial, antidepressant, and antiviral activity, as well as diuretic effects, and they are used in popular medicine as a sedative [13]. A leaf methanol extract of Damask rose with high amounts of (+)-catechin and (+)-epi-catechin had higher antioxidant activities than butylated hydroxytoluene (BHT), Trolox, and butylated hydroxyanisole (BHA) standards.

Agricultural practices, genetic makeup, and environmental factors affect the quality of plant products [11]. In semi-arid and rainfed areas where water is scant, incorporating different shade net houses and mulch types over the soil surface are key to meeting the increasing demand for herbs [3,4].

Despite the economic importance of *Rosa damascena* Mill. var. *trigentipetala* for the livelihood for Saudi smallholder farmers, it is cultivated in a traditional and primitive manner [14]. Yet, no published information is available on (i) its limits of tolerance, or (ii) the underlying mechanisms implicated in its tolerance to water deficit. Therefore, an investigation of its responses to drought and the mechanisms involved may support to know how to improve drought tolerance in Rosa species. This study aimed to (i) determine the limits of tolerance to water deficit of Damask rose, and (ii) identify the main physiological and biochemical mechanisms that are linked to drought resistance. Such information will be crucial in defining culture conditions that optimize biomass, biomolecules production and anti-oxidation efficiency, along with a better valorization of this underused species in water management to improve secondary metabolites production with a global goal of new water-efficient genotype screening programs.

2. Materials and Methods

2.1. Experimental Design and Irrigation Treatment

The experiment was undertaken from January to May 2020 in a greenhouse at the Biology Department, Faculty of Science, Taïf University, Saudi Arabia (21°26′02.4″ N, 40°29′36.9″ E), with a natural photoperiod (approximately 14 h light), temperature (day/night) of 32/22 °C, and relative humidity of 70%.

Two-year-old rooted cuttings of *Rosa damascena* Mill. var. *trigentipetala* were transplanted to Wagner pots (height: 30 cm; diameter: 20 cm) filled with sandy soil and watered daily with half-strength nutrient solution [15]. Uniform cuttings were selected, grouped into ten replicates (main factor), and exposed to one of three irrigation regimes—25%, 50%, or 100% field capacity (FC) as non-stress (WW), moderate water stress (MWS), and severe water stress (SWS) treatments, respectively—for 90 days. The pots were watered to their corresponding weights every two days with quarter-, half-, and full-strength Hewitt nutrient solution, respectively. Soil FC (%) for the 100%, 50%, and 25% FC treatments were 11.5%, 5.75%, and 2.9%, respectively. Evaporation from the soil surface was prevented by enclosing the pots in plastic bags. Ten pots without plants were used to monitor soil evaporation.

2.2. Growth Parameters and Leaf Water Potential (Ψw) Measurement

Leaf water potential (Ψ_{leaf}) was recorded on five mature and fully spread leaves using a Scholander pressure chamber (Soil Moisture Equipment Corp., Santa Barbara, CA, USA) at first light (Ψ_{PD}, 07:00 h) and midday (Ψ_{MD}, 12:00 h).

Fresh material of each plant was then placed in clean paper bags, labeled, and oven-dried at 60 °C for 48 h to determine the respective dry weight (DW) following the protocol of Wasli et al. [8].

2.3. Relative Chlorophyll Content (RCC)

The pigment concentrations in rose leaves were determined by measuring the absorption spectra of frond extracts using a UV spectrophotometer (Spectro UV-VIS Dual Beam 8 auto cell UUS-2700). Two hundred mg of leaf plant material frozen in liquid N_2 was ground to a fine powder (on ice) and immediately immersed in 5 mL acetone (80/20 *v/v*) solution. The total extraction took place after 72 h in darkness at 4 °C, with the absorbance of the extracts measured at 663, 645 and 470 nm for Chl *a* and Chl *b* and carotenoids, respectively [16]. Varian 220Z, Mulgrave, Victoria, Australia

2.4. Characterization and Quantification of Phenolic Pools by Colorimetric and Chromatographic Analysis

Contents of total phenolic compounds and flavonoids (obtained with 3 g dry powder in 30 mL methanol 80%) were determined according to the method of Wasli et al. [17], and the results were expressed as mg of gallic acid or mg catechin per gram of dried residue, respectively using a UV-spectrometer (Varian 220Z, Mulgrave, Victoria, Australia).

In turn, to characterize and quantify individual phenolics dried samples (evaporated with in a rotavap at 40 °C) were hydrolyzed according to the method of Proestos et al. [18] with some modifications. Next, 10 mL of MeOH (80:20 *v/v*) containing butylated hydroxytoluene (0.5 g/L) was added to 250 µg of the dried sample. Then, 5 mL of 1 M HCl was added. The mixture was stirred carefully and sonicated for 15 min and refluxed in a water bath at 90 °C for 2 h. The obtained mixture was injected to RP-HPLC using a system model Agilent 1200 with the UV spectra of standards measured from 200–400 nm. The column was a reversed phase Zorbax SB-C18 of 4.6 mm × 250 mm and 3.5 µm particle size. The column temperature was thermostated at 25 °C. The injected sample volume was 20 µL, and peaks were monitored at 280 nm. The mobile phase comprised acetonitrile (solvent A) and water/sulfuric acid (98:2) (solvent B). The optimized gradient elution occurred as at a flow rate of 0.6 mL/min: 0–5 min, 10–20% A; 5–10 min, 20–30% A; 10–15 min, 30–50% A; 15–20 min, 50–70% A; 20–25 min, 70–90% A; 25–30 min, 90–50% A; 30–35 min, return to initial conditions.

The compounds were identified by comparing the retention times and peak area with pure standards and reported as mg per g sample dry weight. For quantitative analysis, the limits of detection and quantification were calculated from calibration curve parameters obtained by injecting known concentrations of a different standard. The results were expressed in mg per g of dry weight.

Standards with high purities were purchased from Sigma–Aldrich, including catechin hydrate ($\geq$96% purity), chlorogenic acid ($\geq$96% purity), caffeic acid ($\geq$97% purity), *p*-coumaric acid (98% purity), ellagic acid ($\geq$95% purity), epicatechin-3-*O*-gallate ($\geq$96%

purity), ferulic acid ($\geq$95% purity), gallic acid ($\geq$95% purity), rosmarinic acid (95% purity), luteolin-7-*O*-glucoside ($\geq$95% purity), kaempferol ($\geq$97% purity), sinapic acid (98% purity), syringic acid ($\geq$96% purity), and *trans*-cinnamic acid ($\geq$95% purity).

2.5. Biological Activities

The antiradical capacity of rose extract (RE) against the 2,2′-azino-bis (3-ethlybenzothiazoline-6-sulfonic acid) (ABTS) radical was assessed according to Wasli et al. [8]. Briefly, 250 µL of stable radical ABTS (prepared by reacting ABTS stock solution (7 mM) with potassium persulfate (2.45 mM) in a 1:1 ratio) was added to 50 µL of increasing RE concentrations ranging from 25 to 100 µg/mL. After 6 min of incubation at room temperature, the absorbance was read against a blank at 734 nm using an ELX800 microplate reader (Bio-Tek Instruments, Inc.; Winooski, VT, USA). The ABTS scavenging ability was expressed as IC_{50} (mg/mL), the inhibiting concentration of 50% of the synthetic radical.

The inhibition percentage (IP%) of ABTS radical was calculated as follows:

$$IP\ (\%) = [(A_{control} - A_{sample})/A_{control}] \times 100 \tag{1}$$

The NO scavenging assay followed the protocol of Wasli et al. [16]. Briefly, 200 µL of different rose extracts (25–150 µg/mL) were mixed with 200 µL sodium nitroprusside (3.33 mM) in 100 mM PBS (pH 7.4). The reaction was initiated by adding Griess reagent, and the absorbance was measured at 562 nm. The results were expressed as CI_{50} (µg/mL).

For ORAC assessment, AAPH (240 mM), fluorescein (70.30 nM), and Trolox (3.24–130.88 µM) were prepared in 75 mM of phosphate buffer with a pH of 7.4. Fluorescence intensity (an excitation wavelength of 485 nm and emission wavelength of 520 nm) was applied every 90 s over a total measurement period of 120 min. The results were expressed in micromoles of Trolox equivalents (TE) per gram (mmol TE/g).

The FRAP assay (reaction of reductants) was traduced by the altering of the test solution from yellow to green.

One milliliter of RE from different treatments at different concentrations ranging from 50 to 500 µg/mL was mixed with 2.5 mL of Na_3PO_4 buffer (0.2 M, pH 6.6) and 2.5 mL potassium ferricyanide ($K_3Fe\ (CN)_6$; 1% w/v). The mixtures were incubated in a water bath at 50 °C for 20 min. Then 2.5 mL of TCA (10%, w/v) were inserted followed by a vigorous centrifugation for 10 min at 650 g. At the final step, 2.5 mL of the supernatant was blended with 2.5 mL of deionized water and 0.5 mL of $FeCl_3$ solution (0.1%, w/v). The absorbance was assessed at 700 nm against a blank sample and ascorbic acid was used as a positive control. The EC_{50} value (µg/mL) expressed the RP [17].

For the β-carotene test, 20 mg of β-carotene was suspended in 10 mL of chloroform; linoleic acid (50 mg) and Tween 80 (1 g) were then added to 1 mL of this solution [19]. Chloroform was removed using a vacuum at 40 °C, before adding 100 mL of oxygenated water. The resulting β-carotene/linoleic acid emulsion was vigorously shaken before 250 µL was added to each well of 96-well microliter plates along with 50 µL of test samples. The initial absorbance at 470 nm was recorded.

The emulsion system with two controls (one containing BHA as a positive control (Figure 1) and the other with the same volume of distilled water instead of the extracts) was incubated at 50 °C for 2 h, and the absorbance at 470 nm was read using a model ELX800 microplate reader (Bio-Tek Instruments, Inc.; Winooski, VT, USA).

Readings for all samples were taken immediately and after 2 h incubation. Blanching inhibition of the β-carotene was determined as follows:

$$\%\ inhibition = \frac{((C_{t=0} - C_{t=2}) - (E_{t=0} - E_{t=2})}{(C_{t=0} - C_{t=2})} \times 100 \tag{2}$$

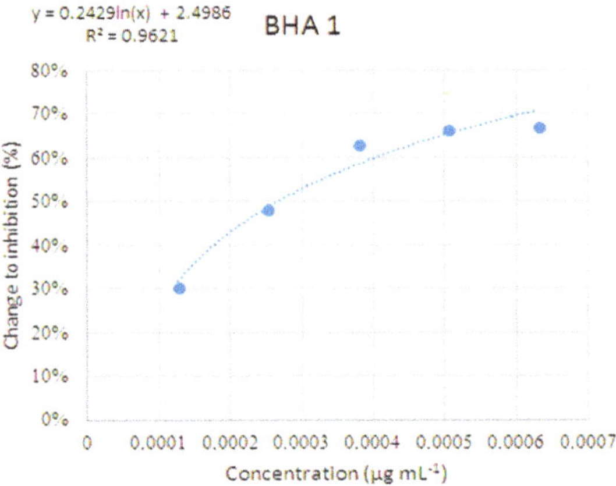

Figure 1. Standard curve of butylated hydroxyanisol (BHA) as a positive control in the β-carotene linoleic acid model system.

2.6. Lipoxygenase (LOX) Inhibitory Activity

LOX activity was assessed according to Wasli et al. [17]. In a 96-well quartz plate, 10 µL *R. damascena* leaf extract (10–100 µg/mL) was added to 5 µL enzyme solution (0.054 g/mL), 50 µL linoleic acid (0.001 M), and borate buffer 937 µL (0.1 M, pH 9), and the absorbance measured at 234 nm. Enzymatic activity was approximated as $(A_0 - A_e/A_0) \times 100$, where A_0 is the absorbance of the control reaction, and A_e is the absorbance of the extract. The results were expressed as IC_{50} values.

2.7. Acetylcholinesterase (AChE) Inhibitory Activity

Sixty microliters of *R. damascena* sample at different concentrations (50–100 µg/mL) was mixed with 425 µL Tris-HCl buffer (0.1 M, pH 8) and 25 µL enzyme (0.28 µM/L), incubated at 37 °C for 15 min, before adding 75 µL substrate (0.005 g iodine acetylcholine in 10 mL buffer) and 475 µL 5,5′-dithiobis (2-nitrobenzoic acid) (DTNB) (0.059 g in 50 mL buffer) to finish the reaction. The absorbance was read at 405 nm, and the results were traduced to IC_{50} values [20].

2.8. Data Analysis

Data were analyzed using one-way ANOVA followed by Tukey's post hoc test. The statistical tests were applied using Graph Pad Prism, version 6, at a $p < 0.05$ significance level. Multivariate data analysis was carried out using Pearson's correlation in XLSTAT, considering variables centered on their means.

3. Results and Discussion

3.1. Effect of Moderate and Severe Drought Stress on Growth Activity and Chlorophyll Content

Biomass production and water status are considered to be the most-used criteria for assessing plant behavior to osmotic stress [14,21]. Both water deficit treatments decreased whole-plant FW and DW, declining by approximately 29% under moderate (50% FC) water stress and 48% and 33% under severe (25% FC) water stress, respectively, relative to the well-watered plants. In turn, water stress increased the root-to-shoot dry weight ratio, relative to the control plants (Table 1, $p \leq 0.05$).

Water stress reduced leaf water potential (Ψ_w) from −1.6 MPa (for control plants) to −2 and −2.4 MPa, respectively, for MWS and SWS (Table 1, $p \leq 0.05$). Leaf water content also declined in response to water stress, more so in the SWS treatment (Table 1, $p \leq 0.05$).

Reduced biomass accumulation in response to drought is mainly due to the reduction in leaf biomass explained by the reduction in leaf area, leaf number, and leaf size [22]. The greater inhibition in shoot growth than root growth (as indicated by the lower shoot-to-root ratio) is explained by the preferential allocation of dry matter to roots, representing a criterion for drought adaptation [23,24]. Changes in root architecture in cereals due to the fact of enhanced cytokinin degradation are a promising strategy for enhancing nutrient uptake, biofortification, and drought tolerance [25].

Table 1. Changes in physiological parameters of *Rosa damascena* Mill. var. *trigentipetala* plants under three watering regimes: well-watered (100% field capacity (FC), WW), moderate water stress (50% FC, MWS), or severe water stress (25% FC, SWS).

Parameters	Control	MWS	SWS
Whole plant FW (g/plant)	20.3 ± 2.6 [a]	14.4 ± 2.2 [b]	10.5 ± 1.0 [c]
Whole plant DW (g/plant)	8.6 ± 0.6 [a]	6.07 ± 1.3 [b]	5.7 ± 0.5 [bc]
Shoot/root ratio	0.81 ± 0.10 [a]	0.55 ± 0.1 [b]	0.54 ± 0.09 [b]
Leaf Ψw (–MPa)	1.6 ± 0.1 [c]	2.00 ± 0.05 [b]	2.40 ± 0.14 [a]
Leaf WC (%)	63.4 ± 1.5 [a]	60.6 ± 1.4 [ab]	52.6 ± 2.4 [b]
Chl *a* (mg/g FW)	0.58 ± 0.01 [a]	0.52 ± 0.01 [a]	0.49 ± 0.05 [b]
Chl *b* (mg/g FW)	0.52 ± 0.02 [a]	0.33 ± 0.04 [b]	0.24 ± 0.07 [c]
Chl (*a* + *b*) (mg/g FW)	1.10 ± 0.03 [a]	0.90 ± 0.05 [b]	0.68 ± 0.02 [c]
Chl *a/b* ratio	1.11 ± 0.04 [c]	1.57 ± 0.17 [b]	2.04 ± 0.56 [a]
CAR (mg/g FW)	0.23 ± 0.04 [a]	0.15 ± 0.01 [b]	0.09 ± 0.01 [c]

Whole plant fresh weight (g/plant), whole plant dry weight (g/plant), shoot/root ratio, water content (WC%), leaf water potential (Ψw), chlorophyll a, chlorophyll b, total chlorophyll (a + b), chlorophyll a/b ratio, and carotenoids (CAR). Values are the mean of six replicates and standard deviations. Values with different superscripts differed significantly at $p < 0.05$ (Tukey's test).

In the same line, total chlorophyll (T Chl), Chl *a*, and Chl *b* concentrations decreased with water stress (Table 1). Chl *b* was the most affected pigment, declining by 37% in the MWS treatment and 54% in the SWS treatment, relative to well-watered plants. The Chl *a/b* ratio increased with water stress by 29% and 46% in the MWS and SWS treatments, respectively (Table 1).

A chlorophyll reduction might be considered an adjustment mechanism to ROS generation from energy absorption by photosynthetic apparatus [7], owing to the presence of antioxidant leaf pigment betalain (betacyanin and betaxanthin), which absorbed a significant amount of radiation and protected drought-stressed chloroplasts from harmful excessive light.

In addition, the increased Chl *a/b* ratio could be correlated with the reduced size of the PSII light-harvesting antenna, ensuring the supply of electrons from PSII to keep pace with the excitation rate of PSI [26].

3.2. Variation in Phenolic Pools under Moderate and Severe Drought Stress

Table 2 showed the TPh and TF values based on the absorbance results of the FC reagent–reactive extract solutions and aluminum chloride method compared with the gallic acid and catechin equivalent standard solutions. Under the control condition (WW; 100% FC), *R. damascena* leaves had 45.63 mg GAE/g DW and 13.44 mg CE/g DW for TPh and TF contents, respectively. At 50% and 25% FC, the levels of TPC significantly increased in leaves by about 31% and 13% respectively for MWS and SWS, relative to the control.

Correspondingly, a higher concentration and stimulation of flavonoids was detected, with TFC values changing from 13.44 to 16.97 mg CE g^{-1} DW (under MWS) and from 13.44 to 15.74 mg CEg^{-1} DW (under SWS) in control and dehydrated leaves, respectively (Table 2).

Phenolic characterization by the RP-HPLC method, showed ten phenolic acids (caffeic chlorogenic, *p*-coumaric, ellagic, ferulic, gallic, syringic, sinapic, rosmarinic, and *trans*-hydroxy-cinnamic acids) with four flavonols/flavonones (catechin hydrate, epicatechin-3-*O*-gallate, luteolin-7-*O*-glucoside, and kaempferol-3-*O*-rutinoside) (Figure 2). Rosmarinic acid was the major detected phenolic acid, with an amount of 33.94 ± 0.05 mg/g DW, followed by syringic (6.05 ± 0.33 mg/g DW) and *trans*-cinnamic acids (6.55 ± 0.45 mg/g

DW). Moderate water stress enhanced cinnamic and benzoic forms by approximately 1.5-fold compared to well-watered plants (Figure 2). It was observed that severe water dehydration induced a significant increase in the biosynthesis of phenolics, in particularly with respect to ferulic acid and its derivatives (Figure 2).

Table 2. Total polyphenol and flavonoid contents in *Rosa damascena* Mill. var. *trigentipetala* plants under three watering regimes: well-watered (100% field capacity (FC), WW), moderate water stress (50% FC, MWS), or severe water stress (25% FC, SWS).

	WW	MWS	SWS
TPC (mg GAE/g DW)	45.63 ± 1.23 [c]	66.02 ± 4.51 [a]	52.83 ± 3.77 [b]
TFC (mg CE/g DW)	13.44 ± 2.04 [c]	16.97 ± 0.98 [a]	15.74 ± 0.44 [b]
Compounds			
Benzoic acids	**7.34 ± 0.19** [c]	**10.84 ± 0.97** [a]	**8.45 ± 0.74** [b]
Gallic acid	0.24 ± 0.02 [b]	0.33 ± 0.05 [a]	0.13 ± 0.03
Ellagic acid	0.93 ± 0.08 [a]	0.50 ± 0.04	0.37 ± 0.06
p-Coumaric acid	0.12 ± 0.00 [b]	0.14 ± 0.04	0.10 ± 0.01 [b]
Syringic acid	6.05 ± 0.09 [c]	9.87 ± 0.84 [a]	7.85 ± 0.64 [b]
Cinnamic acids	**46.95 ± 0.72** [c]	**58.12 ± 1.41** [a]	**55.26 ± 2.5** [b]
Caffeic acid	4.3 ± 0.03 [c]	16.09 ± 0.25 [a]	12.83 ± 0.20 [b]
Rosmarinic acid	33.94 ± 0.05 [a]	30.99 ± 0.60 [a]	33.89 ± 1.4 [a]
trans-Cinnamic acid	6.55 ± 0.48 [c]	9.14 ± 0.31 [a]	7.28 ± 0.66 [b]
Chlorogenic acid	0.45 ± 0.03 [a]	0.43 ± 0.02 [a]	0.08 ± 0.00 [b]
Ferulic acid	0.13 ± 0.01 [a]	0.11 ± 0.02 [a]	0.14 ± 0.01 [a]
Sinapic acid	1.62 ± 0.12 [a]	1.36 ± 0.21 [b]	1.04 ± 0.29 [c]
Flavonoids (Flavonols/flavonones)	**18.42 ± 0.97** [b]	**21.68 ± 1.73** [a]	**18.97 ± 0.50** [b]
Luteolin-7-*O*-glucoside	8.89 ± 0.32 [a]	7.82 ± 0.87 [b]	7.17 ± 0.13 [b]
Epicatechin-3-*O*-gallate	4.51 ± 0.53 [c]	9.80 ± 0.50 [a]	8.30 ± 0.24 [b]
Catechin hydrate	0.47 ± 0.01 [b]	0.54 ± 0.09 [a]	0.29 ± 0.04 [c]
Kaempferol-3-*O*-rutinoside	4.55 ± 0.11 [a]	3.52 ± 0.27 [b]	3.03 ± 0.09 [b]

Values are the mean of three replicates and standard deviations. Values with different superscripts differed significantly at $p < 0.05$ (Tukey's test).

The flavonol and flavanone groups accounted for approximately 32% of the total phenolic constituents in control leaves, which were mostly represented by luteolin-7-*O*-glucoside, epicatechin-3-*O*-gallate, and kaempferol-3-*O*-rutinoside, with minor amounts of catechin hydrate (Table 2). Moderate water stress raised flavonoid levels by 1.2-fold despite the decline in luteolin-7-*O*-glucoside (−12%), and kaempferol (−23%). Although severe water stress increased flavonoid content by about 8.48%.

Higher phenolic contents suggest that rose can efficiently accumulate secondary metabolites in order to adapt to water deficiency [27]. Phenols play a key role in cell protection and osmotic adjustment, either directly by inducing ROS detoxification processes or indirectly by stimulating the antioxidative defense system [28]. Phenols can function as a filter to absorb radiation by limiting chlorophyll excitation in the photosynthetic apparatus during unfavorable conditions [29].

The shielding contribution of flavonoids is ascribed to their OH groups, the omnipresence of double bonds, and their predilection to glycosylation and methylation [30]. Flavonoids with an ortho-dihydroxy pattern in the B-ring in skeleton verge are thought to preserve in higher amounts than mono-hydroxylated in the B-ring [31]. Such biomolecules can also uphold the integrity of the envelope membrane through lipid adjusting during cellular dehydration [32].

Many studies have reported a variation in phenolic composition under abiotic stress. For example, Meot-Duros and Magné [33] reported an accumulation of caffeic acid in the leaves of *Crithmum maritimum* exposed to water stress (due to the nature of sandy substrate) and ionic stress (due to the sea sprays). In *Prunella vulgaris*, moderate drought stress enhanced the production of ursolic, rosmarinic, and oleanolic acids [34]. As recorded by Bettaieb-Rebey et al. [28], ferulic acid was involved in the adaptation of *S. officinalis* to drought stress through the assimilation of UV light and its transformation into blue fluorescence, which sheltered the plants from the destructive effects of UV light. Cinnamic, vanillin, *p*-hydroxybenzoic, and vanillic acids were showed to be involved in drought tolerance of Q8 rice cultivar [35], which could be related to cell wall lignification correlated with the implication of specific amino acids for osmotic adjustment [36]. Al Yasi et al. [21] suggested that the cell wall rigidity of Damask rose exposed to drought allows it to cope with toxic molecules and ROS [37,38].

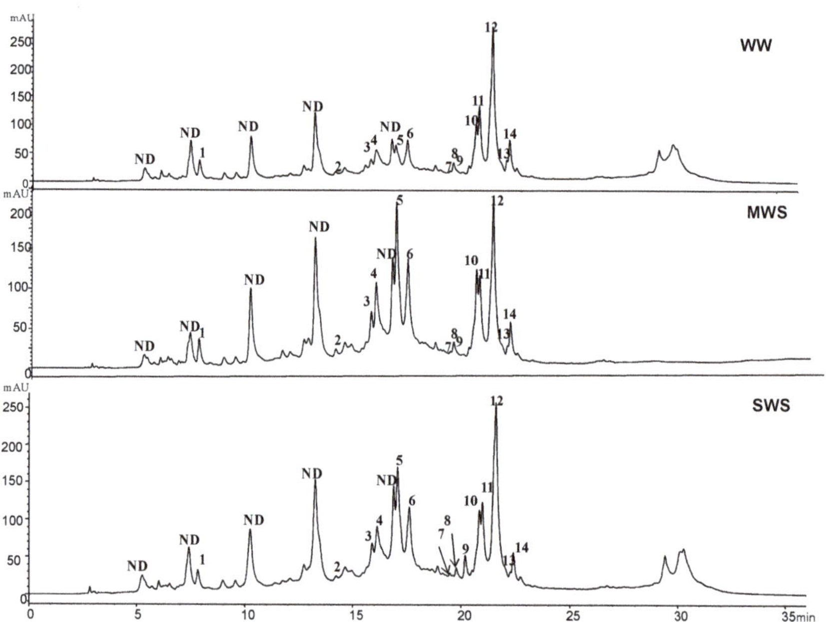

Figure 2. RP-HPLC chromatograms of *Rosa damascena* plants. The signal was monitored at 280 nm. The peak numbers corresponded to (1) gallic acid; (2) catechin hydrate; (3) chlorogenic acid; (4) epicatechin3-*O*-gallate; (5) caffeic acid; (6) syringic acid; (7) *p*-coumaric acid; (8) sinapic acid; (9) ferulic acid; (10) luteolin 7-*O*-glucoside; (11) *trans*-hydroxycinnamic acid; (12) rosmarinic acid; (13) ellagic acid; (14) kaempferol-3-*O*-rutinoside. Well-watered (100% field capacity (FC), WW), moderate water stress (50% FC, MWS), and severe water stress (25% FC, SWS).

Changes in flavonoid groups have occurred in diverse plants under water limitation; for instance, kaempferol and quercetin increased in dehydrated tomato plants [39]. Ding et al. [40] suggested that characteristic catechins (a subgroup of flavan-3-ols) play essential key roles in the stress response of tea plants by minimizing excess ROS production.

Luteolin is an inhibitor of α-amylase activity in plants. Thus, drought-stressed *Achillea pachycephala* species reduced their photosynthetic rate and soluble sugars, producing signals for discriminate gene expression patterns [31].

The metabolic investigation of flavonoid alternation by LC-QTOF-MS, during drought stress in *Arabidopsis thaliana* (wild type, Col-0); proved that glycosides flavonoid forms (kaempferol, quercetin, and cyanidin) were assigned in the mitigation of oxidative and drought stress.

3.3. Moderate and Severe Drought Stress Effects on Antioxidant Activities

Rosa damascena Mill. var. *trigentipetala* extracts were explored for their antioxidant potentialities using distinct in vitro methods—ABTS•+, NO•, ORAC, FRAP, and β-carotene–

linoleic acid model systems—to estimate their quenching ability for distinct radicals and their aptitude for reducing trivalent iron $^+$(Fe III) to its bivalent form (Fe II) and inhibiting the bleaching of the antioxidant pigment β-carotene [41].

The results confirmed an increment in antioxidant activities in both drought stress treatments (Table 3). Indeed, quenching activities against ABTS$^{\bullet+}$ and peroxyl radicals were found to be increased, as shown by their superior ORAC values and decreased IC_{50} values. A similar trend occurred for RPA, with EC_{50} values decreasing from 356 to 189 µg/mL (for MWS) and from 356 to 234 µg/mL (for SWS), respectively. The MWS samples (IC_{50} = 0.42 mg/mL) could also inhibit β-carotene bleaching more than the control (IC_{50} = 0.67 mg/mL) and SWS (IC_{50} = 0.55 mg/mL) samples.

Table 3. Antioxidant activities in *Rosa damascena* Mill. var. *trigentipetala* plants under three watering regimes: well-watered (100% field capacity (FC), WW), moderate water stress (50% FC, MWS), or severe water stress (25% FC, SWS).

	ABTS$^{\bullet+}$ (IC_{50} µg/mL)	NO$^\bullet$ (IC_{50} µg/mL)	ORAC (µmol/TE g)	FRAP (EC_{50} µg/mL)	β-Carotene (IC_{50} µg/mL)
WW	89.11 ± 0.14 [a]	139.28 ± 0.01 [b]	11.53 ± 0.38 [d]	356.24 ± 11.44 [a]	420 ± 0.02 [c]
MWS	54.87 ± 0.83 [c]	79.67 ± 0.07 [c]	25.09 ± 4.17 [b]	189.18 ± 9.10 [c]	670 ± 0.01 [a]
SWS	63.19 ± 1.68 [b]	75.44 ± 0.03 [c]	19.34 ± 0.07 [c]	234.78 ± 34.12 [b]	550 ± 0.08 [b]
Ascorbic acid	1.9 ± 0.01 [d]	213 ± 0.17 [a]	-	-	-
BHA	-	-	-	-	0.024 ± 0.00 [d]
BHT	-	-	-	18.5 ± 0.00 [d]	-
Trolox	-	-	39.14 ± 0.30 [a]	-	-

Values are the means of three replicates and standard deviations. Values with different superscripts differed significantly at $p < 0.05$ (Tukey's test).

Water limitation also affected NO-quenching activity, despite not being completely aligned with the previous tests. The inhibited activity of NO radicals raised by approximately two-fold in both treatments, suggesting that, despite the overall increase in antioxidant activity in water-stressed plants, the effects on leaves vary depending on the specific reaction and/or mechanism involved. Our results showed that *Rosa damascena* Mill. var. *trigentipetala* extracts were more active against NO than the reference compound, ascorbic acid (IC_{50} = 213 µg/mL).

The dependence on antioxidant activity, obtained from various assays, in relation to TPh and TF had a linear correlation between the IC_{50} values for the ABTS$^{\bullet+}$ ($r = -0.87$ and -0.76) scavenging activity, ORAC quantity ($r = -0.75$ and -0.69), EC_{50} values of FRAP ($r = -0.99$ and -0.86) and β-carotene linoleic acid ($r = -0.90$ and -0.84) model systems, respectively (Table 4).

Indeed, the antioxidant capacity and extract phenolic content are always positively correlated, owing to the direct contribution of phenolic compounds in antioxidant activities [42]. Nevertheless, a moderate correlation between TPh (or TFC) and antioxidant activity was observed for the NO-quenching potential with r values ranging from 0.5 to 0.6. Likewise, a high linear correlation occurred between the ABTS$^{\bullet+}$ and NO$^\bullet$ scavenging activities; FRAP and β-carotene assays and caffeic acid; and syringic acid, *trans*-cinnamic acid, and luteolin-7-O-glucoside, indicating the potential role of cinnamic and benzoic acids with flavonol groups in offsetting oxidative stress under water-limited conditions.

Table 4. Correlation coefficients between total phenolic content (TPC)/total flavonoid content (TFC), individual phenolic compounds, and the IC_{50}/EC_{50} values for ABTS$^{\bullet+}$, NO, ORAC, FRAP, and β-carotene activities.

Variables	TPC	TFC	ABTS$^{\bullet+}$	NO$^{\bullet}$	ORAC	FRAP	β-Carotene
TPC	1	0.97	−0.87	−0.54	−0.75	−0.99	−0.90
TFC	0.97	1	−0.76	−0.56	−0.69	−0.86	−0.84
GA	0.26	0.45	−0.17	0.21	−0.11	−0.18	−0.42
Chl A	−0.22	−0.02	0.31	0.65	0.27	0.30	0.05
CA	0.99	0.95	−0.99	−0.91	−0.89	−0.99	−0.97
Sy A	0.97	0.99	−0.95	−0.76	−0.93	−0.95	−0.99
FA	−0.65	−0.79	0.59	0.23	0.49	0.598	0.77
trans-CA	0.88	0.96	−0.84	−0.57	−0.73	−0.85	−0.95
RA	−0.76	−0.87	0.70	0.38	0.64	0.71	0.86
EA	−0.84	−0.71	0.88	0.99	0.79	0.88	0.73
p-CA	0.32	0.51	−0.24	0.14	−0.29	−0.25	−0.48
Sp A	−0.61	−0.43	0.67	0.90	0.60	0.67	0.46
EP-3-*O*-G	0.99	0.96	−0.99	−0.91	−0.99	−0.99	−0.97
Lut-7-*O*-glu	−0.74	−0.59	0.80	0.96	0.65	0.79	0.62
C	0.12	0.33	−0.04	0.34	−0.01	−0.05	−0.29
K-3-*O*-R	−0.99	−0.97	0.99	0.89	0.81	0.99	0.98

GA, gallic acid; Chl A, chlorogenic acid; CA, caffeic acid; SyA, syringic acid; FA, ferulic acid; *trans*-CA, *trans*-cinnamic acid; RA, rosmarinic acid; EA, ellagic acid; CA, *p*-coumaric acid; Sp A, sinapic acid; EP-3-*O*-G, epicatechin-3-*O*-gallate; Lut-7-glu, luteolin-7-*O*-glucoside; C, catechin hydrate; K-3-*O*-R, kaempferol-3-*O*-rutinoside. Blue color reflects a strong correlation between different parameters ($r > 0.5$); red color shows a correlation for the same parameter.

3.4. Moderate and Severe Drought Stress Effects on LOX and AChE Inhibitory Enzyme Activities

LOX isoforms play a pivotal role in the mobilization of storage lipids during the germination process [43], and play a critical role in the generation of protective components, such as jasmonates, divinyl ethers, and leaf aldehydes, which assist plants to recover from biotic (insects and pathogens) and abiotic stress [44,45]. In turn, LOX reactions with unsaturated fatty acids can produce off-flavors/off-odors and cause food spoilage [41].

The inhibitory potential for LOX activity varied depending on the severity of stress. Moderate water stress had higher anti-LOX activity (27 µg/mL) compared to SWS (48 µg/mL), as reflected in the lower IC_{50} values (Table 5). Increased lipid peroxidation under stress conditions is principally induced by higher lipolytic activity in the membrane, stimulating LOX activity [42,45,46]. The ability to reduce LOX activity (either directly or by down-regulating its expression) is considered beneficial for plants, as LOX are oxidative enzymes that can set radicals and ROS free [47,48]. The stimulation of PgLOX3 (LOX3 isoform) gene expression is a possible adaptive strategy under water deficit [48].

Table 5. Inhibitory activities of lipoxygenase (LOX) and acetylcholinesterase (AChE) in hydromethanolic extracts in leaves of rose plants under three watering regimes: well-watered (100% field capacity (FC), WW), moderate water stress (50% FC, MWS), or prolonged water stress (25% FC, SWS).

	LOX (CI$_{50}$ µg/mL)	AChE (CI$_{50}$ µg/mL)
WW	56 ± 3.07 [a]	321 ± 1.50 [a]
MWS	27 ± 0.34 [c]	205 ± 4.63 [c]
SWS	48 ± 2.77 [b]	281 ± 3.87 [b]

Variables	TPC	TFC	LOX	AChE
TPC	1	0.97	−0.74	−0.99
TFC	0.97	1	−0.85	−0.96
LOX	−0.74	−0.85	1	0.93
AChE	−0.99	−0.96	0.93	1

Data represent the mean ± standard deviation of three independent assays. Values with different superscripts significantly differ ($p < 0.05$) for correlation coefficients between total phenolic content (TPC)/total flavonoid content (TFC) and IC_{50} values for LOX and AChE activities. Red color shows a correlation for the same parameters.

The genomic DNA structure of PgLOX3 disclosed a nucleotide sequence with high identity with EST in severely drought-stressed *Populus* encoding the PgPsad2 protein, which suggests the involvement of PgLOX3 genes in drought stress tolerance [48].

The recognition of new AChE inhibitors derived from natural sources with few side effects is required [41]. Our research showed that moderately dehydrated rose leaves had a higher inhibitory AChE capacity (CI_{50} = 205 µg/mL) than severely dehydrated leaves (CI_{50} = 281 µg/mL).

The correlation analysis (Table 5) showed strong correlation coefficients between TPC/TFC and the IC_{50} values of LOX ($r = -0.74$; $r = -0.85$) and AChE ($r = -0.99$; $r = -0.96$), indicating the potential role of non-antioxidant compounds in hampering LOX and AChE enzymes under drought stress. The inhibition of AChE activity was considered a tolerance mechanism in *Pimpinella anisum* leaves exposed to Zn excess, which could be linked to the omnipresence of bioactive molecules [41].

4. Conclusions

In summary, an integrated approach combining biochemical and physiological studies revealed new insights into the mechanisms and processes involved in *Rosa damascena* drought adaptation. When cultivated under water-limiting conditions, *R. damascena* shoot and whole plant biomass production significantly declined, whereas that of shoot/root was not affected. In addition, such constraints resulted in a significant reduction of chlorophyll linked with an alteration in water potential, probably to support its nutrient use efficiency associated with the preservation of an adequate level of chlorophyll in leaves. In turn, the contents of biomolecules under water deficit were positively correlated with antioxidant and inhibitory enzyme activities (LOX, AChE), as evaluated using different test systems, suggesting an adequate protection against oxidative damage, and thus adaptation to water limitation. Moisture deficit can successfully enhance health-promoting phytochemicals in rose, which could be manipulated through agricultural techniques and screening programs to develop drought-tolerant genotypes.

Further omics technologies such as transcriptomics and proteomics could help us to identify pathways/cycles involved in the establishment of enhanced drought tolerance.

Author Contributions: K.H.: performed the experimental work, the analysis and data interpretation, statistical analysis, writing- original draft preparation. H.W.: contributed in data analysis and co-wrote the manuscript. H.M.A.-Y., E.F.A., A.A.I. and F.A.S.H.: helped in experimental work and paper revision. K.H.M.S.: revised the paper. Conceptualization, K.H. and H.W.; methodology, K.H., E.F.A. and H.M.A.-Y.; software, K.H., F.A.S.H. and A.A.I.; validation, K.H., H.W. and H.M.A.-Y.; formal analysis, E.F.A., F.A.S.H., A.A.I., H.W.; data curation, K.H. and H.W.; writing—original draft preparation, H.W. and K.H.; writing—review and editing, K.H. and K.H.M.S., visualization, E.F.A., H.M.A.-Y., A.A.I. and F.A.S.H.; supervision, K.H. and K.H.M.S. All authors have read and agreed to the published version of the manuscript.

Funding: This work was supported by the deputyship for research & Innovation, Ministry of Education in Saudi Arabia: 1-441-131.

Institutional Review Board Statement: Not applicable.

Informed Consent Statement: Not applicable.

Data Availability Statement: Not applicable.

Acknowledgments: The authors extend their appreciation to the Deputyship for Research and Innovation, Ministry of Education in Saudi Arabia for funding this research work through the project number 1-441-131.

Conflicts of Interest: The authors declare no conflict of interest.

References

1. Klein, R.J.T.; Midgley, G.F.; Preston, B.L.; Alam, M.; Berkhout, F.G.H.; Dow, K.; Shaw, M.R. Adaptation opportunities, constraints, and limits. In *Climate Change 2014—Impacts, Adaptation and Vulnerability: Part A: Global and Sectoral Aspects: Working Group II Contribution to the IPCC Fifth Assessment*; Field, C.B., Barros, V.R., Dokken, D.J., Mach, K.J., Mastrandrea, M.D., Bilir, T.E., Chatterjee, M., Ebi, K.L., Estrada, Y.O., Genova, R.C., et al., Eds.; Cambridge University Press: Cambridge, UK, 2014; pp. 899–943.
2. Grafton, R.Q.; Daugbjerg, C.; Qureshi, M.E. Towards food security by 2050. *Food Secur.* **2015**, *7*, 179–183. [CrossRef]
3. Marín-de la Rosa, N.; Lin, C.W.; Kang, Y.J.; Dhondt, S.; Gonzalez, N.; Inzé, D.; FalterBraun, P. Drought resistance is mediated by divergent strategies in closely related Brassicaceae. *New Phytol.* **2019**, *223*, 783–797. [CrossRef] [PubMed]
4. Jafari, S.; Garmdareh, S.E.H.; Azadegan, B. Effects of drought stress on morphological, physiological, and biochemical characteristics of stock plant (*Matthiola incana* L.). *Sci. Hortic.* **2019**, *253*, 128–133. [CrossRef]
5. Caser, M.; D'Angiolillo, F.; Chitarra, W.; Lovisolo, C.; Ruffoni, B.; Pistelli, L.; Pistelli, L.; Scariot, V. Ecophysiological and phytochemical responses of *Salvia sinaloensis* Fern. to drought stress. *Plant Growth Regul.* **2018**, *84*, 383–394. [CrossRef]
6. Mandoulakani, B.A.; Eyvazpour, E.; Ghadimzadeh, M. The effect of drought stress on the expression of key genes involved in the biosynthesis of phenylpropanoids and essential oil components in basil (*Ocimum basilicum* L.). *Phytochemistry* **2017**, *139*, 1–7. [CrossRef]
7. Maleki, M.; Shojaeiyan, A.; Mokhtassi-Bidgoli, A. Genotypic variation in biochemical and physiological responses of fenugreek (*Trigonellafoenum-graecum* L.) landraces to prolonged drought stress and subsequent rewatering. *Sci. Hortic.* **2021**, *287*, 110224. [CrossRef]
8. Wasli, H.; Jelali, N.; Silva, A.M.S.; Ksouri, R.; Cardoso, S.M. Variation of polyphenolic composition, antioxidants and physiological characteristics of dill (*Anethum graveolens* L.) as affected by bicarbonate-induced iron deficiency conditions. *Ind. Crops Prod.* **2018**, *126*, 466–476. [CrossRef]
9. Sreenivasulu, N.; Grimm, B.; Wobus, U.; Weschke, W. Differential response of antioxidant compounds to salinity stress in salt-tolerant and salt-sensitive seedlings of foxtail millet (*Setaria italica*). *Physiol. Plant.* **2002**, *109*, 435–442. [CrossRef]
10. Abreu, I.N.; Mazzafera, P. Effect of water and temperature stress on the content of active constituents of *Hypericum Brasiliense* Choisy. *Plant Physiol. Biochem.* **2005**, *43*, 241–248. [CrossRef]
11. Bettaieb-Rebey, I.; Jabri-Karoui, I.; Hamrouni-Sellami, I.; Bourgou, S.; Limam, F.; Marzouk, B. Effect of drought on the biochemical composition and antioxidant activities of cumin (*Cuminum cyminum* L.) seeds. *Ind. Crops Prod.* **2012**, *36*, 238–245. [CrossRef]
12. Naquvi, K.J.; Ansari, S.H.; Ali, M.; Najmi, A.K. Volatile oil composition of *Rosa damascena* Mill. (Rosaceae). *J. Pharmacogn. Phytochem.* **2014**, *2*, 130–134.
13. Baydar, N.G.; Baydar, H. Phenolic compounds, antiradical activity and antioxidant capacity of oil-bearing rose (*Rosa damascena* Mill.) extracts. *Ind. Crops Prod.* **2014**, *41*, 375–380. [CrossRef]
14. Al-Yasi, H.; Attia, H.; Alamer, K.; Hassan, F.; Ali, E.; Elshazly, S.; Siddique, K.H.; Hessini, K. Impact of drought on growth, photosynthesis, osmotic adjustment, and cell wall elasticity in Damask rose. *Plant Physiol. Biochem.* **2020**, *150*, 133–139. [CrossRef] [PubMed]
15. Hewitt, E.J. *Sand and Water Culture Methods Used in the Study of Plant Nutrition*, 2nd ed.; Commonwealth Agric. Bureaux, East Mailing: Kent, WA, USA, 1966; pp. 431–446.
16. Lichtenthaler, H.K. Chlorophylls and carotenoids: Pigments of photosynthetic biomembranes. *Methods Enzymol.* **1987**, *148*, 350–382.
17. Wasli, H.; Jelali, N.; Ksouri, R.; Cardoso, S.M. Insights on the adaptation of *Foeniculum vulgare* Mill to iron deficiency. *Appl. Sci.* **2021**, *11*, 7072. [CrossRef]
18. Proestos, C.; Boziaris, I.S.; Nychas, G.J.E.; Komaitis, M. Analysis of flavonoids and phenolic acids in Greek aromatic plants: Investigation of their antioxidant capacity and antimicrobial activity. *Food Chem.* **2006**, *95*, 664–671. [CrossRef]
19. Koleva, I.I.; Van Beek, T.A.; Linssen, J.P.H.; de Groot, A.; Evstatieva, L.N. Screening of plant extracts for antioxidant activity: A comparative study on three testing methods. *Phytochem. Anal.* **2002**, *13*, 8–17. [CrossRef]
20. Majdoub, N.; El-Guendouz, S.; Rezgui, M.; Carlier, J.; Costa, C.; Bettaieb-Ben Kaaba, L.; Miguel, M.G. Growth, photosynthetic pigments, phenolic content and biological activities of *Foeniculum vulgare* Mill., *Anethum graveolens* L. and *Pimpinella anisum* L. (Apiaceae) in response to zinc. *Ind. Crops Prod.* **2017**, *109*, 627–636. [CrossRef]
21. Negrão, S.; Schmöckel, S.M.; Tester, M. Evaluating physiological responses of plants to salinity stress. *Ann. Bot* **2017**, *119*, 1–11. [CrossRef]
22. Hessini, K.; Issaoui, K.; Ferchichi, S.; Saif, T.; Abdelly, C.; Siddique, K.H.; Cruz, C. Interactive effects of salinity and nitrogen forms on plant growth, photosynthesis and osmotic adjustment in maize. *Plant Physiol. Biochem.* **2019**, *139*, 171–178. [CrossRef]
23. Mejri, M.; Siddique, K.H.; Saif, T.; Abdelly, C.; Hessini, K. Comparative effect of drought duration on growth, photosynthesis, water relations, and solute accumulation in wild and cultivated barley species. *J. Plant Nutr. Soil Sci.* **2016**, *179*, 327–335. [CrossRef]
24. Farhat, N.; Belghith, I.; Senkler, J.; Hichri, S.; Abdelly, C.; Braun, H.P.; Debez, A. Recovery aptitude of the halophyte *Cakilemaritima* upon water deficit stress release is sustained by extensive modulation of the leaf proteome. *Ecotoxicol. Environ. Saf.* **2019**, *179*, 198–211. [CrossRef] [PubMed]
25. Ramireddy, E.; Hosseini, S.A.; Eggert, K.; Gillandt, S.; Gnad, H.; Von Wirén, N.; Schmülling, T. Root engineering in barley: Increasing cytokinin degradation produces a larger root system, mineral enrichment in the shoot and improved drought tolerance. *Plant Physiol.* **2018**, *177*, 1078–1095. [CrossRef] [PubMed]

26. Kitajima, K.; Hogan, K.P. Increases of chlorophyll a/b ratios during acclimation of tropical woody seedlings to nitrogen limitation and high light. *Plant Cell Environ.* **2003**, *26*, 857–865. [CrossRef] [PubMed]

27. Caser, M.; Chitarra, W.; D'Angiolillo, F.; Perrone, I.; Demasi, S.; Lovisolo, C.; Pistelli, L.; Pistelli, L.; Scariot, V. Drought stress adaptation modulates plant secondary metabolite production in *Salvia dolomitica* Codd. *Ind. Crops Prod.* **2019**, *129*, 85–96. [CrossRef]

28. Bettaieb-Rebey, I.; Jabri-Karoui, I.; Hamrouni-Sellami, I.; Bourgou, S.; Limam, F.; Marzouk, B. Drought effects on polyphenol composition and antioxidant activities in aerial parts of *Salvia officinalis* L. *Acta Physiol. Plant.* **2011**, *33*, 1103–1111. [CrossRef]

29. Nogués, S.; Baker, N.R. Effects of drought on photosynthesis in Mediterranean plants grown under enhanced UV-B radiation. *J. Exp. Bot.* **2000**, *51*, 1309–1317. [CrossRef]

30. Hodaei, M.; Rahimmalek, M.; Arzani, A.; Talebi, M. The effect of water stress on phytochemical accumulation, bioactive compounds and expression of key genes involved in flavonoid biosynthesis in *Chrysanthemum morifolium* L. *Ind. Crops Prod.* **2018**, *120*, 295–304. [CrossRef]

31. Gharibia, S.; Tabatabaei, B.E.S.; Saeidia, G.; Majid Talebi, M.; Matkowsk, A. The effect of drought stress on polyphenolic compounds and expression of flavonoid biosynthesis related genes in *Achillea pachycephala* Rech. f. *Phytochemistry* **2019**, *162*, 90–98. [CrossRef]

32. Jorge, T.F.; Tohge, T.; Wendenburg, R.; Ramalho, J.C.; Lidon, F.C.; Ribeiro-Barros, A.I.; Fernie, A.R.; António, C. Salt-stress secondary metabolite signatures involved in the ability of *Casuarina glauca* to mitigate oxidative stress. *Environ. Exp. Bot.* **2019**, *166*, 103808. [CrossRef]

33. Meot-Duros, L.; Magné, C. Antioxidant activity and phenol content of *Crithmum maritimum* L. leaves. *Plant Physiol. Biochem.* **2009**, *47*, 37–41. [CrossRef] [PubMed]

34. Chen, Y.; Guo, Q.; Liu, L.; Liao, L.; Zhu, Z. Influence of fertilization and drought stress on the growth and production of secondary metabolites in *Prunella vulgaris* L. *J. Med. Plants Res.* **2011**, *5*, 1749–1755.

35. Quan, N.T.; Anh, L.H.; Khang, D.T.; Tuyen, P.T.; Toan, N.P.; Minh, T.N.; Xuan, T.D. Involvement of secondary metabolites in response to drought stress of rice (*Oryza sativa* L.). *Agriculture* **2016**, *6*, 23. [CrossRef]

36. Ayaz, F.A.; Kadioglu, A.R.; Turgut, R. Water stress effects on the content of low molecular weight carbohydrates and phenolic acids in *Ctenanthesetosa* (Rosc.) Eichler. *Can. J. Plant Sci.* **2000**, *80*, 373–378. [CrossRef]

37. Salem, N.; Msaada, K.; Dhifi, W.; Sriti, J.; Mejri, H.; Limam, F.; Marzouk, B. Effect of drought on safflower natural dyes and their biological activities. *Excli J.* **2014**, *13*, 1–18. [PubMed]

38. Okunlola, G.O.; Olatunji, O.A.; Akinwale, R.O.; Tariq, A.; Adelusi, A.A. Physiological response of the three most cultivated pepper species (*Capsicum* spp.) in Africa to drought stress imposed at three stages of growth and development. *Sci. Hortic.* **2017**, *224*, 198–205. [CrossRef]

39. Sanchez-Rodriguez, E.; Moreno, D.A.; Ferreres, F.; Rubio-WilhelmiMdel, M.; Ruiz, J.M. Differential responses of five cherry tomato varieties to water stress: Changes on phenolic metabolites and related enzymes. *Phytochemistry* **2011**, *72*, 723–729. [CrossRef]

40. Ding, C.; Lei, L.; Yao, L.; Wang, L.; Hao, X.; Li, N.; Wang, Y.; Yin, P.; Guo, G.; Yang, Y.; et al. The involvements of calcium-dependent protein kinases and catechins in tea plant [*Camellia sinensis* (L.) O. Kuntze] cold responses. *Plant Physiol. Biochem.* **2019**, *143*, 190–202. [CrossRef]

41. Saada, M.; Wasli, H.; Jallali, I.; Kbouki, R.; Girard-Lalancette, K.; Mshvildadzer, V.; Ksouri, R.; Legault, J.; Cardoso, S.M. Bio-Guided Fractionation of Retama raetam (Forssk.) Webb & Berthel Polar Extracts. *Molecules* **2021**, *26*, 5800.

42. New comer, M.E.; Brash, A.R. The structural basis for specificity in lipoxygenase catalysis. *Protein Sci.* **2015**, *24*, 298–309. [CrossRef]

43. Pallavi, P.C.; Singh, A.K.; Singh, S.; Singh, N.K. In silico structural and functional insights into the lipoxygenase enzyme of legume *Cajanus cajan*. *Int. J. Recent Innov. Trends Comput. Commun.* **2014**, *5*, 87–91.

44. Porta, H.; Rocha-Sosa, M. Plant lipoxygenases. Physiological and molecular features. *Plant Physiol.* **2002**, *130*, 15–21. [CrossRef] [PubMed]

45. Rahman, U.; Uddin, T.; Choudhary, G.; Iqbal, M. Discovery and molecular docking simulation of 7-hydroxy-6-methoxy-2H-chromen-2-one as a LOX Inhibitor. *Pak. J. Pharm. Sci.* **2019**, *32*, 217–220. [PubMed]

46. Bae, K.S.; Rahimi, S.; Kim, Y.J.; Renuka Devi, B.S.; Khorolragchaa, A.; Sukweenadhi, J.; Silva, J.; Myagmarjav, D.; Yang, D.C. Molecular characterization of lipoxygenase genes and their expression analysis against biotic and abiotic stresses in *Panax ginseng*. *Eur. J. Plant Pathol.* **2016**, *145*, 331–343. [CrossRef]

47. Siedow, J.N. Plant lipoxygenase: Structure and function. *Annu. Rev. Plant Physiol. Plant Mol. Biol.* **1991**, *42*, 145–188. [CrossRef]

48. Maccarrone, M.; Veldink, G.A.; Agro, A.F.; Vliegenthart, J.F. Modulation of soybean lipoxygenase expression and membrane oxidation by water deficit. *FEBS Lett.* **1995**, *371*, 223–226.

 *horticulturae*

Article

Interactive Effects of Drought and Saline Aerosol Stress on Morphological and Physiological Characteristics of Two Ornamental Shrub Species

Stefania Toscano [1], Antonio Ferrante [2], Daniela Romano [1,*] and Alessandro Tribulato [1]

[1] Department of Agriculture, Food and Environment (Di3A), Università degli Studi di Catania, Via Valdisavoia 5, 95123 Catania, Italy; stefania.toscano@unict.it (S.T.); atribula@unict.it (A.T.)
[2] Department of Agricultural and Environmental Sciences, Università degli Studi di Milano, Via Celoria 2, 20133 Milan, Italy; antonio.ferrante@unimi.it
* Correspondence: dromano@unict.it

Abstract: Effects of drought and aerosol stresses were studied in a factorial experiment based on a Randomized Complete Design with triplicates on two ornamental shrubs. Treatments consisted of four levels of water container (40%, 30%, 20%, and 10% of water volumetric content of the substrate) and, after 30 days from experiment onset, three aerosol treatments (distilled water and 50% and 100% salt sea water concentrations). The trial was contextually replicated on two species: *Callistemon citrinus* (Curtis) Skeels and *Viburnum tinus* L. 'Lucidum'. In both species, increasing drought stress negatively affected dry biomass, leaf area, net photosynthesis, chlorophyll *a* fluorescence, and relative water content. The added saline aerosol stress induced a further physiological water deficit in plants of both species, with more emphasis on *Callistemon*. The interaction between the two stress conditions was found to be additive for almost all the physiological parameters, resulting in enhanced damage on plants under stress combination. Total biomass, for effect of combined stresses, ranged from 120.1 to 86.4 g plant^{-1} in *Callistemon* and from 122.3 to 94.6 g plant^{-1} in *Viburnum*. The net photosynthesis in *Callistemon* declined by the 70% after 30 days in WC 10% and by the 45% and 53% in WC 20% and WC 10% respectively after 60 days. In *Viburnum* plants, since the first measurement (7 days), a decrease of net photosynthesis was observed for the more stressed treatments (WC 20% and WC 10%), by 57%. The overall data suggested that *Viburnum* was more tolerant compared the *Callistemon* under the experimental conditions studied.

Keywords: *Callistemon citrinus* (Curtis) Skeels; *Viburnum tinus* L. 'Lucidum'; plant biomass; root/shoot ratio; gas exchange; relative water content

Citation: Toscano, S.; Ferrante, A.; Romano, D.; Tribulato, A. Interactive Effects of Drought and Saline Aerosol Stress on Morphological and Physiological Characteristics of Two Ornamental Shrub Species. *Horticulturae* **2021**, *7*, 517. https:// doi.org/10.3390/horticulturae7120517

Academic Editor: Rossano Massai

Received: 21 October 2021
Accepted: 22 November 2021
Published: 23 November 2021

Publisher's Note: MDPI stays neutral with regard to jurisdictional claims in published maps and institutional affiliations.

1. Introduction

The Mediterranean environment is characterized by high summer temperatures often associated with shortage and poor quality of irrigation water. These conditions represent a limit for optimal plant growth and somehow even survival [1]. Along coastal areas, plants in many cases are also affected by salinity stress (soil salinity and spray aerosol) that represents a relevant restriction affecting distribution and survival of native plants [2]. The presence of simultaneous stresses (drought and saline aerosol), which have a cumulative negative effect on plant growth and survival [3]. Álvarez et al. [4] have investigated on drought and saline aerosol stress separately, but only a few studies report the effect of interaction between these stresses. These two abiotic stresses share common physiological, biochemical, and molecular responses in plants. Therefore, plants can undergo synergistic negative effects subjected to both stresses [5].

In coastal areas, the management of landscape and its ornamental value due to the expansion of tourist industry and, as a consequence, the growing of residential uses implies a large interest in green areas realization [6]. Since the end of XX century, tourism has grown by almost 75% in the Mediterranean coasts [7], and projections, before the COVID-19 pandemic, showed a continuing increase in the number of tourists, until to reach 637 million by 2025 [8]. For these reasons the areas destined for both private gardens and public parks are increasing, due to the importance to have recreational areas near homes and hotels. Due to residential use of coastal area, the consumption of water resources has increased, and the irrigation of these area in the summer months for the metropolitan area of Barcelona (Spain), determines a consumption of approximately of 4.28 litres/m^2/day to watering the gardens; this water quantity represents about the 50% of the total domestic water consumption [9]. These quantities of water not always are available; for this reason the expansion of green areas, especially under suboptimal environmental conditions, has motivated number of studies to identify the most suitable species to be used in gardens [10] and their most suitable cultivation methods [11]. Many ornamental plants adopted in green areas present, at inter and intraspecific levels, relevant differences on response to stress conditions [12]. The interest in ornamental species choice is linked to individuate plants that are able to tolerate environmental stresses. The negative effects of abiotic stresses hamper important physiological functions [4] and damage organographic structure and, consequently, aesthetic features of plants.

In nature, all living organisms included plants are continuously exposed to different abiotic and biotic stresses, quite often overlapping at the same time. Abiotic stresses, like drought, salinity, floods, heat, and frost shock and other environmental extreme events, represent the main reasons of losses in plant growth and crop yield [13]. The combination of abiotic stresses, like drought and salinity, often occurs in climatic areas where warm summers are associated with lack of rain/irrigation water, especially along the coasts, resulting in severe yield losses compared to a single stress [14]. Current knowledge indicate that plants are able to manage with a number of overlapping biotic and abiotic stresses through exhibition of tailored responses, which can hardly be understood from data coming from results of a single independently imposed stress [15–17]. Facing these threats, plant resilience results are especially important when stresses could cause negative effects on plant growth and reproduction [18].

The response of plants to both salt and drought stress differs according to the different potentially usable species [19] and, sometimes, within the cultivars of the same species [20]. Salt stress in coastal areas, can occur both at the root and foliage level. At root level, the presence of high levels of ions in the irrigation water or soil reduce the plants' performance. At foliage level, the direct action of saline water spray is due to salt accumulation on leaves. Plants are generally more susceptible to salt damage when the salt directly arrives on leaves than when the saline water arrives to the soil and roots [21]. The salinity stress can be even worse in presence of sandy banks, since marine aerosol carries salts and dust on leaves. The abrasive action of the dust can enhance the salt damage. Moreover, in coastal environments, marine aerosol contains salts and other pollutants that can affect growth and sometimes even survival of indigenous plant species [22]. These two aspects can show an independent trend [23] or, in other words, there is a positive relationship between tolerance to salt spray and to soil salinity [24]. Plants are usually 3 to 4 times more tolerant to salt in the roots than in the leaves [24].

Effects of saline aerosol strongly depend on the intensity and on the duration of the stress [25]. Effects of marine aerosol and pollutants on coastal vegetation have been studied for several decades in the European Mediterranean regions [26]. Several experiments demonstrated that leaves can absorb sodium and chlorine ions from sea water drops accumulated on the surface [27]. The salt experimentally applied on leaves of sensitive species induces symptoms similar to those observed on coastal or roadside vegetation near the sea [28]. The abrasive action of the wind excoriates leaf surface and creates possible ways for salt absorption [29]. The damage observed in plants near the sea is attributed to the excessive absorption of chlorine and sodium ions, facilitated by the presence of surfactants that are present in polluted sea water [26].

As consequence, the efficiency of photosynthesis and gas exchanges is negatively influenced by saline aerosol [30], although differences can be found among species [31] because of the characteristics of their cuticles and the epithelial cells [32]. The salts that lay on the surface of the leaf after the evaporation of water [33] can negatively modify the water balance and compromise the characteristics of the cuticle or guard cells, resulting in incomplete stomatal closure [34]. Although the combined action of drought and salt stress has been a key issue for ornamental plants quality [35,36]. The response to saline aerosol in plants subjected to differentiated conditions of water availability has been little investigated.

Callistemon citrinus (Curtis) Skeels (Myrtaceae) and *Viburnum tinus* L. 'Lucidum' (Adoxaceae) play an important role in European market as ornamental potted shrubs for the showy flowering of *Callistemon* and brilliant color of foliage in *Viburnum* [37,38]; these two species show from moderate to high tolerance to drought and salt stress, allowing their use also in marginal urban areas [39,40]. Although various investigations have been carried out to study the physiological responses of these species to drought and salt stress [4,40], little information is available on the combined effects of two stresses (drought and saline aerosol), which are very frequent in coastal area in the Mediterranean environments.

In this view, it appears quite relevant to elucidate the morphological and physiological mechanisms of adaptation of potted *C. citrinus* and *V. tinus* to drought and saline aerosol stress in order to discriminate effects of single stress or cumulative actions of the two stresses. Responses of the two species to the different stress factors were focusing on growth parameters, leaf gas exchanges, photosystem efficiency, water relations, and leaf functional traits. The information obtained from this experiment could be used to discriminate the effects of drought and saline aerosol stress and to understand how the combination of the two stressors can modify the responses of plants, with the aim to individuate the most functional strategy adopted by plants to overcome these conditions, which are very common in the Mediterranean coastal areas.

2. Materials and Methods

2.1. Experimental Conditions and Plant Materials

The experiment was carried out during 2019 springtime on two ornamental shrubs, *Callistemon citrinus* (Curtis) Skeels and *Viburnum tinus* L. 'Lucidum', grown in a cold greenhouse located in Catania area, Italy (37°41′ N 15°11′ E 80 m a.s.l.). Two-month-old rooted cuttings of both species were transplanted into 3.4 L pots, filled with sand (75%), silt (18%) and clay (7%), and fertilized with 2 g L^{-1} Osmocote Plus (14/13/13, +microelements). At the beginning of the experiment, the dry biomass of the plants was on average 27.8 ± 2.09 g and 33.6 ± 3.28 g for *Callistemon* and *Viburnum*, respectively.

In the experiment two species (*Callistemon* and *Viburnum*), four water regimes (WC 40%, WC 30%, WC 20%, and WC 10%) and three saline aerosol solutions (S0, distilled water; S1, 50% of simulated sea water solution, and S2, 100% of simulated sea water solution) were studied. For the beginning 4 weeks the plants of the two species were subjected to the differentiated water regimes; after this period, three saline aerosol treatments for another 4 weeks were imposed on the same plants differently subject to water stress. To determine the volumetric content of water in the substrate, an automated management system with dielectric sensors EC 5TE (Decagon Devices Inc., Pullman, WA, USA) was used. Two sensors were used for each treatment and for each replicate, and data were recorded using a data acquisition system (data logger) CR1000 (Campbell Scientific Ltd., Loughborough, UK). Sensors were calibrated following the protocol of Starr and Paltineanu [41] and Tribulato et al. [42]. Sensors were calibrated where a series of each measurement is taken in connection with samples of volumetric soil to quantify the relationship between the substrate (measured with the 5TM) and the volumetric water content (WC). The sensors were installed 10 cm below the substrate surface at the center of each sample pot. The use of the probes was preceded by their calibration to determine the real content of water in the substrate samples, which were placed in a thermo-ventilated oven at 70 °C until the constant weight (W cal.) reached (R^2 = 0.9742). The irrigation, scheduled at 8:00 a.m. and at 7:00 p.m., was activated when the water content dropped below the pre-set threshold values of WC 10% (severe drought stress), WC 20% (moderate drought stress), WC 30% (light drought stress), and WC 40% (control, Capacity Control) of the volume of the substrate. The same threshold values were adopted to determine irrigation interruption. Full composition of saline aerosol solution was as follows: NaCl, Na_2SO_4, $MgCl_2$, $CaCl_2$, and KCl at concentrations of 23.48, 3.92, 4.98, 1.10, and 0.66 g L^{-1} respectively, with a concentration of 401.8 mM NaCl [43] while in the treatment S1 it was reduced to 50%. The treatments were performed by spraying the canopies of plants twice a week.

For each species, triplicates of four plants for 12 treatments (144 plants in total for each species) were adopted. The mean air temperatures and relative humidity levels during the experimental periods were registered on a data logger (CR 1000; Campbell Scientific Ltd., Loughborough, UK). The mean temperature was 25.2 °C, the relative humidity levels were 66% (Figure S1) and mean photosynthetic active radiation (PAR) was 13.7 MJ m^{-2} d^{-1}.

2.2. Biomass and Leaf Area

At the end of the experiment, six plants per treatment were separated into stems, leaves, and roots. The dry biomass was determined by drying weighed fresh samples in a thermo-ventilated oven at 70 °C up to constant weight. Total leaf area was measured with leaf area meter (Delta-T Devices Ltd., Cambridge, UK) and the SPAD index of 25 fully expanded leaves per each replicate was registered using a SPAD-502 chlorophyll meter (Minolta Camera Co., Osaka, Japan).

2.3. Leaf Gas Exchanges, Chlorophyll a Fluorescence, and Relative Water Content

At 7, 15, 30, 45, and 60 days from experiment onset, leaf gas exchange was measured using a CO_2/H_2O infrared gas analyzer (LCi, ADC Bioscientific Ltd., Hoddesdon, UK). The measurements were carried out from 09:00 a.m. to 1:00 p.m. For each treatment, net photosynthetic rate (A_N) and stomatal conductance (gs) were measured. The mean irradiance was 224.6, 149.2, 137.0, 170.4, and 290.6 μM m^{-2} s^{-1} of photosynthetically active radiation (PAR) at day 7, 15, 30, 45, and 60 respectively. With a normal concentration of CO_2, the temperature in the measurement chamber was 32.1, 33.3, 32.3, 33.9, and 32.9 °C at day 7, 15, 30, 45, and 60 respectively; the H_2O reference as partial pressure was 20.1, 19.4, 23.6, 20.5, and 22.4 mBar at the day 7, 15, 30, 45, and 60 respectively.

The chlorophyll *a* fluorescence was measured using the modulated chlorophyll fluorimeter OS1-FL (Opti-Sciences Corporation, Tyngsboro, MA, USA). Under abiotic stress conditions, one parameter that is commonly used to identify the presence of photosynthetic plant damage in plants is the measurement of chlorophyll *a* fluorescence. The ratio variable to maximal fluorescence ratio (Fv/Fm) (i.e., the maximum primary photochemical efficiency of the PSII) allows the evaluation of the efficiency of the PSII photosystem, indirectly measuring the physiological state of the plant. Every leaf was dark-adapted using cuvette clips for 15 min. The chlorophyll *a* fluorescence was expressed as the Fv/Fm ratio, which indicates the maximal quantum yield of PSII photochemistry, where Fm = the maximal fluorescence and Fv = the variable fluorescence.

The Relative Water Content (RWC) was measured at the end of the trial between 12:00 a.m. and 2:00 p.m. For the determination of RWC, 30 leaf discs of 10 mm in diameter, per each replicate were collected, and their fresh weights (FW) were registered. Samples were then soaked for 24 h in distilled water in dark conditions and their turgid weight (TW) was determined. The samples later were dried at 75 °C up to constant dry weight (DW). The RWC was measured by using the following formula

$$RWC\% = (FW - DW/TW - DW) \times 100$$

2.4. Statistical Analysis

The experimental design was a completely randomized experiment with three replicates. The CoStat version 6.311 (CoHortSoftware, Monterey, CA, USA) was used for statistical analysis. Data were subjected to three-way ANOVA to compare the effects of drought (D), saline aerosol (A), and species (S). For each species, data were subjected to two-way ANOVA to compare the effects of drought and saline aerosol treatments. The differences between means were conducted by Tukey's test ($p < 0.05$). The data presented in Figures are means $\pm$ standard error (SE) (Graphpad 7.0). The Heat map was realized using Graphpad 7.0. The principal component loading plot and scores of PCA were performed using Minitab 16, LLC.

3. Results

Statistical analysis results, main and interaction effects of drought, salinity, and species on morphometric parameters were reported in Table 1. Three-way ANOVA showed that the morphometric characteristics were affected by drought, saline stress, and species (Table 1). The statistical analysis revealed that total biomass data were statistically different for drought and saline aerosol factors and interactions among all three factors were also significant. Epigeous biomass and leaf number data were statistically significant for all factors and interactions. Significant differences were detected in the root/shoot ratio for the three factors, while the interactions were only significant for DxS and AxS. Total leaf area was statistically significant for the three factors and interactions, except AxS. SPAD data were significantly different for saline aerosol and species, while the only significant interaction was the AxS.

Table 1. Summary of the main and interaction effects of drought, saline aerosol, and species treatments on total and epigeous dry biomass, root/shoot ratio, total leaf area, leaf number, and SPAD of potted *Callistemon* and *Viburnum* plants with the corresponding significance of the F-values.

	TB	EB	R/S	TLA	LN	SPAD
Main Effects						
Drought (D)	F 41.86 $p < 0.001$ ***	F 45.77 $p < 0.001$ ***	F 3.28 $p < 0.001$ ***	F 32.49 $p < 0.001$ ***	F 5.09 $p < 0.01$ **	F 2.34 ns
Saline Aerosol (A)	F 24.25 $p < 0.001$ ***	F 34.18 $p < 0.001$ ***	F 5.63 $p < 0.01$ **	F 46.51 $p < 0.001$ ***	F 3.49 $p < 0.05$ *	F 4.12 $p < 0.05$ *
Species (S)	F 0.85 ns	F 22.91 $p < 0.001$ ***	F 154.95 $p < 0.001$ ***	F 448.94 $p < 0.001$ ***	F 1195.39 $p < 0.001$ ***	F 19.12 $p < 0.001$ ***
Interaction						
D × A	F 9.40 $p < 0.001$ ***	F 10.34 $p < 0.001$ ***	F 1.35 ns	F 4.29 $p < 0.01$ **	F 6.07 $p < 0.001$ ***	F 1.30 ns
D × S	F 1.92 $p < 0.001$ **	F 0.67 $p < 0.01$ **	F 1.85 $p < 0.01$ **	F 4.29 $p < 0.01$ **	F 3.62 $p < 0.05$ *	F 2.73 ns
A × S	F 6.33 $p < 0.001$ **	F 4.23 $p < 0.05$ *	F 3.53 $p < 0.05$ *	F 0.50 ns	F 4.62 $p < 0.05$ *	F 3.43 $p < 0.05$ *
D × A × S	F 10.77 $p < 0.001$ ***	F 14.11 $p < 0.001$ ***	F 1.80 ns	F 6.20 $p < 0.001$ ***	F 5.00 $p < 0.001$ ***	F 1.13 ns

TB = Total dry biomass; EB = Epigeous dry biomass; R/S = Root to shoot ratio; TLA = Total leaf area; LN = leaf number. Significance of differences of parameters: ns = not significant; * $p < 0.05$; ** $p < 0.01$; *** $p < 0.001$ with the corresponding significance of the F-values.

To better understand effects of drought and saline aerosol stress, two species were separately analyzed. The individual and the combined effects of drought and saline aerosol stress on plant growth were reported in Tables 2 and 3.

In *Callistemon*, the total dry biomass was affected by drought and saline treatment, but no interaction was detected among the experimental factors. Drought effect on total dry biomass showed a reduction in WC 30%, WC 20% and WC 10% of about 10%, 22%, and

28% respectively. Saline stress also caused a reduction for this parameter by ~10% for S1 and S2 compared with the S0 (Table 2).

In *Callistemon*, the epigeous dry biomass showed a similar trend as the total dry biomass with a reduction by ~10, 24, and 33% WC 30%, WC 20%, and WC 10% respectively as compared with the control (Table 2). Saline stress as well affected a reduction for this parameter by ~9% for S1 and S2 compared with the S0 (Table 2).

The root-to-shoot ratio increased in *Callistemon* plants grown under higher deficit irrigation (WC 10%) as well as for the saline stress conditions in S1 and S2, with an increase by 36% (Table 2, Figure 1a).

The combination of two abiotic stresses resulted in a greater decrease in total leaf area, as shown by the lower values obtained for the WC 10%, S0, S1, and S2 (Table 2, Figure 1b).

Similarly at the total leaf area, the combination of the two stresses resulted in a greater decrease in the leaf number in WC 10% S0, S1, and S2 (Table 2, Figure 1c).

In *Viburnum* plants, the total dry biomass varied with drought, but not with saline aerosol treatment. Drought affected total dry biomass, which showed a reduction in WC 30%, WC 20%, and WC 10% of ~14%, 18% and 23% respectively, compared with control. The saline stress, instead, did not show any significant reduction (Table 3).

The combined effects of drought and saline aerosol stress showed a significant interaction (Table 3) while the highest decrease was observed from WC 10% S2 (~33%) compared with the control plants (Table 3, Figure 2a).

Table 2. Effects of drought and saline aerosol treatments on total (TB) and epigeous (EB) dry biomass, root/shoot ratio (R/S), total leaf area (TLA), and leaf number (LN) of potted *Callistemon* plants.

Aerosol/Drought	WC 40%	WC 30%	WC 20%	WC 10%	Mean	A × D
TB (g plant^{-1})						ns
S0	130.9 ± 2.9	113.2 ± 2.4	94.2 ± 1.6	94.2 ± 0.9	108.1 ± 4.7 a **	
S1	116.1 ± 0.7	102.3 ± 8.5	96.4 ± 0.9	88.1 ± 4.4	100.7 ± 3.7 b	
S2	113.4 ± 1.9	107.1 ± 6.9	89.9 ± 2.2	77.0 ± 3.5	96.9 ± 4.7 b	
Mean	120.1 ± 2.9 A ***	107.6 ± 3.6 B	93.5 ± 1.3 C	86.4 ± 3.0 C		
EB (g plant^{-1})						ns
S0	113.0 ± 2.4	96.1 ± 0.9	80.2 ± 1.6	76.8 ± 2.0	91.5 ± 4.4 a **	
S1	99.3 ± 1.6	89.7 ± 8.6	80.3 ± 1.8	70.6 ± 2.8	84.9 ± 3.8 b	
S2	97.4 ± 1.8	91.6 ± 6.6	75.8 ± 1.0	58.9 ± 1.2	80.9 ± 4.8 b	
Mean	103.2 ± 2.7 A ***	92.5 ± 3.3 B	78.8 ± 1.0 C	68.8 ± 2.8 D		
R/S						$p < 0.01$
S0	0.16 ± 0.01	0.18 ± 0.02	0.17 ± 0.00	0.20 ± 0.02	0.18 ± 0.01 ns	
S1	0.17 ± 0.01	0.14 ± 0.01	0.20 ± 0.02	0.25 ± 0.01	0.19 ± 0.01	
S2	0.16 ± 0.00	0.17 ± 0.01	0.19 ± 0.02	0.30 ± 0.01	0.20 ± 0.02	
Mean	0.17 ± 0.01 B ***	0.16 ± 0.01 B	0.19 ± 0.01 B	0.25 ± 0.02 A		
TLA (cm^2)						$p < 0.05$
S0	3979.9 ± 139.3	3611.7 ± 14.0	2978.9 ± 116.3	2364.8 ± 60.1	3233.8 ± 190.3 a ***	
S1	3345.9 ± 123.7	2977.2 ± 61.2	2969.4 ± 122.4	2101.6 ± 90.8	2932.1 ± 144.7 b	
S2	3503.9 ± 51.0	3315.5 ± 136.2	2878.1 ± 90.3	2030.8 ± 115.2	2848.5 ± 176.7 b	
Mean	3609.9 ± 110.4 A ***	3301.5 ± 101.4 B	2942.1 ± 57.6 C	2165.7 ± 68.4 D		
LN (n°)						$p < 0.01$
S0	1050.2 ± 59.7	928.1 ± 13.3	908.5 ± 57.2	804.5 ± 43.2	922.7 ± 30.7 a *	
S1	1073.0 ± 40.0	865.5 ± 118.0	905.6 ± 25.1	651.8 ± 39.9	871.7 ± 47.3 ab	
S2	907.8 ± 13.8	931.4 ± 86.1	913.6 ± 160.0	727.7 ± 69.5	861.3 ± 25.7 b	
Mean	1010.3 ± 33.2 A ***	905.3 ± 43.9 B	898.0 ± 10.7 B	727.3 ± 34.3 C		

*: $p < 0.05$; **: $p < 0.01$; ***: $p < 0.001$; ns: $p > 0.05$ Data followed by a different letter were significantly different according to LSD Test. WC 40%: control; WC 30%: light drought stress; WC 20%: moderate drought stress; WC 10%: severe drought stress. S0: Distilled water, S1: 50% Synthetic seawater solution, S2: 100% Synthetic seawater solution. Data are means ± standard error (*n* = 3). Three biological replicates were used for the measurements. TB = Total dry biomass; EB = Epigeous dry biomass; R/S = Root to shoot ratio; TLA = Total leaf area; LN = Leaf number.

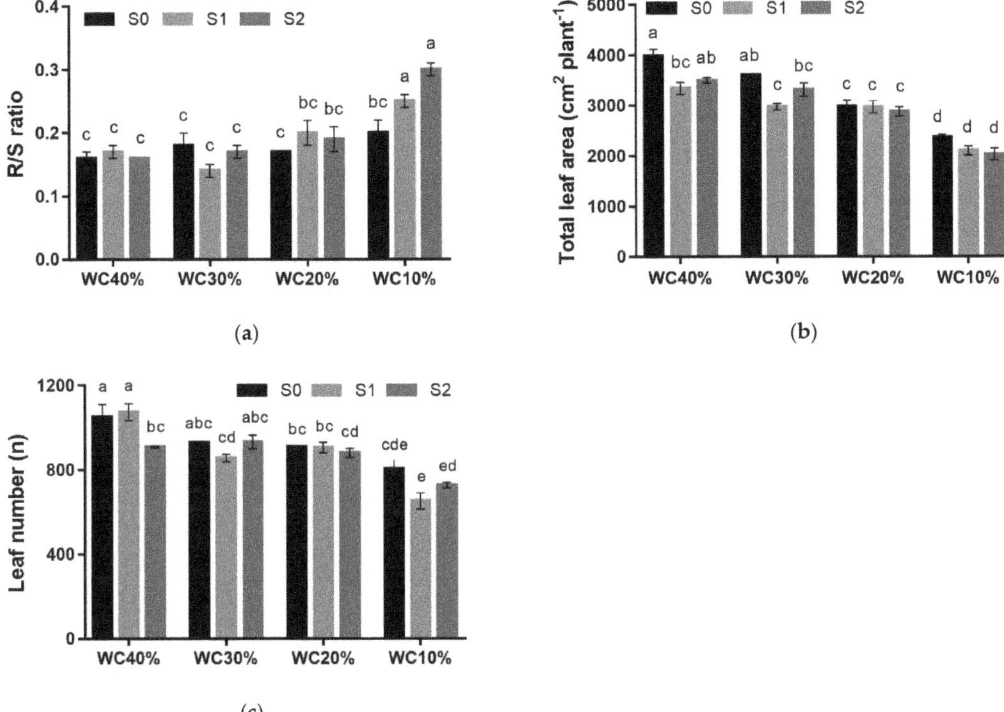

(a)

(b)

(c)

Figure 1. Interaction effect of drought (WC 40%: control; WC 30%: light drought stress; WC 20%: moderate drought stress; WC 10%: severe drought stress) × saline aerosol treatment (S0: Distilled water, S1: 50% Synthetic seawater solution, S2: 100% Synthetic seawater solution) for *Callistemon* plants on root to shoot ratio (**a**), total leaf area (**b**), and leaf number (**c**). Data are means ± standard error ($n = 3$). Three biological replicates were used for the measurements. Data were subjected to two-way ANOVA and differences among means were determined using Tukey's post-test. Different letters indicate statistical differences for $p < 0.05$.

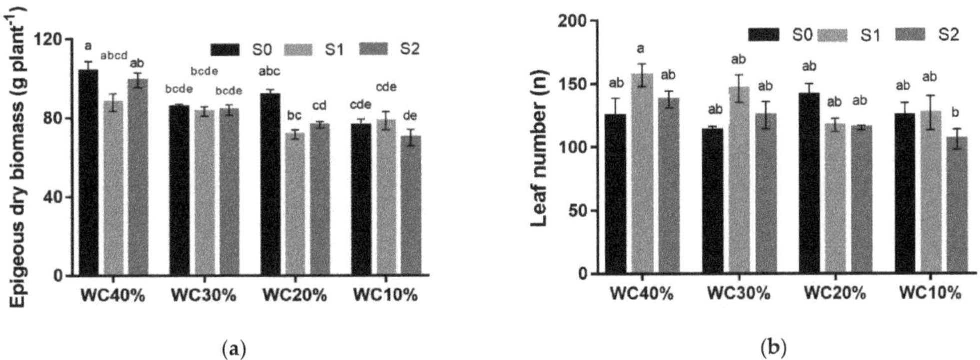

(a)

(b)

Figure 2. Interaction effect of drought (WC 40%: control; WC 30%: light drought stress; WC 20%: moderate drought stress, WC 10% severe drought stress) × saline aerosol treatment (S0: Distilled water, S1: 50% Synthetic seawater solution, S2: 100% Synthetic seawater solution) for *Viburnum* plants on epigeous dry biomass (**a**) and leaf number (**b**). Data are means ± standard error ($n = 3$). Three biological replicates were used for the measurements. Data were subjected to two-way ANOVA and differences among means were determined using Tukey's post-test. Different letters indicate statistical differences for $p < 0.05$.

The root-to-shoot ratio varied with saline aerosol stress but not with drought and the interaction among them (Table 3). An increase in *Viburnum* plants grown under saline aerosol stress (S1) was observed with an increase by ~20% compared with the control plants (Table 3).

In *Viburnum* plants, the total leaf area varied with drought and saline treatment, but no interaction was significant. Drought reduced the total leaf area in WC 30%, WC 20%, and WC 10% by ~16%, 19% and 24% respectively compared with the control plants (Table 3). Saline stress also affected a reduction for this parameter by ~18% for S1 and by 11% for S2 compared with the S0 (Table 3).

At the end of the experiment, with the intensification of water and saline aerosol stress (WC 10% S2), a reduction in leaf number was observed. In particular, leaf number decreased by 33% compared with WC 40% and S1 (Figure 2b).

In *Callistemon* plants, the combination of the two stresses together resulted in a significant decrease for SPAD index, as shown by the lowest values obtained for the WC 10% in S1 and S2 (Figure 3a). No interaction effects were observed in *Viburnum* plants (Figure 3b).

Leaf damage was observed in both species and in particular in *Callistemon* plants, although the percentage of leaf damaged was always lower than 10% (data not shown).

Table 3. Effects of drought and saline aerosol treatments on total (TB) and epigeous (EB) dry biomass, root/shoot ratio (R/S), total leaf area (TLA), and leaf number (LN) of potted *Viburnum* plants.

Aerosol/Drought	WC 40%	WC 30%	WC 20%	WC 10%	Mean	A × D
TB (g plant^{-1})						*p* > 0.05
S0	126.3 ± 8.0	105.2 ± 0.7	111.4 ± 2.8	95.4 ± 2.1	109.6 ± 3.9 ns	
S1	114.7 ± 1.6	103.8 ± 3.6	95.0 ± 4.4	99.4 ± 6.1	103.2 ± 2.9	
S2	126.0 ± 3.9	106.9 ± 2.2	95.6 ± 1.2	88.9 ± 2.2	104.3 ± 3.5	
Mean	122.3 ± 3.2 A ***	105.3 ± 1.3 B	100.7 ± 3.1 BC	94.6 ± 2.8 C		
EB (g plant^{-1})						*p* < 0.05
S0	103.8 ± 4.9	85.3 ± 1.6	91.6 ± 2.8	76.2 ± 3.3	89.2 ± 3.3 a **	
S1	87.9 ± 4.5	83.4 ± 2.4	71.5 ± 2.3	78.5 ± 4.7	80.3 ± 2.4 ab	
S2	99.1 ± 3.8	84.0 ± 2.7	76.4 ± 1.8	69.8 ± 4.2	82.3 ± 3.6 b	
Mean	96.9 ± 3.2 A ***	84.3 ± 1.2 B	79.8 ± 3.3 BC	74.8 ± 2.4 C		
R/S						ns
S0	0.22 ± 0.0	0.23 ± 0.0	0.22 ± 0.0	0.26 ± 0.1	0.23 ± 0.0 b *	
S1	0.31 ± 0.1	0.24 ± 0.0	0.33 ± 0.0	0.27 ± 0.0	0.29 ± 0.0 a	
S2	0.27 ± 0.0	0.27 ± 0.0	0.25 ± 0.0	0.27 ± 0.0	0.27 ± 0.1 ab	
Mean	0.27 ± 0.0 ns	0.25 ± 0.0	0.27 ± 0.0	0.27 ± 0.0		
TLA (cm^2)						ns
S0	6386.0 ± 240.3	5073.1 ± 9.9	5288.2 ± 436.3	4736.0 ± 111.4	5370.8 ± 215.9 a ***	
S1	4890.8 ± 184.6	4584.9 ± 124.1	3963.6 ± 268.0	4202.1 ± 202.5	4401.4 ± 136.3 c	
S2	5805.8 ± 251.1	4732.9 ± 308.6	4637.6 ± 145.8	4008.3 ± 319.6	4796.2 ± 225.3 b	
Mean	5694.2 ± 245.5 A ***	4784.9 ± 123.0 B	4629.8 ± 245.3 BC	4315.5 ± 157.5 C		
LN (n°)						*p* < 0.05
S0	125.0 ± 14.0	113.3 ± 3.2	141.4 ± 8.7	125.0 ± 10.1	126.2 ± 5.2 ab *	
S1	157.1 ± 8.9	146.2 ± 11.0	117.5 ± 5.3	127.1 ± 13.5	137.0 ± 6.4 a	
S2	137.8 ± 6.6	125.2 ± 10.7	115.5 ± 1.8	106.2 ± 8.2	121.2 ± 4.8 b	
Mean	140.0 ± 7.0 ns	128.2 ± 6.6	124.8 ± 5.1	119.4 ± 6.3		

*: *p* < 0.05; **: *p* < 0.01; ***: *p* < 0.001; ns: *p* > 0.05 Data followed by a different letter were significantly different according to LSD Test. WC 40%: control; WC 30%: light drought stress; WC 20%: moderate drought stress; WC 10%: severe drought stress. S0: 0% Distilled water, S1: 50% Synthetic seawater solution, S2: 100% Synthetic seawater solution. TB = Total dry biomass; EB = Epigeous dry biomass; R/S = Root to shoot ratio; TLA = Total leaf area; LN = Leaf number.

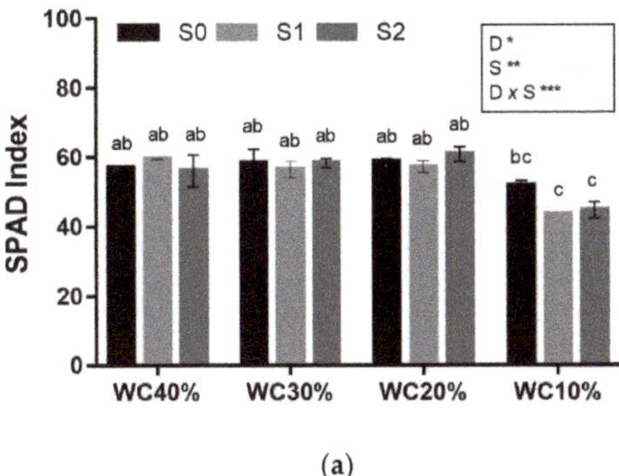
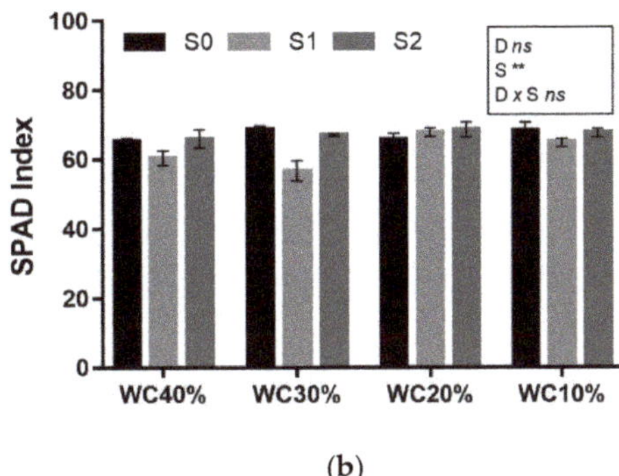

(a)   (b)

Figure 3. SPAD index in *Callistemon* (**a**) and *Viburnum* (**b**) potted plants subjected to drought (WC 40%: control; WC 30%: light drought stress; WC 20%: moderate drought stress; WC 10% severe drought stress) and saline aerosol treatment (S0: Distilled water, S1: 50% Synthetic seawater solution, S2: 100% Synthetic seawater solution). Data are means ± standard error (*n* = 3). Three biological replicates were used for the measurements. Data were subjected to two-way ANOVA and differences among means were determined using Tukey's post-test. Different letters indicate statistical differences for $p < 0.05$; *: $p < 0.05$; **: $p < 0.01$; ***: $p < 0.001$; ns: $p > 0.05$.

Gas exchange measurements in *Callistemon* plants were severely affected under water deficit. The A_N was reduced by water deficit and plants exposed to severe water stress (WC 10%) for 30 days ($p < 0.0026$ **) showed a reduction by 70% compared to control. At the end of the trial, 60 days, with a reduction by 45% and 53% in WC 20% and WC 10% plants ($p < 0.0002$ ***). The addition of saline aerosol stress, after 30 days, differences were amplified comparing treatments with control plants, and at the end of the trial, *Callistemon* showed significant differences ($p < 0.0001$ ***) in WC 20% S1 and S2 and WC 10% S0, S1, and S2 treatments compared with the control plants (Figure 4a, Table 4).

The gs in *Callistemon* plants was reduced in WC 10% after 7 days ($p < 0.0068$ **). At 30 days significant differences ($p < 0.0000$ ***) were observed in the more stressed treatments (WC 20% and WC 10%). No significant differences were observed at the end of the trial ($p > 0.5225$ ns) (Figure 4b, Table 4).

Table 4. Summary of the main and interaction effects of days (T) and drought (D) in the first 30 days, and of days (T), drought (D), and saline aerosol (A) for the other 30 days on net photosynthesis (A_N) and stomatal conductance (gs) of potted *Callistemon*.

Factor	A_N	gs
Main effect		
Days (T)	$p > 0.1360$ ns	$p > 0.9479$ ns
Drought (D)	$p < 0.001$ ***	$p < 0.001$ ***
Interaction		
T × D	$p > 0.9195$ ns	$p > 0.3881$ ns
Main effect		
Days (T)	$p < 0.000$ ***	$p > 0.9335$ ns
Drought (D)	$p < 0.000$ ***	$p > 0.7446$ ns
Aerosol Saline (A)	$p < 0.0223$ *	$p > 0.7487$ ns
Interaction		
T × D	$p < 0.02$ *	$p > 0.1859$ ns
T × A	$p > 0.2517$ ns	$p > 0.4371$ ns
D × A	$p < 0.0492$ *	$p > 0.1965$ ns
T × D × A	$p > 0.695$ ns	$p > 0.4665$ ns

Data were subjected to two-way ANOVA to compare the effects of days (T) and drought (D), while data were subjected to three-way ANOVA to compare the effects of days (T), drought (D), saline aerosol (A). Significance of differences of parameters: ns = not significant; * $p < 0.05$; *** $p < 0.001$.

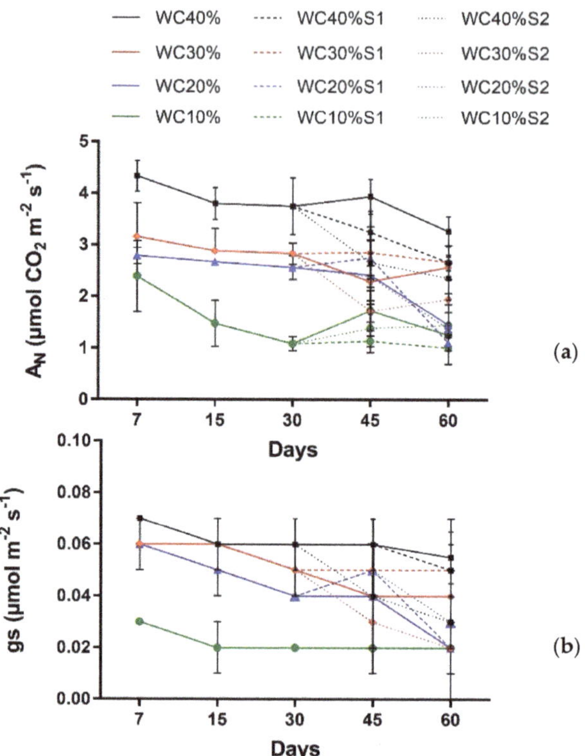

Figure 4. Trend of net photosynthesis (A_N) (**a**) and stomatal conductance (gs) (**b**) in *Callistemon* plants affected by drought (WC 40%: control; WC 30%: light drought stress; WC 20%: moderate drought stress; WC 10% severe drought stress) and saline aerosol treatment (S0: Distilled water, S1: 50% Synthetic seawater solution, S2: 100% Synthetic seawater solution). The plants were subjected of drought stress for four weeks; subsequently the plants were also treated with saline aerosol treatment. Data are means $\pm$ standard error ($n = 3$). Three biological replicates were used for the measurements.

In *Viburnum* plants, as observed for *Callistemon*, since the first measurement (7 days) a decrease of net photosynthesis was observed for the more stressed treatments (WC 20% and WC 10%, by 57%, $p < 0.0068$ **) and remained thereafter significant for the entire experimental period. With the addition of saline aerosol stress after 30 days, also for *Viburnum* plants, the differences with the control plants, at the end of the trial, were more pronounced in WC 20% S1 and S2, and WC 10% S0, S1, and S2 ($p < 0.0000$ ***) (Figure 5a, Table 5).

The gs in *Viburnum* plants was reduced in WC 10% since the first measurement ($p < 0.0068$ **). At 30 days significant differences ($p < 0.003$ **) were observed in the more stressed treatments (WC 20% and WC 10%). With the addition of saline aerosol stress the differences with the control plants, at the end of the trial, were more pronounced in WC 10% S0, S1, and S2 ($p < 0.0009$ ***) (Figure 5b, Table 5).

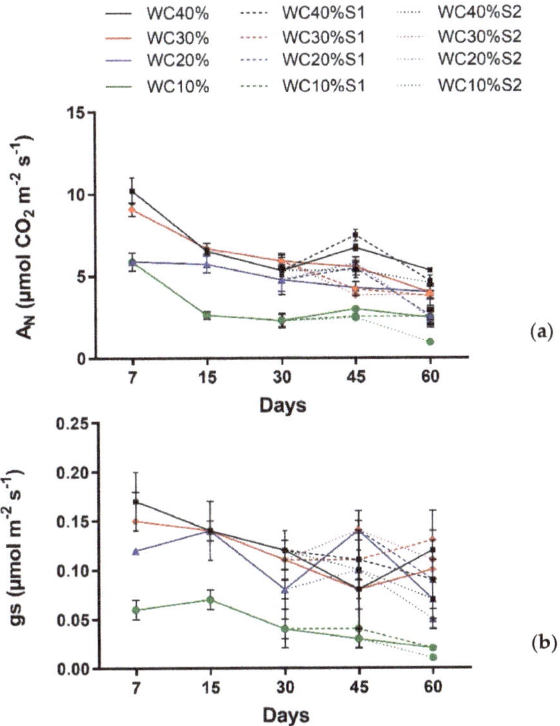

Figure 5. Trend of net photosynthesis (A_N) (**a**) and stomatal conductance (gs) (**b**) in *Viburnum* plants affected by drought (WC 40%: control; WC 30%: light drought stress; WC 20%: moderate drought stress; WC 10% severe drought stress) and saline aerosol treatment (S0: Distilled water, S1: 50% Synthetic seawater solution, S2: 100% Synthetic seawater solution). The plants were subjected of drought stress for four weeks; subsequently the plants were also treated with saline aerosol treatment. Data are means ± standard error (*n* = 3). Three biological replicates were used for the measurements.

Table 5. Summary of the main and interaction effects of days (T) and drought (D) in the first 30 days, and of days (T), drought (D), and saline aerosol (A) for the other 30 days on (A_N) and stomatal conductance (gs) of potted *Viburnum*.

Factor	A_N	gs
Main effect		
Days (T)	$p < 0.001$ ***	$p < 0.0024$ **
Drought (D)	$p < 0.001$ ***	$p < 0.000$ ***
Interaction		
T × D	$p < 0.0117$ *	$p > 0.6831$ ns
Main effect		
Days (T)	$p < 0.0000$ ***	$p < 0.0123$ *
Drought (D)	$p < 0.0000$ ***	$p < 0.0000$ ***
Aerosol Saline (A)	$p < 0.0001$ ***	$p > 0.1899$ ns
Interaction		
T × D	$p < 0.0002$ ***	$p < 0.0298$ *
T × A	$p > 0.1567$ ns	$p > 0.2644$ ns
D × A	$p > 0.0677$ ns	$p > 0.1738$ ns
T × D × A	$p < 0.0000$***	$p > 0.5243$ ns

Data were subjected to two-way ANOVA to compare the effects of days (T) and drought (D), while data were subjected to three-way ANOVA to compare the effects of days (T), drought (D), saline aerosol (A). Significance of differences of parameters: ns = not significant; * $p < 0.05$; ** $p < 0.01$; *** $p < 0.001$.

Chlorophyll *a* fluorescence is a good non-destructive marker of stress in plants. In our experiment, the maximum quantum efficiency of PSII (Fv/Fm ratio) decreased at the end of the experiment in both species. The Fv/Fm ratio in *Callistemon* showed significant differences ($p < 0.0034$ **) at the end of the experiment in WC 20% and WC 10% S0, S1, and S2 treatments compared with the control plants (Figure 6a, Table 6). No significant differences for Fv/Fm were observed between WC 30% and control plants (Figure 6a). Significant differences was observed in *Viburnum* at the end of the trial in WC 20% S0, S1, and S2 and WC 10% S0 and S1 ($p < 0.0000$ ***) compared with the control plants (Figure 6b, Table 6).

In *Callistemon* the RWC was only influenced by drought treatments and, in the most stressed treatment (WC 10%), the decrease from the control (WC 40%) was by 37%; in *Viburnum*, instead, the RWC was influenced by saline aerosol stress with a decrease in the more stressed treatments by 8% (Figure 7a,b).

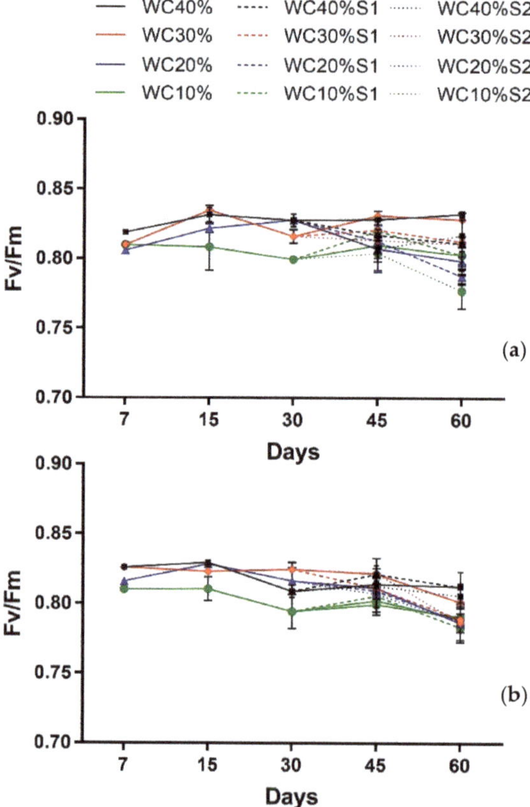

Figure 6. Trend of maximum quantum efficiency of PSII (Fv/Fm) in plants of *Callistemon* (**a**) and *Viburnum* (**b**) subjected to drought (WC 40%: control; WC 30%: light drought stress; WC 20%: moderate drought stress; WC 10% severe drought stress) and saline aerosol treatment (S0: Distilled water, S1: 50% Synthetic seawater solution, S2: 100% Synthetic seawater solution). Data are means ± standard error ($n = 3$). Three biological replicates were used for the measurements.

Table 6. Summary of the main and interaction effects of days (T) and drought (D) in the first 30 days, and of days (T), drought (D), and saline aerosol (A) for the other 30 days on of maximum quantum efficiency of PSII (Fv/Fm) of potted *Callisemon* and *Viburnum*.

Factor	Fv/Fm *Callistemon*	Fv/Fm *Viburnum*
Main effect		
Days (T)	$p < 0.0001$ ***	$p < 0.0002$ ***
Drought (D)	$p < 0.0000$ ***	$p < 0.0000$ ***
Interaction		
T × D	$p < 0.0018$ **	$p < 0.0259$ *
Main effect		
Days (T)	$p > 0.4355$ ns	$p < 0.0000$ ***
Drought (D)	$p < 0.0000$ ***	$p < 0.0005$ ***
Aerosol Saline (A)	$p < 0.0000$ ***	$p > 0.7511$ ns
Interaction		
T × D	$p > 0.4146$ ns	$p < 0.0157$ *
T × A	$p < 0.0418$ *	$p > 0.1429$ ns
D × A	$p < 0.0000$ ***	$p < 0.4412$ *
T × D × A	$p < 0.0099$ **	$p > 0.6064$ ns

Data were subjected to two-way ANOVA to compare the effects of days (T) and drought (D), while data were subjected to three-way ANOVA to compare the effects of days (T), drought (D), saline aerosol (A). Significance of differences of parameters: ns = not significant; * $p < 0.05$; ** $p < 0.01$; *** $p < 0.001$

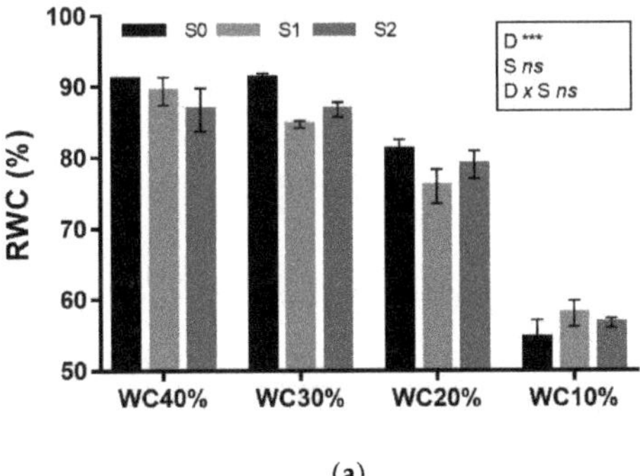

(a)

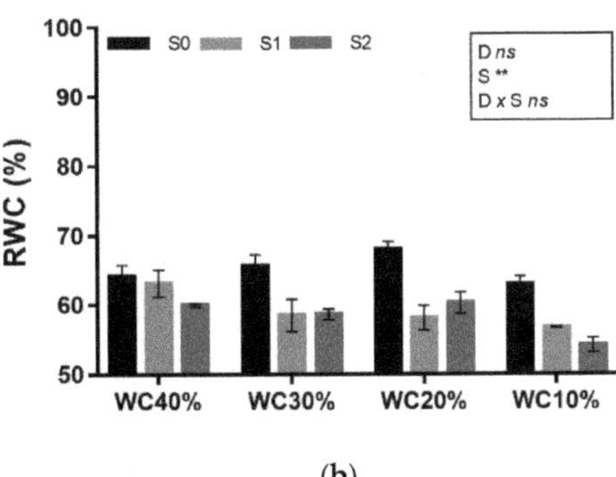

(b)

Figure 7. Relative water content (RWC) at the end of the experiment in plants of *Callistemon* (a) and *Viburnum* (b) subjected of drought (D: WC 40%: control; WC 30%: light drought stress; WC 20%: moderate drought stress; WC 10% severe drought stress) and saline aerosol treatment (S: S0: Distilled water, S1: 50% Synthetic seawater solution, S2: 100% Synthetic seawater solution). Data are means ± standard error ($n = 3$). Three biological replicates were used for the measurements. Data were subjected to two-way ANOVA and differences among means were determined using Tukey's post-test. Significance of differences of parameters: ns = not significant; ** $p < 0.01$; *** $p < 0.001$.

To visualize the effects of drought and saline aerosol stress on the relationships among the measured parameters, a correlation-based heat map was displayed (Figure 8). The heat map clearly revealed a considerable variation among the lines in their responses to progressive drought stress and interaction with saline aerosol stress. Cumulative effects of increasing the drought quantity and salinity levels have resulted in higher decreases in vegetative growth. Under severe treatments, the total and epigeous dry biomass decreased in all lines, while the R/S ratio increased in both species. The reduction of the total leaf area and the total leaf number was specifically observed in *Callistemon*. In *Callistemon* the combination of the two stresses together resulted in a significant decrease in SPAD values while effects were observed in *Viburnum*. Irrigation and saline aerosol treatments noticeably affected the leaf relative water content (RWC), chlorophyll a fluorescence (Fv/Fm) in the most stressed *Callistemon* plants; the differences in net photosynthetic activity (A_N) and stomatal conductance (gs) were more prominently in *Viburnum* plants.

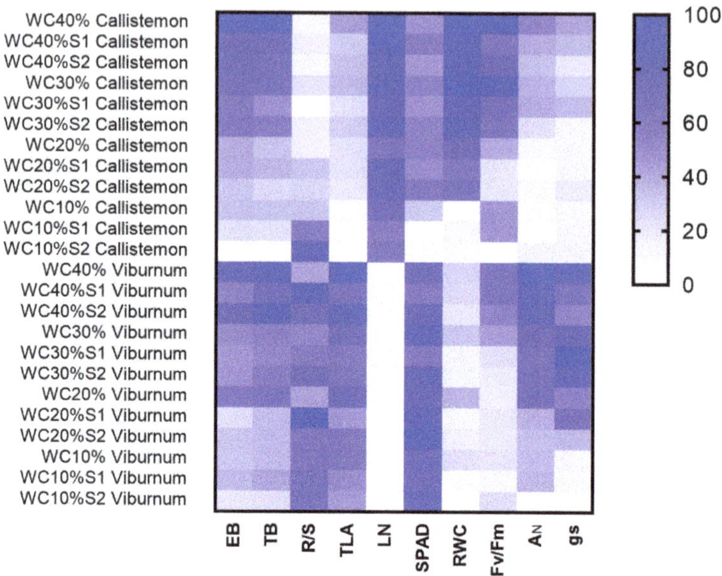

Figure 8. Heat map analysis summarizing the morphological and physiological changes of potted *Callistemon* and *Viburnum* plants responses to drought (WC 40%: control; WC 30%: light drought stress; WC 20%: moderate drought stress; WC 10% severe drought stress) and saline aerosol treatment (S0: Distilled water, S1: 50% Synthetic seawater solution, S2: 100% Synthetic seawater solution). Blue color indicates higher and white color indicates lower value. The mean values were normalized. TB = total dry biomass; EB = epigeous dry biomass; R/S = root to shoot ratio; TLA = total leaf area; LN = leaf number; SPAD = SPAD index; RWC = relative water content; Fv/Fm = maximum quantum efficiency of PSII; A_N = net photosynthesis; gs = stomatal conductance.

From the PCA analysis, effects were summary score plot, similar response of the two species to drought and saline aerosol stress treatments were observed (Figure 9a,b). In *Callistemon*, the first two PCs were related with eigen values >1 and explained more than 80% of the total variance, with PC1 and PC2 accounting for 75.2% and 10.0%; in *Viburnum* the values were 57.7% and 18.2% respectively for PC1 and PC2. The PCA showed that the most morphological and physiological parameters were associated in *Callistemon* with the WC 40% and WC 30% and related saline treatments, whereas the Spad index and R/S ratio, were associated with the WC 10% S1 and WC 10% S2 (Figure 9a).

In *Viburnum*, the most important morphological and physiological parameters were also associated with the WC 40% and WC 30% and saline treatments, the SPAD index was associated with WC 20% S2, WC 10%, and WC 10% S2, whereas the R/S ratio with WC 20%, WC 20% S1, and WC 10% S1 (Figure 9b).

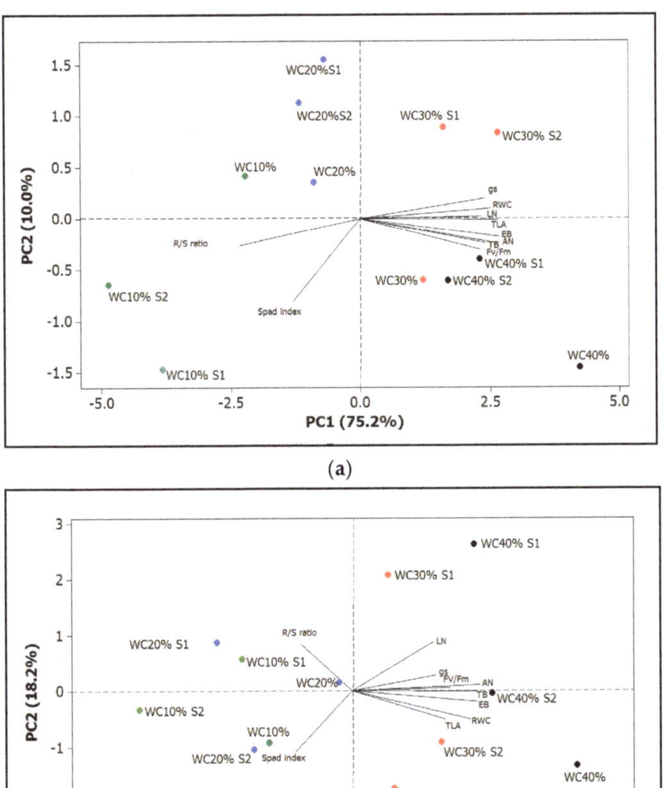

Figure 9. Principal component loading plot and scores of PCA epigeous and total dry biomass, R/S ratio, total leaf area, leaf number, Spad index, gas exchange, chlorophyll a fluorescence, and RWC for *Callistemon* (**a**) and *Viburnum* (**b**) as drought and saline aerosol stress treatments. TB = total dry biomass; EB = epigeous dry biomass; R/S = root to shoot ratio; TLA = total leaf area; LN = leaf number.

4. Discussion

The selection of ornamental plants tolerant/resistant to number of stresses is a priority to increasing the private or public gardens, in urban and peri-urban coastal areas, but information in literature to discriminate the response of plant species to abiotic stress and hence help the plant species choice is lacking [2]. In our study, two ornamental shrubs exposed to drought and subsequently saline aerosol for 4 weeks showed different responses. The analysis of physiological and morphological traits presented in this study may help to clarify the different strategies adopted by these species. In our study, the interaction between drought and saline aerosol stress was responsible for plant growth parameters in both shrubs. Dry biomass in *Callistemon* plants decreased under drought and saline aerosol treatments without interaction effects. In *Viburnum*, the effects of the two stresses appeared to be more evident in epigeous dry biomass; in fact, the differences due to the drought and saline aerosol stresses were more pronounced than in total biomass. Plants subjected to drought stress did not show the same amount of total biomass compared to plants grown under optimal irrigation, and these reductions also increased when saline aerosol stress was imposed. Significant interactions were observed between the water regime and saline aerosol treatment: plants, under the WC 10% treatment, were the most stressed due to the lack of water and were the most affected by the adoption of the saline solution. Species' response to stresses in terms of growth is the final manifestation of several interacting physiological and biochemical parameters and has been often used to characterize salinity or drought tolerance [44].

When plants are subjected to drought and/or saline stresses, the reduction of leaf area is considered an avoidance mechanism leading to the reduction of water losses by regulating the stomata closure, which is the most common defense strategy of many species under osmotic stress [45,46]. In our trial, the total leaf area of *Callistemon* showed interaction effects with differences due to saline aerosol in the control and light drought stress, while in *Viburnum* plants no interaction effects were observed. Previous attempts to study environmental stresses, such as drought, reported that plants can shift biomass allocation and change root/shoot ratios to cope with various environmental conditions [47]. It is proved that increasing R/S ratio is one of the avoidance mechanisms that enable plants to optimize water uptake under drought condition [48]. In presence of drought stress, plants need a wider root surface, while in saline aerosol stress conditions in some cases this surface can be even reduced to minimize toxic ions uptake and their consequent accumulation in the shoots, inducing in both situations a different distribution of the roots [49–51]. Our results reported that the plants of *Callistemon* in presence of severe drought stress and saline aerosol (10% WC S1 and S2), showed a higher R/S ratio thus optimizing water uptake.

Drought and salinity stresses induce the generic response of creating a physiological water deficit in plants. So far, interaction between drought and saline aerosol stress was reported to be detrimental for a number of physiological parameters, with the result of an increase of damage in plants under combined stress [52]. Both drought and salinity stresses induce a very complex photosynthetic response in plants, which occur in different leaf cell sites during the different steps of plants growth and development. The strength, length, and rate of development of the stress affect plant reactions to water shortage and salinity, because these issues dictate whether mitigation processes occur or not [53]. As a result, both single and overlapped water shortage and saline aerosol stress imposed to plants lead to drastic inhibition of net photosynthetic rate, stomatal conductance, and increased oxidative damage [52].

In our study, the photosynthetic rate and the stomatal conductance in both shrubs significantly decreased under both drought and saline aerosol treatments. As reported by Toscano et al., [54] in a study regarding two ornamental shrubs subjected by drought stress, a severe water deficit also had a negative effect on the photosynthetic rate and stomatal conductance of both species; however, these parameters were more affected, particularly in early summer, when these plants had very low gs values. Other studies reported that both stresses can adversely affect the photosynthetic activity in plants [55–58]. Good photosynthesis activity, even under salinity aerosol stress, can contribute to the biosynthesis of different related primary metabolism as an antioxidant defense mechanism, which help plants in detoxifying free radicals induced by stress conditions [43].

In this work, as expected, drought and saline aerosol stress resulted in reduced photosynthesis as a consequence of reduced CO_2 assimilation following stomatal closure. Achou et al. [59] reported the same results in a study on tomato plants. Moreover, stomatal closure minimizes water loss by transpiration, which leads to a decrease in photosynthetic carbon assimilation [60–62] and affects osmotic regulation [63,64]. Beside the gas exchange analyses, the stress conditions can be also monitored using chlorophyll *a* fluorescence that is a non-destructive method, largely used in studying plant response and adaptation to stressful environments [43]. The reduction of Fv/Fm can help to assess the tolerance of different species to the different stresses.

The combination of drought stress and saline aerosol treatments determined significant reduction in the values of parameter Fv/Fm at the end of the experiment in both species in plants treated with severe water deficit (20% WC and 10% WC). This indicates that no photo-damage of PSII reaction centers or relaxing, developing slowly quenching excitation energy, occurred [65]. Instead, in the combination of the two stresses, the lower Fv/Fm values indicated that PSII had been damaged. In a study conducted by Umar et al. [66] the maximum quantum yield of PSII (Fv/Fm) decreased in all sunflower cultivars when exposed to combined stress as compared with control while the Fv/Fm was not much affected under salt and drought stress alone. A study on the effect of marine aerosol on ornamental plants showed that *Callistemon citrinus* has been considered as intermediate saline spray tolerant species [6]. The tolerance of the species was evaluated using chlorophyll *a* fluorescence and relative derived indexes, classifying this species as medium tolerant [6].

RWC is a strong parameter to detect water status in plant tissues and is more reliable than cell water potential since, through a direct connection with cell volume, it better demonstrates the balance between the plant water content and transpiration rate. In general, by increasing drought stress intensity, the reduction in RWC can be due to a decrease in water potential of leaves [67]. RWC is an indicator of the water status of plant tissues during drought stress. It decreases with the water deficit increase, although this reduction is genotype specific [68]. In particular, in *Callistemon* the reduction for effect of drought stress was of about 37%; in *Viburnum* differences were observed only in correspondence of saline aerosol. In our study, the lowest RWC was obtained at 10% WC treatment in *Callistemon* and at 10% WC, S1 and S2 in *Viburnum*.

Gas exchanges are closely related to the status of leaf water, which could be also considered indicators of stress under drought and saline conditions. In different studies it was reported that gas exchanges had a close relationship to leaf water status, in fact, the net photosynthesis of the plants decreased with the relative water content and leaf water potential [69–71]. Similar results were found in our experiment, where the more stressing treatments, showed a major reduction of RWC and gas exchanges.

To better understand the tolerance mechanisms to environmental stresses it is interesting to study and interpret the morphological and physiological parameters collected from plants grown under stressed and not-stressed conditions. Correlation analysis, PCA, and clustering give useful indication for evaluating the relationships between the parameters and their principal components for stress tolerance [72,73]. In our study, heat map and PCA showed that the differences in stress tolerance between *Callistemon* and *Viburnum* were largely linked to variations in physiological parameters, especially in *Callistemon* plants. The PCA analysis confirmed that the drought and saline aerosol treatments with higher water capacity did not differ between both shrubs analyzed, demonstrating that the level of stress did not influence the plant morphology and physiology. To overcome stressful conditions, plants implement a different change such as the increase in the R/S ratio and the increase in the SPAD index. A heat map is a visual method that can be useful to show the intensity of variation between multiple parameters measured from different treatments. After drought stress, the saline aerosol amplified the negative effects, especially in plants previously damaged by water shortage.

From the visual appearance and ornamental quality point of view, *Viburnum* showed lower changes compared with *Callistemon* at severe stress conditions. In mild stress conditions, *Callistemon* can also be considered as ornamental plants for garden and urban green areas.

5. Conclusions

The experiment demonstrates the combined effects of the two stresses investigated—drought and saline aerosol—on morphological and physiological parameters of two ornamental shrubs largely adopted in Mediterranean green areas near the sea. Both species showed a tolerance to saline aerosol, probably for the sclerophyll leaf characteristics, as demonstrated by small presence of leaf necrosis that can be associated to higher tolerance. Plants exposed to severe drought stress were more sensitive to saline aerosol, as demonstrated by marked reductions in plant morphological and physiological parameters. Results of this experiment can be useful for the utilization of these two species in coastal seaside green areas with or without irrigation systems and subjected to different degrees of salinity stress. The overall data analyses at morphological and physiological levels suggest that *Viburnum tinus* was the more tolerant species compared to *Callistemon citrinus*.

Supplementary Materials: The following are available online at https://www.mdpi.com/article/10.3390/horticulturae7120517/s1, Figure S1: Mean air temperature (°C) and relative humidity (%) during the trial.

Author Contributions: Conceptualization, S.T. and D.R.; Methodology, S.T.; Software, S.T.; Validation, D.R., A.T. and A.F.; Formal analysis, S.T.; Investigation, S.T.; Resources, D.R.; Data curation, S.T. and A.F.; Writing—original draft preparation, S.T., A.T. and D.R.; Writing—review and editing, S.T., D.R., A.F. and A.T.; Visualization, A.F.; Supervision, D.R. and A.T.; Funding acquisition, D.R. All authors have read and agreed to the published version of the manuscript.

Funding: This research received no external funding.

Institutional Review Board Statement: Not applicable.

Informed Consent Statement: Not applicable.

Data Availability Statement: Main data are contained within the article; further data presented in this study are available on request from the corresponding author.

Conflicts of Interest: The authors declare no conflict of interest.

References

1. Savé, R.; Castell, C.; Terradas, J. Gas exchange and water relations. In *Ecology of Mediterranean Evergreen Oak Forest*; Rodá, F., Retama, J., Gracia, A., Bellot, J., Eds.; Ecological Studies; Springer: Berlin, Geramny, 1999; pp. 135–147.
2. Farieri, E.; Toscano, S.; Ferrante, A.; Romano, D. Identification of ornamental shrubs tolerant to saline aerosol for coastal urban and peri-urban greening. *Urban For Urban Green* **2016**, *18*, 9–18. [CrossRef]
3. Glenn, E.P.; Nelson, S.G.; Ambrose, B.; Martinez, R.; Soliz, D.; Pabendinskas, V.; Hultine, K. Comparison of salinity tolerance of three Atriplex spp. in well-watered and drying soils. *Environ. Exp. Bot.* **2012**, *83*, 62–72. [CrossRef]
4. Álvarez, S.; Sánchez-Blanco, M.J. Comparison of individual and combined effects of salinity and deficit irrigation on physiological, nutritional and ornamental aspects of tolerance in *Callistemon laevis* plants. *J. Plant Physiol.* **2015**, *185*, 65–74. [CrossRef] [PubMed]
5. Abdelraheem, A.; Esmaeili, N.; O'Connell, M.; Zhang, J. Progress and perspective on drought and salt stress tolerance in cotton. *Ind. Crops Prod.* **2019**, *130*, 118–129. [CrossRef]
6. Ferrante, A.; Trivellini, A.; Malorgio, F.; Carmassi, G.; Vernieri, P.; Serra, G. Effect of seawater aerosol on leaves of six plant species potentially useful for ornamental purposes in coastal areas. *Sci. Hortic.* **2011**, *128*, 332–341. [CrossRef]
7. EEA-UNEP/MAP Action Plan, 2014. In *Horizon 2020 Mediterranean Report toward Shared Environmental Information Systems (Technical Report No 6/2014)*; Publications Office of the European Union: Luxembourg, 2014. Available online: https://www.eea.europa.eu/publications/horizon-2020-mediterranean-report/file (accessed on 12 November 2021).
8. Pinna, M.S.; Bacchetta, G.; Cogoni, D.; Fenu, G. Is vegetation an indicator for evaluating the impact of tourism on the conservation status of Mediterranean coastal dunes? *Sci. Total Environ.* **2019**, *674*, 255–263. [CrossRef]
9. Domene, E.; Saurí, D. Urbanisation and water consumption: Influencing factors in the metropolitan region of Barcelona. *Urban Stud.* **2006**, *43*, 1605–1623. [CrossRef]
10. Toscano, S.; Ferrante, A.; Romano, D. Response of Mediterranean ornamental plants to drought stress. *Horticulturae* **2019**, *5*, 6. [CrossRef]
11. Araújo-Alves, J.P.L.; Torres-Pereira, J.M.; Biel, C.; De Herralde, F.; Savé, R. Effects of minimum irrigation technique on ornamental parameters of two Mediterranean species used in xerigardening and landscaping. *Acta Hortic.* **1999**, *541*, 353–358. [CrossRef]
12. Giordano, M.; Petropoulos, S.A.; Rouphael, Y. Response and defence mechanisms of vegetable crops against drought, heat and salinity stress. *Agriculture* **2021**, *11*, 463. [CrossRef]
13. Acquaah, G. *Principles of Plant Genetics and Breeding*, 2nd ed.; John Wiley & Sons, Ltd.: Chichester, UK, 2012; p. 740.
14. Mittler, R. Abiotic stress, the field environment and stress combination. *Trend Plant Sci.* **2006**, *11*, 15–19. [CrossRef]
15. Atkinson, N.J.; Urwin, P.E. The interaction of plant biotic and abiotic stresses: From genes to the field. *J. Exp. Bot.* **2012**, *63*, 3523–3543. [CrossRef]
16. Kissoudis, C.; van de Wiel, C.; Visser, R.G.; van der Linden, G. Enhancing crop resilience to combined abiotic and biotic stress through the dissection of physiological and molecular crosstalk. *Front. Plant Sci.* **2014**, *5*, 207. [CrossRef] [PubMed]

17. Ramegowda, V.; Senthil-Kumar, M. The interactive effects of simultaneous biotic and abiotic stresses on plants: Mechanistic understanding from drought and pathogen combination. *J. Plant Physiol.* **2015**, *176*, 47–54. [CrossRef]
18. Suzuki, N.; Rivero, R.M.; Shulaev, V.; Blumwald, E.; Mittler, R. Abiotic and biotic stress combinations. *New Phytol.* **2014**, *203*, 32–43. [CrossRef] [PubMed]
19. Sánchez-Blanco, M.J.; Rodriguez, P.; Morales, M.A.; Torrecillas, A. Comparative growth and water relations of *Cistus albidus* and *Cistus monspeliensis* plants during water deficit conditions and recovery. *Plant Sci.* **2002**, *162*, 107–113. [CrossRef]
20. Cai, X.; Starman, T.; Niu, G.; Hall, C.; Lombardini, L. Response of selected garden roses to drought stress. *HortScience* **2012**, *47*, 1050–1055. [CrossRef]
21. Lumis, G.P.; Hofstra, G.; Hall, R. Sensitivity of roadside trees and shrubs to aerial drift of deicing salts in soils around trees. *J. Arboric.* **1973**, *20*, 196–200.
22. Sánchez-Blanco, M.J.; Rodríguez, P.; Morales, M.A.; Torrecillas, A. Contrasting physiological responses of dwarf sea-lavender and marguerite to simulated sea aerosol deposition. *J. Environ. Qual.* **2003**, *32*, 3338–3344. [CrossRef]
23. Ashraf, M.; McNeilly, T.; Bradshaw, A.D. Tolerance of *Holcus lanatus* and *Agrostis stolonifera* to sodium chloride in soil solution and saline spray. *Plant Soil* **1986**, *96*, 77–84. [CrossRef]
24. Wu, L.; Guo, X.; Harivandi, A. Salt tolerance and salt accumulation of landscape plants irrigated by sprinkler and drip irrigation systems. *J. Plant Nutr.* **2001**, *24*, 1473–1490. [CrossRef]
25. Niinemets, Ü. Responses of forest trees to single and multiple environmental stresses from seedlings to mature plants: Past stress history, stress interactions, tolerance and acclimation. *For. Ecol. Manag.* **2010**, *260*, 1623–1639. [CrossRef]
26. Bussotti, F.; Grossoni, P.; Pantani, F. The role of marine salt and surfactants in the decline of Tyrrhenian coastal vegetation in Italy. *Ann. For. Sci.* **1995**, *52*, 251–261. [CrossRef]
27. Waisel, Y. *Biology of Halophytes*; Academic Press: New York, NY, USA, 1972; p. 410.
28. Shannon, M.C. Adaptation of plants to salinity. *Adv. Agron.* **1997**, *60*, 75–120. [CrossRef]
29. Boyce, S.G. The salt spray community. *Ecol. Monogr.* **1954**, *24*, 29–67. [CrossRef]
30. Bussotti, F.; Bottacci, A.; Grossoni, P.; Mori, B.; Tani, C. Cytological and structural changes in *Pinus pinea* L. needles following the application of an anionic surfactant. *Plant Cell Environ.* **1997**, *20*, 513–520. [CrossRef]
31. Schönherr, J.; Bauer, H. Analysis of effects of surfactants on permeability of plant cuticles. In *Adjuvant and Agrichemicals*; Foy, C.L., Ed.; CRC: Boca Raton, FL, USA, 1992; pp. 17–35.
32. Jefree, E.J. The cuticle, epicuticular waxes and trichomes of plants, with reference to their structure, functions and evolution. In *Insects and the Plant Surface*; Juniper, B., Southwood, R., Eds.; Edward Arnold: London, UK, 1986; pp. 23–64.
33. Esch, A.; Mengel, K. Combined effects of acid mist and frost drought on the water status of young spruce trees (*Picea abies*). *Environ. Exp. Bot.* **1998**, *39*, 57–65. [CrossRef]
34. Burkhardt, J.; Basi, S.; Pariyar, S.; Hunsche, M. Stomatal penetration by aqueous solutions–an update involving leaf surface particles. *New Phytol.* **2012**, *196*, 774–787. [CrossRef]
35. Rodríguez, P.; Torrecillas, A.; Morales, M.A.; Ortuño, M.F.; Sánchez-Blanco, M.J. Effects of NaCl salinity and water stress on growth and leaf water relations of *Asteriscus maritimus* plants. *Environ. Exp. Bot.* **2005**, *53*, 113–123. [CrossRef]
36. Navarro, A.; Bañón, S.; Conejero, W.; Sánchez-Blanco, M.J. Ornamental characters, ion accumulation and water status in *Arbutus unedo* seedlings irrigated with saline water and subsequent relief and transplanting. *Environ. Exp. Bot.* **2008**, *62*, 364–370. [CrossRef]
37. Vernieri, P.; Mugnai, S.; Borghesi, E.; Petrognani, L.; Serra, G. Non-chemical growth control of potted *Callistemon laevis*. *Agric. Med.* **2006**, *160*, 85–90.
38. Gori, R.; Lubello, C.; Ferrini, F.; Nicese, F.P.; Coppini, E. Reuse of industrial wastewater for the irrigation of ornamental plants. *Water Sci. Technol.* **2008**, *57*, 883–889. [CrossRef] [PubMed]
39. Álvarez, S.; Sánchez-Blanco, M.J. Long-term effect of salinity on plant quality, water relations, photosynthetic parameters and ion distribution in *Callistemon citrinus*. *Plant Biol.* **2014**, *16*, 757–764. [CrossRef] [PubMed]
40. Cirillo, C.; De Micco, V.; Arena, C.; Carillo, P.; Pannico, A.; De Pascale, S.; Rouphael, Y. Biochemical, Physiological and Anatomical Mechanisms of Adaptation of *Callistemon citrinus* and *Viburnum lucidum* to NaCl and $CaCl_2$ Salinization. *Front. Plant Sci.* **2019**, *10*, 742. [CrossRef]
41. Starr, J.L.; Paltineanu, I.C. Methods for Measurement of SoilWater Content: Capacitance Devices. In *Methods of Soil Analysis*; United States Department of Agriculture, Agricultural Research Service: Washington, DC, USA, 2002.
42. Tribulato, A.; Toscano, S.; Di Lorenzo, V.; Romano, D. Effects of water stress on gas exchange, water relations and leaf structure in two ornamental shrubs in the Mediterranean area. *Agronomy* **2019**, *9*, 381. [CrossRef]
43. Toscano, S.; Branca, F.; Romano, D.; Ferrante, A. An evaluation of different parameters to screen ornamental shrubs for salt spray tolerance. *Biology* **2020**, *9*, 250. [CrossRef] [PubMed]
44. Sidari, M.; Mallamaci, C.; Muscolo, M. Drought, salinity and heat differently affect seed germination of *Pinus pinea*. *J. For. Res.* **2008**, *13*, 326–330. [CrossRef]
45. Savé, R.; Biel, C.; Domingo, R.; Ruiz-Sánchez, M.C.; Torrecillas, A. Some physiological and morphological characteristics of citrus plants for drought resistance. *Plant Sci.* **1995**, *110*, 167–172. [CrossRef]
46. Ruiz-Sánchez, M.C.; Domingo, R.; Torrecillas, A.; Pérez-Pastor, A. Water stress preconditioning to improve drought resistance in young apricot plants. *Plant Sci.* **2000**, *156*, 245–251. [CrossRef]

47. Zhou, G.; Zhou, X.; Nie, Y.; Bai, S.H.; Zhou, L.; Shao, J.; Cheng, W.; Wang, J.; Hu, F.; Fu, Y. Drought-induced changes in root biomass largely result from altered root morphological traits: Evidence from a synthesis of global field trials. *Plant Cell Environ.* **2018**, *41*, 2589–2599. [CrossRef]
48. Chaves, M.M.; Maroco, J.P.; Pereira, J.S. Understanding plant responses to drought—From genes to the whole plant. *Funct. Plant Biol.* **2003**, *30*, 239–264. [CrossRef] [PubMed]
49. Munns, R. Comparative physiology of salt and water stress. *Plant Cell Environ.* **2002**, *25*, 239–250. [CrossRef] [PubMed]
50. Alarcón, J.J.; Morales, M.A.; Ferrández, T.; Sánchez-Blanco, M.J. Effects of water and salt stresses on growth, water relations and gas exchange in *Rosmarinus officinalis*. *J. Hortic. Sci. Biotechnol.* **2006**, *81*, 845–853. [CrossRef]
51. Álvarez, S.; Castillo, M.; Acosta, J.; Navarro, A.; Sánchez-Blanco, M. Photosynthetic response, biomass distribution and water status changes in *Rhamnus alaternus* plants during drought. *Acta Hortic.* **2012**, *937*, 853–860. [CrossRef]
52. Pandey, P.; Ramegowda, V.; Senthil-Kumar, M. Shared and unique responses of plants to multiple individual stresses and stress combinations: Physiological and molecular mechanisms. *Front. Plant Sci.* **2015**, *6*, 723. [CrossRef]
53. Chaves, M.M.; Flexas, J.; Pinheiro, C. Photosynthesis under drought and salt stress: Regulation mechanisms from whole plant to cell. *Ann. Bot.* **2009**, *103*, 551–560. [CrossRef]
54. Toscano, S.; Ferrante, A.; Tribulato, A.; Romano, D. Leaf physiological and anatomical responses of Lantana and Ligustrum species under different water availability. *Plant Physiol. Biochem.* **2018**, *127*, 380–392. [CrossRef]
55. Farooq, M.; Basra, S.M.A.; Wahid, A.; Rehman, H. Exogenously applied nitric oxide enhances the drought tolerance in fine grain aromatic rice (*Oryza sativa* L.). *J Agron Crop Sci.* **2009**, *195*, 254–261. [CrossRef]
56. Xu, C.; Leskovar, D.I. Growth, physiology and yield responses of cabbage to deficit irrigation. *Hortic. Sci.* **2014**, *41*, 138–146. [CrossRef]
57. Ors, S.; Ekinci, M.; Yildirim, E.; Sahin, U. Changes in gas exchange capacity and selected physiological properties of squash seedlings (*Cucurbita pepo* L.) under well-watered and drought stress conditions. *Arch. Agron. Soil Sci.* **2016**, *62*, 1700–1710. [CrossRef]
58. Ekinci, M.; Ors, S.; Yildirim, E.; Turan, M.E.T.İ.N.; Sahin, U.; Dursun, A.; Kul, R. Determination of physiological indices and some antioxidant enzymes of chard exposed to nitric oxide under drought stress. *Russ. J Plant Physiol.* **2020**, *67*, 740–749. [CrossRef]
59. Achuo, E.A.; Prinsen, E.; Höfte, M. Influence of drought, salt stress and abscisic acid on the resistance of tomato to *Botrytis cinerea* and *Oidium neolycopersici*. *Plant Pathol.* **2006**, *55*, 178–186. [CrossRef]
60. Adams, H.D.; Germino, M.J.; Breshears, D.D.; Barron-Gafford, G.A.; Guardiola-Claramonte, M.; Zou, C.B.; Huxmanet, T.E. Nonstructural leaf carbohydrate dynamics of *Pinus edulis* during drought-induced tree mortality reveal role for carbon metabolism in mortality mechanism. *New Phytol.* **2013**, *197*, 1142–1151. [CrossRef] [PubMed]
61. Chen, L.; Zhang, S.; Zhao, H.; Korpelainen, H.; Li, C. Sex-related adaptive responses to interaction of drought and salinity in *Populus yunnanensis*. *Plant Cell Environ.* **2010**, *33*, 1767–1778. [CrossRef] [PubMed]
62. Choat, B.; Brodribb, T.J.; Brodersen, C.R.; Remko, A.; Duursma, R.A.; Rosana López, R.; Medlyn, B.E. Triggers of tree mortality under drought. *Nature* **2018**, *558*, 531–539. [CrossRef]
63. Hartmann, H.; Trumbore, S. Understanding the roles of nonstructural carbohydrates in forest trees—From what we can measure to what we want to know. *New Phytol* **2016**, *211*, 386–403. [CrossRef] [PubMed]
64. Karst, J.; Gaster, J.; Wiley, E.; Simon, M.; Landhäusser, S.M. Stress differentially causes roots of tree seedlings to exude carbon. *Tree Physiol.* **2017**, *37*, 154–164. [CrossRef] [PubMed]
65. Umar, M.; Uddin, Z.; Siddiqui, Z.S. Responses of photosynthetic apparatus in sunflower cultivars to combined drought and salt stress. *Photosynthetica* **2019**, *57*, 627–639. [CrossRef]
66. Nogués, S.; Baker, N.R. Effects of drought on photosynthesis in Mediterranean plants grown under enhanced UV-B radiation. *J. Exp. Bot.* **2000**, *51*, 1309–1317. [CrossRef] [PubMed]
67. Sharif, P.; Seyedsalehi, M.; Paladino, O.; Van Damme, P.; Sillanpää, M.; Sharifi, A.A. Effect of drought and salinity stresses on morphological and physiological characteristics of canola. *Int. J. Environ. Sci. Technol.* **2018**, *15*, 1859–1866. [CrossRef]
68. Zhu, Y.; Luo, X.; Nawaz, G.; Yin, J.; Yang, J. Physiological and biochemical responses of four cassava cultivars to drought stress. *Sci. Rep.* **2020**, *10*, 6968. [CrossRef] [PubMed]
69. Chaves, M. Effects of water deficits on carbon assimilation. *J. Exp. Bot.* **1991**, *42*, 1–16. [CrossRef]
70. Lawlor, D.W. Limitation to photosynthesis in water-stressed leaves: Stomata vs. metabolism and the role of ATP. *Ann. Bot.* **2002**, *89*, 871–885. [CrossRef] [PubMed]
71. Yan, W.; Zhong, Y.; Shangguan, Z. A meta-analysis of leaf gas exchange and water status responses to drought. *Sci. Rep.* **2016**, *6*, 20917. [CrossRef]
72. Dehbalaei, S.; Farshadfar, E.; Farshadfar, M. Assessment of drought tolerance in bread wheat genotypes based on resistance/tolerance indices. *Int. J. Agric. Crop Sci.* **2013**, *5*, 2352–2358.
73. Sun, J.; Luo, H.; Fu, J.; Huang, B. Classification of genetic variation for drought tolerance in Tall Fescue using physiological traits and molecular markers. *Crop Sci.* **2013**, *53*, 647–654. [CrossRef]

horticulturae

Article

Morphological, Physiological, and Biochemical Responses of Zinnia to Drought Stress

Stefania Toscano and Daniela Romano *

Department of Agriculture, Food and Environment (Di3A), Università Degli Studi di Catania, Via Valdisavoia 5, 95123 Catania, Italy; stefania.toscano@unict.it
* Correspondence: dromano@unict.it

Abstract: Bedding plants in the nursery phase are often subject to drought stress because of the small volume of the containers and the hydraulic conductivity of organic substrates used. To analyse the morphological, physiological, and enzymatic responses of zinnia (*Zinnia elegans* L.) plants at different irrigation levels, four treatments were performed: irrigated at 100% (100% field capacity, FC); light deficit irrigation (75% FC), medium deficit irrigation (50% FC), and severe deficit irrigation (25% FC). The growth of zinnia was significantly influenced by drought stress treatments. Different morphological parameters (dry biomass, leaf number, root to shoot ratio (R/S)) were modified only in the more severe drought stress treatment (25% FC). The stomata density increased in 50% FC and 25% FC, while the stomata size was reduced in 25% FC. The net photosynthesis, stomatal conductance, and transpiration were reduced in 50% FC and 25% FC. The relative water content (RWC) was reduced in 25% FC. Severe drought stress (25% FC) increased proline content up to seven-fold. Catalase (CAT), peroxidase (GPX), and superoxide dismutase (SOD) activity significantly increased in 50% FC and 25% FC. Principal component analysis (PCA) showed that the morphological and physiological parameters were mostly associated with the 100% FC and 75% FC treatments of the biplot, whereas the stomata density, R/S ratio, and antioxidant enzymes (GPX, CAT) were associated with 50% FC, and proline and DPPH were associated with 25% FC, respectively.

Keywords: *Zinnia elegans* L.; bedding plants; deficit irrigation; stomata characteristics; gas exchange; proline; enzyme activity

Citation: Toscano, S.; Romano, D. Morphological, Physiological, and Biochemical Responses of Zinnia to Drought Stress. *Horticulturae* **2021**, 7, 362. https://doi.org/10.3390/horticulturae7100362

Academic Editor: Othmane Merah

Received: 8 September 2021
Accepted: 30 September 2021
Published: 4 October 2021

Publisher's Note: MDPI stays neutral with regard to jurisdictional claims in published maps and institutional affiliations.

1. Introduction

Bedding plants play a relevant role in public green areas and private gardens. These plants can suffer from drought stress because they are not always properly watered, especially when grown in pots or show small root systems [1]. Bedding plants are in fact at greater risk of undergoing drought stress during the nursery phase because they are cultivated in small pots that can limit root growth, making the plants subject to greater levels of drought stress. Furthermore, the hydraulic conductivity of the substrates, which are often organic, used in the production of bedding plants decreases rapidly with small changes in the substrate water content [2], making the extraction of water very difficult for plants when the water content in the substrate is low. However, limited research exists regarding the physiological mechanisms that allow bedding plants to tolerate drought stress.

Drought stress results in damage to the plant's physiological and biochemical processes and represents one of the most relevant environmental factors that impair plant growth and performance [3]. Plants may exhibit numerous drought stress response mechanisms at the morphological and physiological levels [4,5]. At the level of the whole plant, some species increase their root biomass to enhance water uptake [6] and hence maintain the water state of the plant and ensure photosynthesis in drought conditions. In the nursery stage of bedding plant cultivation, this acclimatization response may not be possible because both root growth and available water are limited by the small pots.

Exposure to drought stress causes morphological changes in shoots; the plants produce smaller leaves and drop the older leaves to reduce transpiration and hence water loss [7]. The reduction in leaf area, if it can help maintain a favourable water status of the plant, reduces plant photosynthesis and plant carbon gain. Photosynthesis, which is essential for plant growth, markedly declines in plants in drought conditions because this process is highly sensitive to drought stress [8]. Although several studies have analysed the influence of drought conditions in the modification of the photosynthesis in bedding plants [9], there is limited research-based information linking the morphological, physiological, and biochemical acclimatization to drought in bedding plants. This information can be important for deepening the knowledge on the physiological responses of plants to drought, individuating guidelines to mitigate drought stress, and for selecting and developing suitable species to resist water shortages [10].

In drought conditions, water loss reduction at the leaf level is determined by transiently lowering stomatal conductance (gs) [11]; this can help maintain the level of foliar photosynthesis in drought conditions, albeit at a lower rate, for a longer time. In plants exposed to drought stress, a good correlation between gs and leaf water potential was observed [12].

Osmotic adaptation on a cellular scale is a drought acclimatization response to the concentration of compatible solutes within cells [13]. This reduces leaf water and maintains the potential gradient necessary for root water uptake from the substrate and allows the maintenance of a positive turgor potential in drought conditions [14]. Although light-harvesting mechanisms, including photosystems, are generally tolerant to drought stress, severe stress levels can impair photosystem II [15]. Drought stress causes lipid peroxidation and causes irreversible damage to the structural and functional integrity of the membrane [16]. For this reason, the accumulation of malondialdehyde (MDA) in the cell and the stability of the cell membrane are widely used as indicators of plant tolerance to drought stress [17].

Plants show various physiological and biochemical responses to drought stress. The accumulation of osmotic compounds, such as proline, is one of the most common plant responses to drought stress [18]. Proline is a compatible solute involved in cellular osmotic regulation and the protection of cellular components during dehydration [19].

Activation of the antioxidant defence and osmoprotection systems are two main drought resistance mechanisms in plants [20,21]. The overproduction of reactive oxygen species (ROS) under stress conditions is a typical tolerance response [22]. ROS assure a key role in the process of acclimatization to various abiotic stresses [23]. Antioxidant mechanisms, both enzymatic and non-enzymatic, are known to be involved in plant protection against ROS. A physiological mechanism for mitigating the negative effects of ROS on plant cells are antioxidant enzymes; among these are catalase (CAT), superoxide dismutase (SOD), and ascorbate peroxidase (APX), which determine the protection of plant cells against oxidative damage [24]. Studies have shown antioxidant enzymatic activity is positively associated with plant stress tolerance, which has been found in various field crops, such as pea [25], maize [26], and wheat [27].

Proline also works as a free radical scavenger and suppresses free radical-mediated damage during drought stress. Several studies have demonstrated that, during drought stress conditions, proline content increases, and proline buildup is associated with more efficient drought tolerance in tall fescue and other plants [19].

Among the ornamental summer flowering plants, zinnia (*Zinnia elegans* L.) is rightly appreciated for its spectacular display of colourful flowers [28]. It belongs to the Compositae family (Asteraceae) and it is native to Central America and Mexico. Zinnia flowers have a long vase life and present uniform and bright colours and sturdy stems [29]. It is one of the suitable bedding and cut flower plants grown during the summer season in hot climates [30].

The application of deficient irrigation strategies to floriculture can make a significant contribution to the conservation of irrigation water. In the near future, the warming climate will enhance the frequency and severity of drought [31]. Therefore, in a changing climate, studying the main physiological limits to productivity in drought conditions will be crucial for enhancing yield stability.

Because the increased drought frequency strongly negatively affects plant growth and development [32], analyzing the effects of water deficit on plants is relevant to hypothesise the influence of future climate changes on the growth of a particular plant species [33]. Among bedding plants, studying the response to drought stress in various species and/or cultivars is strategic to individuate genotypes able to improve landscape performance and expand the use of these plants in drought areas [34].

The aim of this research was to determine the morphological, physiological, and enzymatic responses of zinnia plants at different irrigation levels and to evaluate the response to different intensities of drought and hence the possibility to reduce the water quantity used in the nursery phase.

2. Materials and Methods

2.1. Experimental Conditions, Plant Material, and Irrigation Treatments

The trial was realised in a nursery near Catania, Italy (37°41′ N 15°11′ E 89 m a.s.l.) in April 2021 on zinnia (*Zinnia elegans* L.). Seeds of zinnia 'Limette', (Fratelli Ingegnoli, Milan, Italy) were sown in cellular trays on the substrate Brill® Semina Bio (Geotec, Bolzano, Italy). At the fourth leaf stage, the seedlings were transplanted into 10 cm Ø pots (one plant per pot), filled with peat and soil (2/1, *v/v*), and fertilised with 2 g L^{-1} of Osmocote Plus (14/13/13, N,P,K + microelements).

Plants were grouped into three repetitions of nine plants per treatment and irrigated every day for 30 days. Four treatments were performed: irrigated at 100% (100% field capacity, FC), light deficit irrigation (75% FC), irrigated at 75% from the 100% FC treatment, medium deficit irrigation (50% FC), irrigated at 50% from the 100% FC treatment and severe deficit irrigation (25% FC), and irrigated at 25% from the 100% FC treatment. Water loss was determined following the methodology of Toscano et al. [35] through the gravimetric method; during the experimental period, the differences in weights (weight after irrigation, weight when drainage stopped, and weight before reirrigating) were calculated.

The treatments started when the plants showed four leaves and ended after 30 days when approximately 50% of plants showed the beginning of inflorescence emergence.

To determine the maximum water-holding capacity of the substrate, pots were mixed and submerged in water to 50% of their height, and the substrate was left to imbibe overnight. To avoid water evaporation, aluminium foil was placed on the upper surfaces of the containers. The next day, the containers were removed from the water bath and left to drain until reaching a constant weight. The weight of each container was then determined and considered as the weight at volumetric water content. Then, the substrate was dried in a thermo-ventilated oven at 105 °C until reaching a constant weight to determine the dry weight and calculate the volumetric water content. The difference between the fresh and dry weight was calculated, and the volumetric water content was determined (75%) and used as the substrate's container capacity [36].

The mean air temperature, relative humidity, and global radiation were determined on a data logger CR1000 (Campbell Scientific Ltd., Loughborough, UK) during the experimental periods. The minimum, maximum, and mean temperatures were 16.9, 54.6, and 26.2 °C, respectively. The mean relative humidity (RH) was 58.8%.

2.2. Biomass and Leaf Area

After the experimental period ended, for six plants per treatment, the roots were separated from the substrate with tap water, and the aerial parts were divided into stems and leaves. Drying the weighed samples in a thermo-ventilated oven at 70 °C until reaching constant weight allowed the determination of dry biomass. The total leaf area was measured using a leaf area meter (Delta-T Devices Ltd., Cambridge, UK).

2.3. Stomata Characteristics

Unfolded and mature leaves (4 leaves per treatment and for each repetition) were detached from the plants and used for stomata characteristics. A microscope (Nikon Eclipse E200, Japan) was utilised for determining the number and size of the stomata. On each slide, along a diagonal transect of the peel, four stomata for three leaves for each repetition were measured for pore lengths at 40×. Stomatal size (S) was defined by guard cell length and width.

2.4. Leaf Gas Exchange, Chlorophyll a Fluorescence, and RWC

At the end of the trial, the gas exchange in six plants per treatment (two plants for each repetition and three leaves per plant) was measured with a CO_2/H_2O infrared gas analyser (LCi, ADC Bioscientific Ltd., Hoddesdon, UK). The reliefs were carried out in the morning (from 09:00 to 13:00). For each drought stress treatment, the net photosynthetic rate (A_N), stomatal conductance (gs), transpiration rate (E), and water use efficiency (WUE) were determined.

The chlorophyll *a* fluorescence (Fv/Fm) was recorded in the same leaves using a modulated chlorophyll fluorimeter OS1-FL (Opti-Sciences Corporation, Tyngsboro, MA, USA). The leaf was dark-adapted using cuvette clips for 15 min (Opti-Sciences Corporation, Tyngsboro, MA, USA). The chlorophyll *a* fluorescence was reported as the Fv/Fm ratio, where Fm = the maximum fluorescence and Fv = the variable fluorescence.

The relative water content (RWC) was evaluated on fully opened leaves. For each replicate, 30 discs 10 mm in diameter were taken, and the fresh weights (FW) were determined. Then, the samples were immersed in distilled water for 24 h and re-weighed to measure the turgid weight (TW). Subsequently, the samples were dried at 75 °C for 24 h to measure the dry weights (DW). The RWC was expressed according to the formula:

$$RWC\% = (FW - DW/TW - DW) * 100$$

2.5. Determination of Chl and Carotenoid Content

After ending the experiment, chlorophyll *a* (Chl *a*), chlorophyll *b* (Chl *b*), total chlorophyll, and carotenoids were determined. For the extraction, 100 mg of fresh material was extracted with 5 mL of 99% methanol and incubated in the dark for 24 h at 4 °C. Quantification was performed by spectrophotometry (7315 Spectrophotometer, Jenway, Staffordshire, UK) at 665.2 nm, 652.4 nm, and 470 nm. The calculation of chlorophylls was performed following the formula reported by Lichtenthaler et al. [37]:

$$Chl\ a = 16.75_{A665.2} - 9.16_{A652.4}.$$
$$Chl\ b = 34.09_{A652.4} - 15.28_{A665.2}.$$
$$Carotenoids = (1000_{A470} - 1.63Chla - 104.96Chlb)/221.$$

2.6. Estimation of Proline Content

Proline content was determined according to Ahmad et al. [38] using L-proline as the standard. Leaf samples (1 g) were homogenised in 5 mL of 3% aqueous sulfosalicylic acid and centrifuged at 14,000× *g* for 15 min (Neya 10R, REMI, Mumbai, India). The homogenate (2 mL) was added at the same quantity of acetic acid and ninhydrin, mixed and incubated for 1 h at 100 °C. Then, the reaction was stopped in an ice bath, and the supernatant was extracted with 4 mL of toluene. The absorbance of the extract was read at 525 nm (7315 Spectrophotometer, Jenway, Staffordshire, UK).

2.7. Estimation of MDA Content

Malondialdehyde (MDA) content was determined according to Li et al. [39]. Leaf samples (0.5 g) were extracted in 1.5 mL of 5% trichloroacetic acid (*w/v*). The homogenate was centrifuged at 5000× *g* for 10 min, and then the extract was diluted to 10 mL. A quantity of 2 mL of the diluted extract was homogenised with the same quantity of 0.67%

thiobarbituric acid. The mixture was incubated at 95 °C for 30 min and then centrifuged at $5000 \times g$ for 10 min. The MDA content was calculated using the following formula: $C\ (\mu mol/L) = 6.45 \times (A_{532} - A_{600}) - 0.56 \times A_{450}$.

2.8. Extraction and Assay of Antioxidant Enzymes

Leaf samples (0.5 g) were extracted with 4 mL of extraction buffer (50 mM potassium phosphate, 1 mM ethylenediaminetetraacetic acid [EDTA], 1% polyvinylpyrrolidone [PVP], 1 mM dithiothreitol [DTT], and 1 mM phenylmethylsulfonyl [(PMSF), pH 7.8). The samples were centrifuged ($15,000 \times g$ for 30 min, 4 °C), and the supernatant was used for the enzyme assay [40].

The catalase activity (CAT; EC 1.11.1.6) was evaluated according to Aguilera et al. [41] with modifications; 20 µL of the extract was homogenised to 830 µL potassium phosphate buffer (50 mM, pH 7). The reaction started with the addition of 150 µL of H_2O_2, and the decrease was monitored at 240 nm for 2 min. The unit of CAT was expressed as units mg^{-1} protein.

The glutathione peroxidase activity (GPX) was determined according to Ruley et al. [42]. The same amount of extract and 17 mM H_2O_2 was homogenised with 2% guaiacol to obtain a final volume of 1 mL. The increase in absorbance was monitored at 510 nm for 3 min. The activity of GPX was defined as units mg^{-1} protein.

The superoxide dismutase activity (SOD; EC 1.15.1.1) was evaluated following Giannopolitis and Ries [43]. The SOD activity was read at 560 nm; the unit of SOD was defined as the amount of enzyme required to cause 50% inhibition of the reduction of NBT. The unit of SOD was expressed as units mg^{-1} protein.

Using Bradford's method [44], the protein content was quantified.

All samples were read using a spectrophotometer (7315 Spectrophotometer, Jenway, Staffordshire, UK).

2.9. 2,2-Diphenyl-1-picrylhydrazyl (DPPH) Radical Scavenging Activity

The scavenging activity against the DPPH radical was evaluated using DPPH. A quantity of 1 g of fresh weight was homogenised with 1.5 mL of 80% methanol, sonicated for 30 min, and centrifuged for 10 min at 5 °C and $5000 \times g$. Subsequently, 0.01 mL of the supernatant was homogenised with 1.4 mL of DPPH solution (150 µM) and incubated for 30 min in the dark. Later, the samples were read at 517 nm. The DPPH activity was expressed as Trolox equivalent antioxidant activity (mg TE g^{-1}).

2.10. Statistical Analysis

The trial was conducted as a randomised complete design with three replicates. The statistical analyses were conducted with CoStat version 6.311 (CoHort Sofware, Monterey, CA, USA); one-way ANOVA was adopted. The differences between the means were determined using Tukey's test ($p < 0.05$). The data presented in the figures are the means $\pm$ standard error (SE) (Graphpad 7.0). The principal component loading plot and scores of PCA were performed using Minitab 16, LLC.

3. Results

3.1. Evapotranspiration

Figure 1 shows the trend of evapotranspiration (L day^{-1}) in the 100% FC treatment during the experimental period. The amounts of water manually added to each pot were 1.88, 1.41, 0.92 and 0.47 L, respectively, for 100% FC, 75% FC, 50% FC and 25% FC. The electrical conductivity of the water was 0.005 dS m^{-1}.

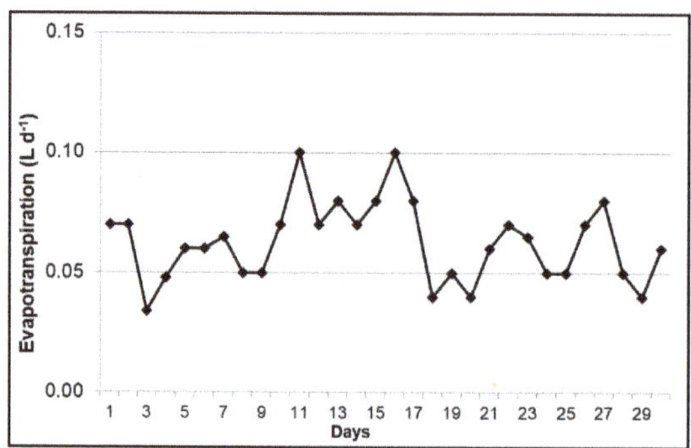

Figure 1. Evapotranspiration (L d^{-1}) in zinnia 100% FC during the experimental period (30 days).

3.2. Biomass and Leaf Area

The growth of zinnia was significantly modified by drought stress treatments. Plant height was reduced because of the different irrigation regimes (Table 1). By increasing the level of drought stress from control to 25% FC, the height was reduced by ~17% and ~38%, respectively, for 50% FC and 25% FC ($p < 0.001$). No significant differences were observed in 75% FC compared with unstressed plants (Table 1). A similar trend was observed for the stem diameter, with a reduction in more stressed treatments by ~23 and ~37%, respectively, for 50% FC and 25% FC ($p < 0.001$). Leaf number was significantly reduced under 25% FC (by 39%) compared with the other treatments ($p < 0.001$) (Table 1); the total leaf area showed a decrease from the moderate deficit irrigation with a reduction of 21% in 50% FC and by 50% in 25% FC. ($p < 0.001$). No significant differences were observed in 75% FC (Table 1).

Table 1. Effects of irrigation treatments on plant height (cm), stem diameter (mm), leaf number (n.), total leaf area (cm^2), leaf, stem, root fresh weight (g plant^{-1}), total dry biomass (g plant^{-1}), and root/shoot ratio (R/S) of potted zinnia plants at the end of the experimental period. Plants were irrigated every day. Four treatments were performed: irrigated at 100% (100% field capacity, FC); light deficit irrigation (75% FC), irrigated at 75% from the 100% FC treatment; medium deficit irrigation (50% FC), irrigated at 50% from the 100% FC treatment and severe deficit irrigation (25% FC), and irrigated at 25% from the 100% FC treatment.

Drought Stress	Plant Height (cm)	Stem Diameter (mm)	Leaf Number (n.)	Total Leaf Area (cm^2)	Leaf Fresh Biomass (g plant^{-1})	Stem Fresh Biomass (g plant^{-1})	Root Fresh Biomass (g plant^{-1})	Total Dry Biomass (g plant^{-1})	R/S
100% FC	17.2 ± 0.3 [a]	4.6 ± 0.1 [a]	15.8 ± 0.4 [a]	210.4 ± 4.5 [a]	7.6 ± 0.2 [a]	2.7 ± 0.2 [a]	7.8 ± 0.6 [a]	2.1 ± 0.1 [a]	0.5 ± 0.0 [b]
75% FC	15.8 ± 0.8 [ab]	4.6 ± 0.0 [a]	14.8 ± 0.4 [a]	218.5 ± 6.4 [a]	8.1 ± 0.7 [a]	2.8 ± 0.2 [a]	7.7 ± 0.6 [a]	2.6 ± 0.0 [a]	0.7 ± 0.0 [b]
50% FC	14.2 ± 0.3 [b]	3.6 ± 0.2 [b]	13.8 ± 0.6 [a]	166.2 ± 9.2 [b]	5.5 ± 0.3 [b]	1.7 ± 0.1 [b]	6.0 ± 0.3 [a]	1.9 ± 0.3 [a]	1.1 ± 0.2 [a]
25% FC	10.7 ± 0.2 [c]	2.9 ± 0.2 [b]	9.7 ± 0.4 [b]	104.4 ± 5.5 [c]	3.6 ± 0.3 [c]	1.1 ± 0.1 [c]	3.0 ± 0.5 [b]	1.0 ± 0.1 [b]	0.4 ± 0.0 [b]
Significance	***	***	***	***	***	**	***	**	*

The values are the means ± standard error (SE). The statistical analysis was one-way ANOVA; * significant at $p < 0.05$; ** significant at $p < 0.01$; *** significant at $p < 0.001$. The values in the same column followed by the same letter are not significantly different at $p < 0.05$ (Tukey's test).

A similar trend was detected for the leaf and stem fresh weight, with a reduction of 28% (leaf) and 15% (stem) and 56% (leaf) and 44% (stem), respectively, for 50% FC and 25% FC ($p < 0.001$ and $p < 0.01$). The light deficit treatments (75% FC) showed the same trend as unstressed plants. Root fresh biomass was only modified in 25% FC, with a reduction of 54% compared with the other treatments ($p < 0.001$) (Table 1). In addition, the the total dry biomass decreased only in 25% FC (by ~48%) ($p < 0.01$) compared with the other treatments (Table 1). The root-to-shoot ratio increased in plants grown under moderate deficit irrigation treatments (50% FC) ($p < 0.05$) (Table 1).

Stomata density showed significant differences among the irrigation treatments, with an increase in the more severe drought stress (50% FC and 25% FC) (Table 2). The drought stress influenced the stomata size only in 25% FC. The latter showed a significant reduction of 31% compared with 100% FC and 75% FC. The 50% FC treatment did not show a significant change among treatments (Table 2 and Figure 2).

Table 2. Effects of irrigation treatments on stomata density (n mm^{-2}) and size (μm) of potted zinnia plants at the end of the experimental period. Plants were irrigated every day. Four treatments were performed: irrigated at 100% (100% FC); light deficit irrigation (75% FC), irrigated at 75% from the 100% FC treatment; medium deficit irrigation (50% FC), irrigated at 50% from the 100% FC treatment and severe deficit irrigation (25% FC), and irrigated at 25% from the 100% FC treatment.

Drought Stress	Stomata	
	Density (n mm^{-2})	Size (μm)
100% FC	321.8 ± 25.2 [b]	84.8 ± 1.7 [a]
75% FC	253.8 ± 23.9 [b]	83.8 ± 4.8 [a]
50% FC	376.2 ± 9.7 [a]	72.4 ± 0.6 [ab]
25% FC	396.2 ± 8.0 [a]	55.7 ± 0.1 [b]
Significance	*	**

The values are the means ± standard error (S.E). The statistical analysis was one-way ANOVA; * significant at $p < 0.05$; ** significant at $p < 0.01$. The values in the same column followed by the same letter are not significantly different at $p < 0.05$ (Tukey's test).

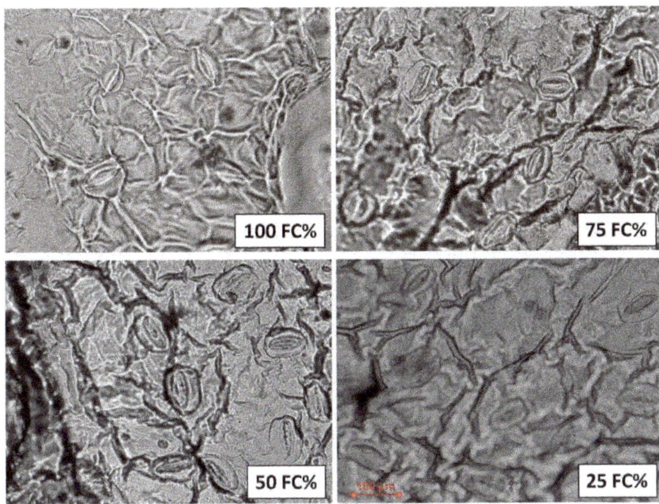

Figure 2. Light microscopy of leaf portions showing stomata traits in different drought stress treatments.

Significant effects of drought stress treatments for gas exchange were observed in zinnia plants (Figure 3). The net photosynthesis, stomatal conductance, and transpiration rate showed similar behaviours. Significant differences for A_N were observed in the 50% FC and 25% FC treatments compared with the control plants and 75% FC. In particular, the plants irrigated at 50% FC and 25% FC showed a reduction of ~30% and 66% compared with control plants (Figure 3a). A similar trend was observed for the stomatal conductance, with a reduction of ~51% and 80%, respectively, for 50% FC and 25% FC (Figure 3b) compared with the control and 75% FC. The transpiration rate showed a significant reduction only in the severe drought stress treatment, with a reduction of 61% compared with 100% FC and the other stress treatments. (Figure 3c). Water use efficiency (WUE) showed significant differences among the more stressed treatment (25% FC) and the other treatments with an increase of ~45% (Figure 3d).

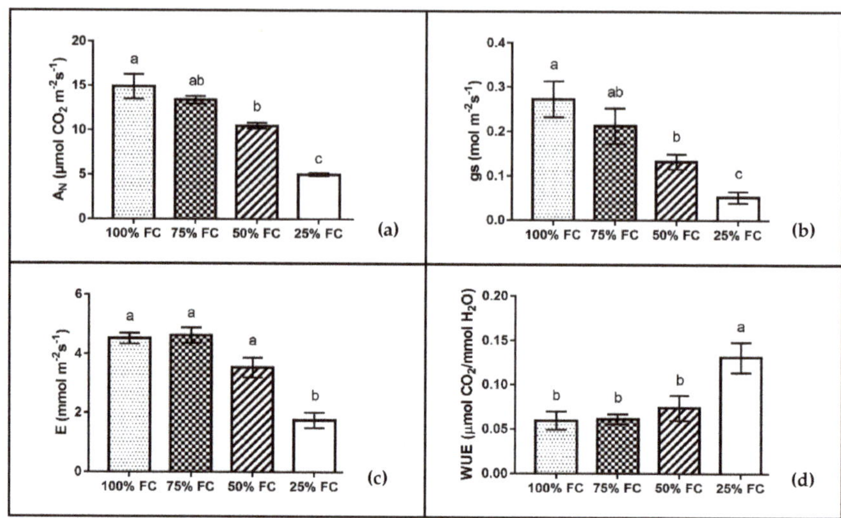

Figure 3. Net photosynthesis (A_N) (**a**), leaf conductance (gs) (**b**), transpiration rate (E) (**c**), and water use efficiency (WUE) (**d**) in zinnia. Plants were irrigated at field capacity (100% FC) or subjected to drought stress (75% FC, 50% FC, and 25% FC). Mean values ± standard error (S.E) (n = 6). Different letters indicate significant differences among the treatments as determined by Tukey's test.

Significant differences for Chl *a* and *b*, total Chl, and carotenoids were observed in the 25% FC treatment. In particular, the plants irrigated at 25% FC showed reductions of ~36%, 39%, 37%, and 40%, respectively, for Chl *a* and *b*, total Chl, and carotenoids compared with control plants (Figure 4a–d). Among the other treatments, no significant differences were observed (Figure 4a–d).

The RWC under light and moderate drought stress (75% FC and 50% FC) did not show significant differences compared with control plants and the other treatments. In the condition of severe drought stress (25% FC), a significant reduction of ~32% was observed (Table 3).

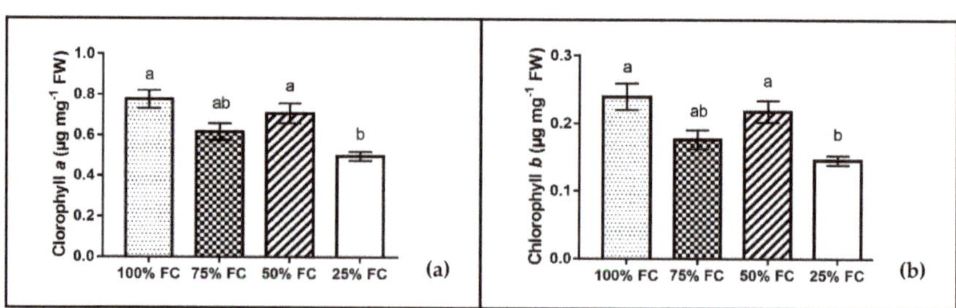

Figure 4. *Cont.*

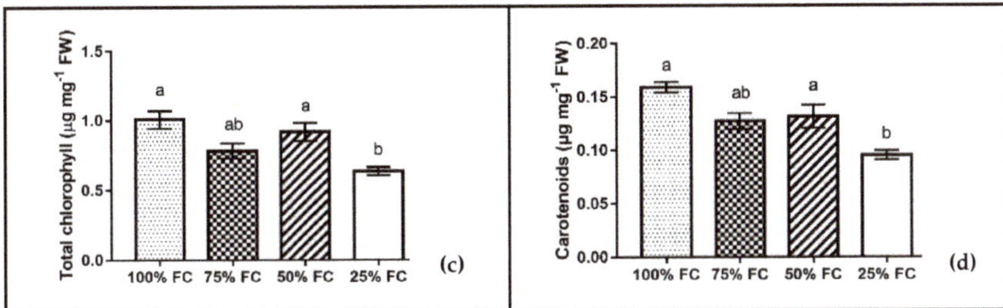

Figure 4. Chlorophyll *a* (**a**), chlorophyll *b* (**b**), total chlorophyll (**c**), and carotenoids (**d**) in zinnia plants. Plants were irrigated at field capacity (100% FC) or subjected to drought stress (75% FC, 50% FC, and 25% FC). Mean values ± standard error (S.E) (n = 6). Different letters indicate significant differences among the treatments as determined by Tukey's test.

Table 3. Effects of irrigation treatments on relative water content (RWC, %) and chlorophyll *a* fluorescence (Fv/Fm) of potted zinnia plants at the end of the experimental period. Plants were irrigated every day. Four treatments were performed: irrigated at 100% (100% FC); light deficit irrigation (75% FC), irrigated at 75% from the 100% FC treatment; medium deficit irrigation (50% FC), irrigated at 50% from the 100% FC treatment and severe deficit irrigation (25% FC), irrigated at 25% from the 100% FC treatment.

Drought Stress	RWC (%)	Fv/Fm
100% FC	73.8 ± 3.1 [a]	0.74 ± 0.01 [a]
75% FC	66.7 ± 2.7 [a]	0.73 ± 0.00 [a]
50% FC	66.7 ± 2.9 [a]	0.72 ± 0.02 [ab]
25% FC	50.7 ± 1.5 [b]	0.68 ± 0.02 [b]
Significance	**	*

The values are the means ± standard error (SE). The statistical analysis was one-way ANOVA; * significant at $p < 0.05$; ** significant at $p < 0.01$. The values in the same column followed by the same letter are not significantly different at $p < 0.05$ (Tukey's test).

In zinnia plants, a reduction in the maximum quantum efficiency (Fv/Fm) was observed only in severe drought stress conditions (10% FC), with a value of 0.68 (Table 3).

The amount of leaf proline content increased in 25% FC. Severe drought stress (25% FC) increased proline content up to seven-fold compared with the control plants (Figure 5a). Among the other treatments, no significant differences were observed (Figure 5a).

The MDA content did not show a significant change among treatments (Figure 5b).

The results showed that drought stress significantly affected enzyme activity (Table 4). The CAT activity significantly increased in moderate and severe drought stress (50% FC and 25% FC). The increment was 41% for CAT and 33% for GPX. In addition, SOD activity showed a similar trend: an increase of 42% in 50% FC and 25% FC was observed (Table 4).

The results showed that drought stress significantly affected DPPH activity ($p \leq 0.01$). DPPH activity significantly increased with severe drought stress treatment. The increment from control to 25% FC was ~23%, while among the other treatments, no significant differences were observed (Table 4).

In order to visualise congruence among 100% FC, 75% FC, 50% FC, and 25% FC plants based on all of the morphological, physiological, and biochemical variables, the whole dataset was subjected to a principal component analysis (PCA; Figure 6). The PCA showed that the morphological and physiological parameters were mostly associated with the 100% FC and 75% FC treatments of the biplot, whereas the stomata density, R/S ratio, and antioxidant enzymes (GPX, CAT) were associated with the 50% FC treatment, and proline and DPPH were associated with the 25% FC treatment (Figure 6).

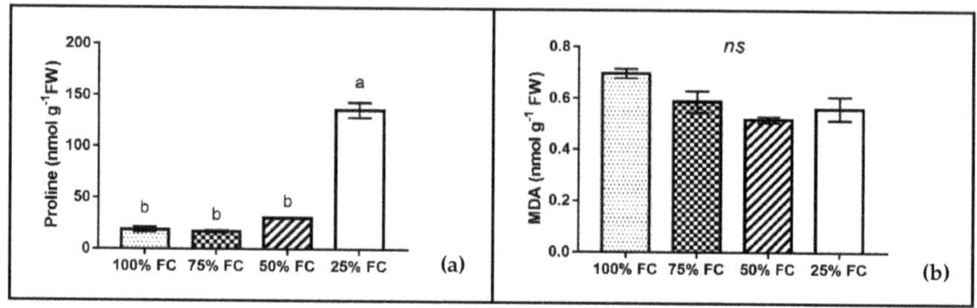

Figure 5. Proline content (**a**) and malondialdehyde content (MDA) (**b**) in zinnia plants. Plants were irrigated at field capacity (100% FC) or subjected to drought stress (75% FC, 50% FC, and 25% FC). Mean values ± standard error (S.E) (n = 6). Different letters indicate significant differences among the treatments as determined by Tukey's test. ns: no significant.

Table 4. Effects of irrigation treatments on catalase (CAT), peroxidase (GPX), and superoxide dismutase (SOD) activity of potted zinnia plants at the end of the experimental period. Plants were irrigated every day. Four treatments were performed: irrigated at 100% (100% FC), light deficit irrigation (75% FC), irrigated at 75% from the 100% FC treatment; medium deficit irrigation (50% FC), irrigated at 50% from the 100% FC treatment and severe deficit irrigation (25% FC), irrigated at 25% from the 100% FC treatment.

Drought Stress	Enzyme Activity			
	CAT (U mg^{-1} Protein)	GPX (U mg^{-1} Protein)	SOD (U mg^{-1} Protein)	DPPH (mg TE g^{-1} FW)
100% FC	0.0040 ± 0.0005 [b]	5.6 ± 0.6 [b]	29.5 ± 0.2 [b]	6.7 ± 0.1 [b]
75% FC	0.0042 ± 0.0008 [b]	5.9 ± 0.4 [b]	23.4 ± 0.1 [b]	6.9 ± 0.1 [b]
50% FC	0.0069 ± 0.0003 [a]	8.0 ± 0.9 [a]	45.8 ± 3.3 [a]	6.7 ± 0.2 [b]
25% FC	0.0072 ± 0.0001 [a]	9.2 ± 0.5 [a]	44.0 ± 1.5 [a]	8.6 ± 0.4 [a]
Significance	**	**	***	***

The values are the mean ± standard error (S.E). The statistical analysis was one-way ANOVA; ** significant at $p < 0.01$; *** significant at $p < 0.001$. The values in the same column followed by the same letter are not significantly different at $p < 0.05$ (Tukey's test).

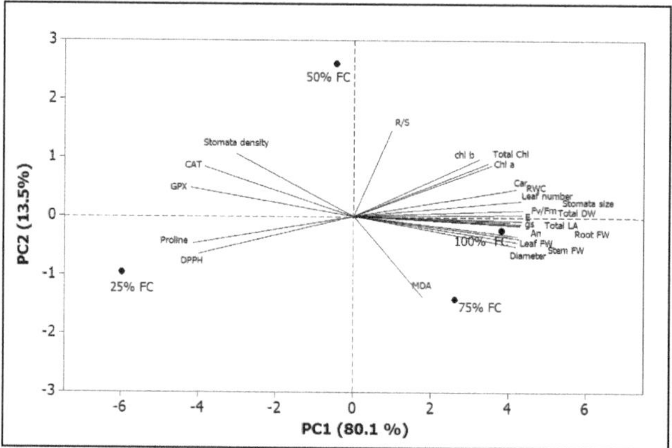

Figure 6. Principal component loading plot and scores of PCA fresh weight and dry biomass, R/S, total leaf area, leaf number, photosynthetic pigments (Chl *a* and *b*, total Chl, and carotenoids), stomata characteristics, gas exchange, RWC, Fv/Fm, MDA, proline (Pro), enzyme activity (GPX, CAT. and SOD), and DPPH for zinnia plants with different drought stress treatments (100% FC, 75% FC, 50% FC, and 25% FC) according to the first two principal components.

4. Discussion

Drought tolerance is the capacity of plants to continue to be functional at lower tissue water potentials. Drought stress determines considerable changes in the physiological and biochemical activity of plants, including photosynthesis, respiration, transpiration, hormone metabolism, and enzyme activity [45]. The mechanisms of drought tolerance involve the maintenance of turgor (by the accumulation of solutes) and/or desiccation tolerance (by protoplasmic resistance) [46]. In urban areas, it is important to select species for planting that are able to tolerate water shortages without losing their aesthetic appearance [47]. Water availability is one of the principal factors that limit bedding and landscaping plant cultivation, particularly for seasonal and annual garden flowers with shallow roots [1]. In the nursery phase, the generalised use of containers, often of small volume, determines root restriction effects [35].

Because of the showiness of flowering herbaceous species, it is difficult to convince people not to use them in urban landscaping. These plants are particularly sensitive to water deficiency. Annual plants are vulnerable to increasing temperatures and decreasing rainfall, as they must complete their life cycle in a single season, and the persistence of the population is highly dependent on the reproductive capacity of a season [48]. As for woody plants, herbaceous plants also adopt different physiological and biochemical mechanisms to overcome drought stress. One of the first responses to drought stress is a reduction in plant growth. In our study, the drought stress treatments significantly reduced the height, shoot and root dry biomass, leaf number, and leaf area in the more stressed treatments. A moderate reduction in the amount of water applied to container-grown zinnias did not reduce the morpho-biometric characteristics. Previous studies have shown that dry biomass reduced significantly with increasing drought stress treatments in *Trachyspermum ammi* L. [49] and *Ocimum basilicum* L. [50]. At the end of the drought treatment, the plant biomass reduction could be linked to a reduction in cell elongation and expansion; lower water absorption led to a reduction in the production of metabolites to maintain normal cellular activity [51].

A lower water amount can have positive effects on growth, so light drought stress can be a useful water-saving strategy. In our study, the plants grown at 50% FC showed no reduction in growth parameters. In a study on parsley, Borges et al. [52] demonstrated that water reduction significantly modified the development of the aerial parts. According to Zulfiqar et al. [53], a reduction in biomass is a good strategy in *Tagetes erecta* to improve water uptake under drought stress in the long term. Drought stress influences the aerial parts of the plants more than the roots; furthermore, the growth of the aerial parts of the plant decreases earlier than the roots, causing an increased root-to-shoot ratio. This ratio is not always modified by drought stress. Sánchez-Blanco et al. [54] found that the R/S in *Pistacia lentiscus* and *Phillyrea angustifolia* was not influenced under water stress treatments. In our study, except for 50% FC, the same trend was observed. Eziz et al. [55], in a meta-analysis to explain the patterns of plant biomass allocation related to drought stress, found that roots of woody plants were more resistant than those of herbaceous plants to drought. Furthermore, drought stress in herbaceous plants had a more negative influence on the leaf mass fraction of woody plants. In our study, leaf fresh biomass reduced as water stress increased. A similar response was observed for the total leaf area and leaf number. The reduction of the leaf area is due to a reduction in the leaf number [36] or leaf size (unit leaf area) [56]. Thus, plants decrease water limitation by reducing the transpiration area. This is a typical mechanism of stress avoidance used by plants to overcome water stress conditions [57].

The gas exchange rates are most affected by water shortage. The principal site of stress in the plants is the photosynthetic apparatus, which is very sensitive to drought; as a consequence, the photosynthetic activity is reduced because of stomatal closure and complex non-gassing effects [58]. In the present study, A_N, gs, and E significantly decreased in the leaves of zinnia under moderate and severe drought stress, and A_N decreased substantially in 25% FC, indicating that the reduction in A_N in the first stage might be determined by the stomatal closure inhibition of CO_2 uptake. Drake et al. [59] suggested that many small stomata lead to greater stomatal resistance and stomatal control, as well as a rapid response in water stress conditions. In our study, in the more stressed treatments (50% FC and 25% FC), an increase in stomata density and a decrease in size were observed. With the reduction of transpiration, the water use efficiency generally increases. Different studies have reported that stressed plants are more able to utilise the energy obtained by photosynthesis because of higher WUE [57]. This is in accordance with our results, which showed that the most stressed plants activated this mechanism by increasing the WUE values. Another typical avoidance strategy during drought stress conditions is a change in the size and density of the stomata [57].

A high RWC during drought stress is a key mechanism for maintaining the metabolic activity in plants; it is a well-known mechanism for inducing drought tolerance in breeding activity [60]. Babaei et al. [61], in a study on *Tagetes minuta*, noted that the low RWC values in this species during drought showing a low recovery capacity. A higher RWC is generally maintained by plants under moderate drought stress in relation to plants under severe drought stress. Furthermore, the growth parameters were more negatively influenced by the imposed water stress than they were in zinnia with higher RWC values.

Chlorophylls are responsible for the correct functioning of the photosynthetic apparatus because they are essential pigments of the superior assimilating tissues of plants. Under salt or water stress, the pigment photosynthetic (chlorophyll *a* + *b*) content can be indicative of stressful conditions [62]. Severe drought stress can also impair the concentrations of photosynthetic pigments [63]. A decrease in chlorophyll content under severe water stress has been noted in different species, including *Catharanthus roseus* [64], *Helianthus annuus* [65], *Tagetes erecta* [66], and *Viola x witthrockiana* [67]. In our study, the effects of the severe drought stress (25% FC), were highlighted in the leaves by a significant lowering of the chlorophyll *a* and *b*, total chlorophyll, and carotenoid content. A physiological adaptation mechanism of plants under drought stress conditions could be the maintenance of high chlorophyll content, even if this is correlated with the biological characteristics of the plant [68]. Indeed, Lu et al. [69] observed that increasing the chlorophyll content was a more effective strategy to avoid photooxidation damage and ROS induction by drought. Plants under drought stress typically accumulate osmotic compounds, such as proline, and change phytohormones [18]. Proline has been found to enhance cell turgor, maintain cell osmotic adjustment, and defend cells during dehydration [67]. In our study, it was noted that proline concentration was higher in zinnia plants under severe drought conditions compared with control plants. These results agree with the results of other studies [70] and underline the relevance of proline as a protective element in the stress response [71]. In addition, Oraee et al. [72] showed that the proline concentration was higher in plants under drought treatments compared with well-watered plants. In drought stress conditions, proline accumulation was inversely proportional to the water status of plants [73]; this was confirmed in our study, in which the highest values of proline in 25% FC corresponded to the lowest values of RWC. This suggests the contribution of this solute in plant osmotic adjustment. Similarly, in Indian grain soybean cultivars exposed to mild water stress, an increased proline accumulation corresponded to a smaller reduction in the relative water content and shoot and root fresh/dry weight [74].

Plants, when synthesizing antioxidants and increasing the activity of antioxidative enzymes, activate a protective mechanism against abiotic stress [75]. Antioxidative enzymes that control ROS level in cells—superoxide dismutase (SOD), catalase (CAT), and peroxidase (POX)—are commonly produced in response to drought. Numerous studies have found that drought-tolerant plants have strong scavenging systems that allow them to maintain low levels of ROS and reduce membrane lipid peroxidation during stress [22].

Among the antioxidative enzymes that act in scavenging ROS species, APX, SOD, and GPX have a key role. Ascorbate peroxidase (APX) acts directly on the H_2O_2 molecules and reduces them to water. SOD conducts the dismutation reaction by reducing the O^{-2} molecule to H_2O_2; APX and GPX convert H_2O_2 to water, thus assuring its removal [76]. In this study, plants under 50% FC and 25% FC showed a significant increment in antioxidant activity at the end of the trial. CAT, SOD, and GPX activity was significantly enhanced at moderate and severe drought stress (50% FC and 25% FC). A similar trend was reported by Amiri et al. [77]. In a study by Tian et al. [78], the results indicated that the increase in enzyme activity in marigold was a tolerant response to drought treatment, while this self-regulation level was lower with the improvement of water deficit. The decrease in CAT activity leads to the accumulation of H_2O_2 and an enhancement in lipid peroxidation, thus increasing MDA and causing damage in plants [78]. Recent studies have observed that proline may also play a role as an antioxidant and not only as an osmotic protectant. Thus, the accumulation of proline plays a key role under drought stress as an antioxidant or through stabilizing macromolecules during water stress [79]. The enhancement of DPPH activity observed in the more stressed plants is in line with other studies [61,80]. The total antioxidant activity is the combined results of all enzymatic and non-enzymatic antioxidant activity in plants subjected to abiotic stresses. Tolerant plants generally show a higher antioxidant content to defend against oxidative stress by keeping high antioxidant enzyme and antioxidant molecule activity and contents under stress conditions [81].

Through the general analysis of the data, we observed an interesting result that the activity of antioxidant enzymes is strongly linked to the increase of secondary metabolites; this is similar to the findings of Catola et al. [82] that secondary metabolites work as substrates for enzymes involved in enzymatic antioxidant reactions [51]. The PCA analysis reported in the present study could help to better understand the influence of drought stress on morphological, physiological, and biochemical characteristics. The PCA results demonstrated the boundary between the drought treatments and the control group was clear. Correlation analysis demonstrated that there were complex and close relationships in the more stressed treatments (50% FC and 25% FC) between the antioxidant enzyme activity, DPPH, proline accumulation, and stomata density, which cooperated to tolerate drought stress. The sensitivity of zinnia plants to drought stress is different, so the response time is also different.

5. Conclusions

The changes in growth parameters, stomata characteristics, gas exchange, osmotic regulators, and antioxidant activity were measured to study the response mechanism of *Z. elegans* to drought stress. During drought exposure in the nursery phase, zinnia showed adaptive changes to water limitation. At the physiological level, zinnia responded to the drought stress by reducing the RWC and biomass and increasing the levels of osmotic regulators (proline) and antioxidant activity. In severe deficit irrigation, the strategies adopted by the plants were not able to resist drought stress. With light deficit irrigation, the plants could perform as well as fully irrigated plants. In medium deficit irrigation, the mechanisms were not always are suitable to overcome drought stress.

Author Contributions: Conceptualization, D.R. and S.T.; methodology, D.R. and S.T.; software, S.T.; validation, D.R.; formal analysis, S.T.; investigation, S.T.; resources, D.R.; data curation, S.T.; writing—original draft preparation, D.R. and S.T.; writing—review and editing, D.R. and S.T.; supervision, D.R.; funding acquisition, D.R. Both authors have read and agreed to the published version of the manuscript.

Funding: This research was funded by PON RICERCA E INNOVAZIONE 2014–2020, Azione II—Obiettivo Specifico 1b—Progetto "Miglioramento delle produzioni agroalimentari mediterranee in condizioni di carenza di risorse idriche—WATER4AGRIFOOD", cod. progetto ARS01_00825.

Institutional Review Board Statement: Not applicable.

Informed Consent Statement: Not applicable.

Data Availability Statement: Main data are contained within the article; further data presented in this study are available on request from the corresponding author.

Conflicts of Interest: The authors declare no conflict of interest.

References

1. Heidari, Z.; Nazarideljou, M.J.; Rezaie Danesh, Y.; Khezrinejad, N. Morphophysiological and biochemical responses of *Zinnia elegans* to different irrigation regimes in symbiosis with *Glomus mosseae*. *Int. J. Hortic. Sci.* **2016**, *3*, 19–32. [CrossRef]
2. Da Silva, F.F.; Wallach, R.; Chen, Y. Hydraulic properties of sphagnum peat moss and tuff (scoria) and their potential effects on water availability. In *Optimization of Plant Nutrition*; Springer: Berlin/Heidelberg, Germany, 1993; pp. 569–576.
3. Talbi, S.; Rojas, J.A.; Sahrawy, M.; Rodríguez-Serrano, M.; Cárdenas, K.E.; Debouba, M.; Sandalio, L.M. Effect of drought on growth, photosynthesis and total antioxidant capacity of the Saharan plant *Oudeneya africana*. *Environ. Exp. Bot.* **2020**, *176*, 104099. [CrossRef]
4. Larkunthod, P.; Nounjan, N.; Siangliw, J.L.; Toojinda, T.; Sanitchon, J.; Jongdee, B.; Theerakulpisut, P. Physiological responses under drought stress of improved drought-tolerant rice lines and their parents. *Not. Bot. Horti Agrobot. Cluj Napoca* **2018**, *46*, 679–687. [CrossRef]
5. Cal, A.J.; Sanciangco, M.; Rebolledo, M.C.; Luquet, D.; Torres, R.O.; McNally, K.L.; Henry, A. Leaf morphology, rather than plant water status, underlies genetic variation of rice leaf rolling under drought. *Plant. Cell Environ.* **2019**, *42*, 1532–1544. [CrossRef]
6. Jaramillo, R.E.; Nord, E.A.; Chimungu, J.G.; Brown, K.M.; Lynch, J.P. Root cortical burden influences drought tolerance in maize. *Ann. Bot.* **2013**, *112*, 429–437. [CrossRef]
7. Nemali, K.S.; Bonin, C.; Dohleman, F.G.; Stephens, M.; Reeves, W.R.; Nelson, D.E.; Castiglioni, P.; Whitsel, J.E.; Sammons, B.; Silady, R.A.; et al. Physiological responses related to increased grain yield under drought in the first biotechnology-derived drought tolerant maize. *Plant Cell Environ.* **2015**, *38*, 1866–1880. [CrossRef]
8. Tezara, W.M.V.J.; Mitchell, V.J.; Driscoll, S.D.; Lawlor, D.W. Water stress inhibits plant photosynthesis by decreasing coupling factor and ATP. *Nature* **1999**, *401*, 914–917. [CrossRef]
9. Kim, J.; van Iersel, M.W. Slowly developing drought stress increases photosynthetic acclimation of *Catharanthus roseus*. *Physiol. Plant.* **2011**, *143*, 166–177. [CrossRef]
10. Nemali, K.; van Iersel, M.W. Relating whole-plant photosynthesis to physiological acclimations at leaf and cellular scales under drought stress in bedding plants. *J. Am. Soc. Hortic. Sci.* **2019**, *144*, 201–208. [CrossRef]
11. Chaves, M.M.; Maroco, J.P.; Pereira, J.S. Understanding plant responses to drought from genes to the whole plant. *Funct. Plant Biol.* **2003**, *30*, 239–264. [CrossRef]
12. Boyle, R.K.; McAinsh, M.; Dodd, I.C. Stomatal closure of *Pelargonium× hortorum* in response to soil water deficit is associated with decreased leaf water potential only under rapid soil drying. *Physiol. Plant.* **2016**, *156*, 84–96. [CrossRef] [PubMed]
13. Sanders, G.J.; Arndt, S.K. Osmotic adjustment under drought conditions. In *Plant Responses to Drought Stress*; Aroca, R., Ed.; Springer: Berlin/Heidelberg, Germany, 2012; pp. 199–229. [CrossRef]
14. Nemali, K.S.; Stephens, M. Plant abiotic stress: Water. In *Encyclopedia of Agriculture and Food Systems*; Van Alfen, N.K., Ed.; Elsevier: London, UK, 2014; pp. 335–342.
15. Flexas, J.; Escalona, J.M.; Medrano, H. Water stress induces different levels of photosynthesis and electron transport rate regulation in grapevines. *Plant Cell Environ.* **1999**, *22*, 39–48. [CrossRef]
16. Fu, J.; Huang, B. Involvement of antioxidants and lipid peroxidation in the adaptation of two cool-season grasses to localized drought stress. *Environ. Exp. Bot.* **2001**, *45*, 105–114. [CrossRef]
17. Dias, M.C.; Correia, S.; Serôdio, J.; Silva, A.M.S.; Freitas, H.; Santos, C. Chlorophyll fluorescence and oxidative stress endpoints to discriminate olive cultivars tolerance to drought and heat episodes. *Sci. Hortic.* **2018**, *231*, 31–35. [CrossRef]
18. Hare, P.D.; Cress, W.A.; Van Staden, J. The involvement of cytokinins in plant responses to environmental stress. *Plant Growth Regul.* **1997**, *23*, 79–103. [CrossRef]
19. Zhang, X.; Ervin, E.H.; Evanylo, G.K.; Haering, K.C. Impact of biosolids on hormone metabolism in drought-stressed tall fescue. *Crop. Sci.* **2009**, *49*, 1893–1901. [CrossRef]
20. Zali, A.G.; Ehsanzadeh, P. Exogenously applied proline as a tool to enhance water use efficiency: Case of fennel. *Agric. Water Manag.* **2018**, *197*, 138–146. [CrossRef]
21. Ghaffari, H.; Tadayon, M.R.; Nadeem, M.; Cheema, M.; Razmjoo, J. Proline-mediated changes in antioxidant enzymatic activities and the physiology of sugar beet under drought stress. *Acta Physiol. Plant.* **2019**, *41*, 23. [CrossRef]
22. Toscano, S.; Farieri, E.; Ferrante, A.; Romano, D. Physiological and biochemical responses in two ornamental shrubs to drought stress. *Front. Plant. Sci.* **2016**, *7*, 645. [CrossRef]
23. Sharma, P.; Jha, A.B.; Dubey, R.S. Oxidative stress and antioxidative defense system in plants growing under abiotic stresses. In *Handbook of Plant and Crop Stress*, 4th ed.; Pessarakli, M., Ed.; CRC Press: Boca Raton, FL, USA, 2019; pp. 93–136.
24. Caverzan, A.; Passaia, G.; Rosa, S.B.; Ribeiro, C.W.; Lazzarotto, F.; Margis-Pinheiro, M. Plant responses to stresses: Role of ascorbate peroxidase in the antioxidant protection. *Genet. Mol. Biol.* **2012**, *35*, 1011–1019. [CrossRef]
25. Moran, J.F.; Becana, M.; Iturbe-Ormaetxe, I.; Frechilla, S.; Klucas, R.V.; Aparicio-Tejo, P. Drought induces oxidative stress in pea plants. *Planta* **1994**, *194*, 346–352. [CrossRef]
26. Wang, Y.J.; Wang, L. Characterization of acetylated waxy maize starches prepared under catalysis by different alkali and alkaline-earth hydroxides. *Starch-Stärke* **2002**, *54*, 25–30. [CrossRef]
27. Keles, Y.; Öncel, I. Growth and solute composition in two wheat species experiencing combined influence of stress conditions. *Russ. J. Plant Physiol.* **2004**, *51*, 203–209. [CrossRef]

28. Dole, J.M.; Fanelli, F.L.; Fonteno, W.C.; Harden, B.; Blankenship, S.M. Post harvest handling of cut linaria, trachelium, and zinnia. *HortScience* **2005**, *40*, 1123B. [CrossRef]
29. Pallavi, B.; Nivas, S.K.; D'souza, L.; Ganapathi, T.R.; Hegde, S. Gamma rays induced variations in seed germination, growth and phenotypic characteristics of *Zinnia elegans* var. Dreamland. *Adv. Hortic. Sci.* **2017**, *31*, 267–274. [CrossRef]
30. Nau, J. Zinnia. In *Ball Red Book Greenhouse Growing*, 15th ed.; Ball, V., Ed.; J. Ball Publishing: West Chicago, IL, USA, 1991; pp. 785–787.
31. Raza, A.; Razzaq, A.; Mehmood, S.S.; Zou, X.; Zhang, X.; Lv, Y.; Xu, J. Impact of climate change on crops adaptation and strategies to tackle its outcome: A review. *Plants* **2019**, *8*, 34. [CrossRef]
32. Lobell, D.B.; Gourdji, S.M. The influence of climate change on global crop productivity. *Plant Physiol.* **2012**, *160*, 1686–1697. [CrossRef]
33. Farooq, M.; Wahid, A.; Kobayashi, N.; Fujita, D.; Basra, S.M.A. Plant drought stress: Effects, mechanisms and management. In *Sustainable Agriculture*; Lichtfouse, E., Navarrete, M., Debaeke, P., Véronique, S., Alberola, C., Eds.; Springer: Dordrecht, The Netherlands, 2009; pp. 153–158. [CrossRef]
34. Niu, G.; Rodriguez, D.S.; Wang, Y.T. Impact of drought and temperature on growth and leaf gas exchange of six bedding plant species under greenhouse conditions. *HortScience* **2006**, *41*, 1408–1411. [CrossRef]
35. Toscano, S.; Ferrante, A.; Tribulato, A.; Romano, D. Leaf physiological and anatomical responses of Lantana and Ligustrum species under different water availability. *Plant Physiol. Biochem.* **2018**, *127*, 380–392. [CrossRef]
36. Alvarez, S.; Bañón, S.; Sánchez-Blanco, M.J. Regulated deficit irrigation in different phenological stages of potted geranium plants: Water consumption, water relations and ornamental quality. *Acta Physiol. Plant.* **2013**, *35*, 1257–1267. [CrossRef]
37. Lichtenthaler, H.K. Chlorophylls and carotenoids: Pigments of photosynthetic biomembranes. *Methods Enzymol.* **1987**, *148*, 350–382. [CrossRef]
38. Ahmad, P.; John, R.; Sarwat, M.; Umar, S. Responses of proline, lipid peroxidation and antioxidative enzymes in two varieties of *Pisum sativum* L. under salt stress. *Int. J. Plant Prod.* **2008**, *2*, 353–366.
39. Li, G.; Wan, S.; Zhou, J.; Yang, Z.; Qin, P. Leaf chlorophyll fluorescence, hyperspectral reflectance, pigments content, malondialdehyde and proline accumulation responses of castor bean (*Ricinus communis* L.) seedlings to salt stress levels. *Ind. Crop. Prod.* **2010**, *31*, 13–19. [CrossRef]
40. Bian, S.; and Jiang, Y. Reactive oxygen species, antioxidant enzyme activities and gene expression patterns in leaves and roots of Kentucky bluegrass in response to drought stress and recovery. *Sci. Hortic.* **2009**, *120*, 264–270. [CrossRef]
41. Aguilera, J.; Bischof, K.; Karsten, U.; Hanelt, D.; Wiencke, C. Seasonal variation in ecophysiological patterns in macroalgae from an Arctic fjord. II. Pigment accumulation and biochemical defence systems against high light stress. *Mar. Biol.* **2002**, *140*, 1087–1095. [CrossRef]
42. Ruley, A.T.; Sharma, N.C.; Sahi, S.V. Antioxidant defense in a lead accumulating plant, *Sesbania drummondii*. *Plant Physiol. Biochem.* **2004**, *42*, 899–906. [CrossRef]
43. Giannopolitis, C.N.; Ries, S.K. Superoxide occurrence in higher plants. *Plant Physiol.* **1977**, *59*, 309–314. [CrossRef]
44. Bradford, M.M. A rapid and sensitive method for the quantitation of microgram quantities of protein utilizing the principle of protein-dye binding. *Anal. Biochem.* **1976**, *72*, 248–254. [CrossRef]
45. Okunlola, G.O.; Olatunji, O.A.; Akinwale, R.O.; Tariq, A.; Adelusi, A.A. Physiological response of the three most cultivated pepper species (*Capsicum* spp.) in Africa to drought stress imposed at three stages of growth and development. *Sci. Hortic.* **2017**, *224*, 198–205. [CrossRef]
46. Jones, M.M.; Turner, N.C.; Osmond, C.B. Mechanisms of drought resistance. In *The Physiology and Biochemistry of Drought Resistance in Plants*; Paleg, L.G., Aspinal, D., Eds.; Academic Press: Sydney, Australia, 1981; pp. 15–35.
47. Chyliński, W.K.; Łukaszewska, A.J.; Kutnik, K. Drought response of two bedding plants. *Acta Physiol. Plant.* **2007**, *29*, 399–406. [CrossRef]
48. Heschel, M.S.; Sultan, S.E.; Glover, S.; Sloan, D. Population differentiation and plastic responses to drought stress in the generalist annual *Polygonum persicaria*. *Int. J. Plant Sci.* **2004**, *165*, 817–824. [CrossRef]
49. Azhar, N.; Hussain, B.; Ashraf, M.Y.; Abbasi, K.Y. Water stress mediated changes in growth: Physiology and secondary metabolites of desi ajwain (*Trachyspermum ammi* L.). *Pak. J. Bot.* **2011**, *43*, 15–19.
50. Khalid, K.A. Influence of water stress on growth essential oil and chemical composition of herbs (*Ocimum* sp.). *Int. Agrophys.* **2006**, *20*, 289–296.
51. Gao, S.; Wang, Y.; Yu, S.; Huang, Y.; Liu, H.; Chen, W.; He, X. Effects of drought stress on growth, physiology and secondary metabolites of two *Adonis* species in Northeast China. *Sci. Hortic.* **2020**, *259*, 108795. [CrossRef]
52. Borges, I.B.; Cardoso, B.K.; Silva, E.I.S.; de Oliveira, J.E.S.; da Silva, R.F.; de Rezende, C.A.M.; Gonçalves, J.E.; Junior, R.P.; de Souza, S.G.H.; Gazim, Z.C. Evaluation of performance and chemical composition of *Petroselinum crispum* essential oil under different conditions of water deficit. *Afr. J. Agric. Res.* **2016**, *11*, 480–486. [CrossRef]
53. Zulfiqar, F.; Younis, A.; Riaz, A.; Mansoor, F.; Hameed, M.; Akram, N.A.; Abideen, Z. Morpho-anatomical adaptations of two *Tagetes erecta* L. cultivars with contrasting response to drought stress. *Pak. J. Bot.* **2020**, *52*, 801–810. [CrossRef]
54. Sánchez-Blanco, M.J.; Álvarez, S.; Ortuño, M.F.; Ruiz-Sánchez, M.C. Root system response to drought and salinity: Root distribution and water transport. In *Root Engineering*; Soil Biology Series; Morte, A., Varma, A., Eds.; Springer: Berlin/Heidelberg, Germany, 2014; Volume 40, pp. 325–352. [CrossRef]

55. Eziz, A.; Yan, Z.; Tian, D.; Han, W.; Tang, Z.; Fang, J. Drought effect on plant biomass allocation: A meta-analysis. *Ecol. Evol.* **2017**, *7*, 11002–11010. [CrossRef] [PubMed]

56. Toscano, S.; Scuderi, D.; Giuffrida, F.; Romano, D. Responses of Mediterranean ornamental shrubs to drought stress and recovery. *Sci. Hortic.* **2014**, *178*, 145–153. [CrossRef]

57. Toscano, S.; Ferrante, A.; Romano, D. Response of Mediterranean ornamental plants to drought stress. *Horticulturae* **2019**, *5*, 6. [CrossRef]

58. Flexas, J.; Medrano, H. Drought-inhibition of photosynthesis in C3 plants: Stomatal and non-stomatal limitations revisited. *Ann. Bot.* **2002**, *89*, 183–189. [CrossRef]

59. Drake, P.L.; Froend, R.H.; Franks, P.J. Smaller, faster stomata: Scaling of stomatal size, rate of response, and stomatal conductance. *J. Exp. Bot.* **2013**, *64*, 495–505. [CrossRef] [PubMed]

60. Soltys-Kalina, D.; Plich, J.; Strzelczyk-Żyta, D.; Śliwka, J.; Marczewski, W. The effect of drought stress on the leaf relative water content and tuber yield of a half-sib family of 'Katahdin'-derived potato cultivars. *Breed. Sci.* **2016**, *66*, 328–331. [CrossRef]

61. Babaei, K.; Moghaddam, M.; Farhadi, N.; Pirbalouti, A.G. Morphological, physiological and phytochemical responses of Mexican marigold (*Tagetes minuta* L.) to drought stress. *Sci. Hortic.* **2021**, *284*, 110116. [CrossRef]

62. Ueda, A.; Kanechi, M.; Uno, Y.; Inagaki, N. Photosynthetic limitations of a halophyte sea aster (*Aster tripolium* L.) under water stress and NaCl stress. *J. Plant Res.* **2003**, *116*, 63–68. [CrossRef] [PubMed]

63. Merwad, A.M.A.; Desoky, E.M.; Rady, M.M. Response of water deficit-stressed *Vigna unguiculata* performances to silicon, proline or methionine foliar application. *Sci. Hortic.* **2018**, *228*, 132–144. [CrossRef]

64. Jaleel, C.A.; Manivannan, P.; Lakshmanana, G.M.A.; Gomathinayagam, M.; Panneerselvam, R. Alterations in morphological parameters and photosynthetic pigment responses of *Catharanthus roseus* under soil water deficits. *Colloids Surf. B Biointerfaces* **2008**, *61*, 298–303. [CrossRef]

65. Kiani, S.P.; Maury, P.; Sarrafi, A.; Grieu, P. QTL analysis of chlorophyll fluorescence parameters in sunflower (*Helianthus annuus* L.) under well-watered and water stressed conditions. *Plant. Sci.* **2008**, *175*, 565–573. [CrossRef]

66. Asrar, A.W.A.; Elhindi, K.M. Alleviation of drought stress of marigold (*Tagetes erecta*) plants by using arbuscular mycorrhizal fungi. *Saudi J. Biol. Sci.* **2011**, *18*, 93–98. [CrossRef]

67. Oraee, A.; Tehranifar, A. Evaluating the potential drought tolerance of pansy through its physiological and biochemical responses to drought and recovery periods. *Sci. Hortic.* **2020**, *265*, 109225. [CrossRef]

68. Rosales-Serna, R.; Kohashi-Shibata, J.; Acosta-Gallegos, J.A.; Trejo-López, C.; Ortiz Cereceres, J.; Kelly, J.D. Biomass distribution, maturity acceleration and yield in drought stressed common bean cultivars. *Field Crop. Res.* **2004**, *85*, 203–211. [CrossRef]

69. Lu, F.; Bu, Z.; Lu, S. Estimating chlorophyll content of leafy green vegetables from adaxial and abaxial reflectance. *Sensors* **2019**, *19*, 4059. [CrossRef]

70. Abid, M.; Ali, S.; Qi, L.K.; Zahoor, R.; Tian, Z.; Jiang, D.; Snider, J.L.; Dai, T. Physiological and biochemical changes during drought and recovery periods at tillering and jointing stages in wheat (*Triticum aestivum* L.). *Sci. Rep.* **2018**, *8*, 4615–4630. [CrossRef] [PubMed]

71. Aghaie, P.; Tafreshi, S.A.H.; Ebrahimi, M.A.; Haerinasab, M. Tolerance evaluation and clustering of fourteen tomato cultivars grown under mild and severe drought conditions. *Sci. Hortic.* **2018**, *232*, 1–12. [CrossRef]

72. Oraee, A.; Tehranifar, A.; Nezami, A.; Shoor, M. Effects of drought stress on cold hardiness of non-acclimated viola (*Viola × wittrockiana* 'Iona Gold with Blotch') in controlled conditions. *Sci. Hortic.* **2018**, *238*, 98–106. [CrossRef]

73. Zegaoui, Z.; Planchais, S.; Cabassa, C.; Djebbar, R.; Belbachir, O.A.; Carol, P. Variation in relative water content, proline accumulation and stress gene expression in two cowpea landraces under drought. *J. Plant. Physiol.* **2017**, *218*, 26–34. [CrossRef]

74. Devi, M.A.; Giridhar, P. Variations in physiological response, lipid peroxidation, antioxidant enzyme activities, proline and isoflavones content in soybean varieties subjected to drought stress. *Proc. Natl. Acad. Sci. USA* **2015**, *85*, 35–44. [CrossRef]

75. Egert, M.; Tevini, M. Influence of drought on some physiological parameters symptomatic for oxidative stress in leaves of chives (*Allium schoenoprasum*). *Environ. Expt. Bot.* **2002**, *48*, 43–49. [CrossRef]

76. Gupta, D.K.; Palma, J.M.; Corpas, F.J. (Eds.) *Redox State as a Central Regulator of Plant-Cell Stress Responses*; Springer International Publishing: Berlin/Heidenberg, Germany, 2016; p. 386. [CrossRef]

77. Amiri, R.; Nikbakht, A.; Etemadi, N. Alleviation of drought stress on rose geranium [*Pelargonium graveolens* (L.) Herit.] in terms of antioxidant activity and secondary metabolites by mycorrhizal inoculation. *Sci. Hortic.* **2015**, *197*, 373–380. [CrossRef]

78. Tian, Z.; Wang, F.; Zhang, W.; Liu, C.; Zhao, X. Antioxidant mechanism and lipid peroxidation patterns in leaves and petals of marigold in response to drought stress. *Hortic. Environ. Biotechnol.* **2012**, *53*, 183–192. [CrossRef]

79. Seki, M.; Umezawa, T.; Urano, K.; Shinozaki, K. Regulatory metabolic networks in drought stress responses. *Curr. Opin. Plant Biol.* **2007**, *10*, 296–302. [CrossRef]

80. Chaeikar, S.S.; Marzvan, S.; Khiavi, S.J.; Rahimi, M. Changes in growth, biochemical, and chemical characteristics and alteration of the antioxidant defense system in the leaves of tea clones (*Camellia sinensis* L.) under drought stress. *Sci. Hortic.* **2020**, *265*, 109257. [CrossRef]

81. Espinoza, A.; Martína, A.S.; Lopez-Climentb, M.; Ruiz-Laraa, S.; Gomez-Cadenasb, A.; Casarettoa, J. Engineered drought-induced biosynthesis of α-tocopherol alleviates stress-induced leaf damage in tobacco. *J. Plant Physiol.* **2013**, *170*, 1285–1294. [CrossRef] [PubMed]
82. Catola, S.; Marino, G.; Emiliani, G.; Huseynova, T.; Musayev, M.; Akparov, Z.; Maserti, B.E. Physiological and metabolomic analysis of *Punica granatum* (L.) under drought stress. *Planta* **2016**, *243*, 441–449. [CrossRef] [PubMed]

horticulturae

Article

Estimation of Water Stress in Potato Plants Using Hyperspectral Imagery and Machine Learning Algorithms

Julio Martin Duarte-Carvajalino [1,*], Elías Alexander Silva-Arero [1], Gerardo Antonio Góez-Vinasco [1], Laura Marcela Torres-Delgado [2], Oscar Dubán Ocampo-Paez [2] and Angela María Castaño-Marín [1]

[1] Corporación Colombiana de Investigación Agropecuaria—Agrosavia, Centro de Investigación Tibaitatá—Km 14 Vía Mosquera, Bogotá D.C. 250047, Cundinamarca, Colombia; esilva@agrosavia.co (E.A.S.-A.); ggoez@agrosavia.co (G.A.G.-V.); amcastano@agrosavia.co (A.M.C.-M.)
[2] Universidad Nacional de Colombia, Carrera 45 N° 26-85–Edificio Uriel Gutiérrez, Bogotá D.C. 111321, Cundinamarca, Colombia; lamtorresde@unal.edu.co (L.M.T.-D.); odocampop@unal.edu.co (O.D.O.-P.)
* Correspondence: jmduarte@agrosavia.co; Tel.: +57-3132495914

Abstract: This work presents quantitative detection of water stress and estimation of the water stress level: none, light, moderate, and severe on potato crops. We use hyperspectral imagery and state of the art machine learning algorithms: random decision forest, multilayer perceptron, convolutional neural networks, support vector machines, extreme gradient boost, and AdaBoost. The detection and estimation of water stress in potato crops is carried out on two different phenological stages of the plants: tubers differentiation and maximum tuberization. The machine learning algorithms are trained with a small subset of each hyperspectral image corresponding to the plant canopy. The results are improved using majority voting to classify all the canopy pixels in the hyperspectral images. The results indicate that both detection of water stress and estimation of the level of water stress can be obtained with good accuracy, improved further by majority voting. The importance of each band of the hyperspectral images in the classification of the images is assessed by random forest and extreme gradient boost, which are the machine learning algorithms that perform best overall on both phenological stages and detection and estimation of water stress in potato crops.

Keywords: water stress; potato; hyperspectral image; machine learning; band importance

Citation: Duarte-Carvajalino, J.M.; Silva-Arero, E.A.; Góez-Vinasco, G.A.; Torres-Delgado, L.M.; Ocampo-Paez, O.D.; Castaño-Marín, A.M. Estimation of Water Stress in Potato Plants Using Hyperspectral Imagery and Machine Learning Algorithms. *Horticulturae* **2021**, *7*, 176. https://doi.org/10.3390/horticulturae7070176

Academic Editors: Sara Álvarez, Stefania Toscano and Giulia Franzoni

Received: 18 May 2021
Accepted: 28 June 2021
Published: 2 July 2021

Publisher's Note: MDPI stays neutral with regard to jurisdictional claims in published maps and institutional affiliations.

1. Introduction

Potato (*Solanum tuberosum* L.) is the third most important food crop in the world [1]. The potato provides an economic and rich source of carbohydrates and it is included in the diet of both developed and undeveloped countries. Water deficit is the most important abiotic stress affecting the development, productivity, and quality of potato cultivars [2]. Hence, it is important to detect, as early as possible, signs of water stress in potato plants avoiding production and quality losses. Due to climate change, crops worldwide are suffering from unexpected and longer severe weather changes such as droughts, which are becoming increasingly more intense [3]. Specifically in Colombia, a good portion of areas suitable for potato production are vulnerable to increased aridity, soil erosion, desertification, and variations in the hydrological system as a consequence of climate change [4]. Therefore, there is a need to map water stress in potato crops using non-destructive technologies such as remote sensing.

Recently, a spectroradiometer (350–2500 nm) was used to explore the effect of water stress on the spectral reflectance of bermudagrass and five vegetation indexes were studied [5]. In the case of potato crops, 12 vegetation indexes including four Normalized Water Indexes (NWIs), have been studied to detect water stress in potato leaves under different watering conditions using also a spectroradiometer (350–2500 nm) [4]. The results indicate clear differences in the spectrum of water-stressed leaves in the 700–1300 nm range [4].

Remote sensing technologies using unmanned aerial vehicles (UAVs) acquiring visible and thermal images were used to map water stress in barley crops [6]. The detection of water stress in plants using aerial imagery has focused on thermal imagery to estimate plant temperature relative to the air temperature computing NWIs. Since stomata close under water stress, the temperature of the leaves relative to the air increases [6–8]. More recently, remote sensing imaging technologies using visible, near-infrared (NIR), short wave infrared (SWIR), and thermography have been proposed to detect water stress in potato crops [9]. Rather than using broadband multispectral images, hyperspectral imagery and machine learning algorithms have been proposed to determine the quality of food products [10]. Hyperspectral imagery (400–1000 nm) has also been proposed to detect water stress in potato crops using spectral indexes [11]. Hyperspectral imagery (400–2500 nm) was used in combination with partial least squares–discriminant analysis (PLS-DA) and partial least squares–support vector machine (PLS-SVM) classification to detect abiotic and biotic drought stress in tomato canopies [12]. Hyperspectral imagery (450–1000 nm) in combination with machine learning algorithms (random forest and extreme gradient boost) has been also used to detect water stress in vine canopies [13]. Another possibility for detecting water stress in plants is to use radar remote sensing technologies [14,15] with the advantage of penetrating the clouds, a limitation of visible and thermal imagery. Finally, ultrasound wave spectroscopy has also been used to estimate the water content of plant leaves using convolutional neural networks and random forest algorithms [16].

As previously indicated, work on detecting water stress in potato cultivars has been based on vegetation indexes (*NDVI*, the Simple Ratio, the Photochemical Reflectance Index, the pigment-specific simple ratio of Chlorophyll-a, the reflectance water index, the Normalized Water Indexes and the dry Zea N index). Here we use a hyperspectral camera (400–1000 nm) and several well-known machine learning algorithms to detect water stress in potato hyperspectral images and to estimate the degree of water stress: none, light, moderate and severe, using all images bands. The use of machine learning algorithms allows us to determine which regions in the spectral signature of the leaves are more influential to better estimate water stress from remote sensing using images in the visible (400–700 nm) and near-infrared (*NIR*) (700–1000 nm) bands.

2. Material and Methods

2.1. Plant Material and Experimental Design

The experiment was developed in greenhouse number 17 of AGROSAVIA (Corporación Colombiana de Investigación Agropecuaria), Tibaitatá research center, Colombia (4°41′25.7064″ N, 74°12′08.23″ W) at 2543 m above the sea level. Certified seeds of *Solanum tuberosum* L., variety Diacol Capiro were planted in the greenhouse. The experiment consisted of a randomized complete blocks design in a factorial 2 × 4 arrangement. The first factor considered was the level of plant development (phenological stage), this was fixed according to [17]: tubers differentiation (TD) and maximum tuberization (MT) (Appendix A). The second factor was the level of water stress severity, determined by the hydric potential of the leaves, measured using a Scholander pressure chamber in Mega Pascals (Mpa). Control plants have a hydric potential in the 0–−0.49 Mpa range, light (L) water stress has a hydric potential in the −0.5–−0.59 Mpa range, moderate (M) water stress has a hydric potential in the −0.6–−0.89 Mpa, and severe (S) water stress has a hydric potential equal to or lower than −0.9 Mpa. These hydric potential ranges were selected based on [18,19], and previous research experience of AGROSAVIA in greenhouses containing potato crops.

Potato plants were sown in a greenhouse in a loamy soil that was kept at field capacity (soil water potential did not decline below −0.033 MPa) by drip irrigation from sowing until the 9th and 13th week after sowing, when each stage of development was reached (TD and TF, respectively). At that time, the water supply was suspended, and the water potential in the leaf was measured daily until reaching each level of stress (L, M, S). Control plants had a water supply throughout the experiment.

2.2. Hyperspectral Imagery

The hyperspectral images were acquired using a 710-VP Surface Optics Corporation camera with 520 × 696 pixels and 128 spectral bands in the 400–1000 nm range, using the Environment for Visualizing Images (ENVI) format. The images were taken at 3 m above the plant's canopy level and the camera looking downwards. The image acquisition campaigns were done at around the same hour of the day. Figure 1 shows a false-color image of the canopy of a plant loaded and visualized with MultiSpec [20]. As can be seen from this image a Spectralon reflectance white panel is also used on each image to convert the hyperspectral intensity images to reflectance. It is easy to segment the white Spectralon panel from the hyperspectral image by computing the average of the red, green, blue, and *NIR* bands and dividing that image by the maximum intensity. Figure 2 shows this normalized average, where the Spectralon reflectance panel can be segmented from the image using a threshold above 0.5.

Figure 1. Hyperspectral image taken at 3 m above the canopy. False-color image showing red as band 90, green as band 60, and blue as band 40.

The reflectance of each hyperspectral image can be computed using:

$$\rho(x,y,\lambda) = \frac{I(x,y,\lambda)\rho_S(\lambda)}{Is(\lambda)} \tag{1}$$

where $\rho(x,y,\lambda)$ is the reflectance image at pixel coordinates x,y and waveband λ, $I(x,y,\lambda)$ is the raw intensity image at pixel coordinates x,y, and waveband λ, $\rho_S(\lambda)$ the known reflectance of the Spectralon panel at λ wavelength (0.99 at visible and *NIR* ranges) and $Is(\lambda)$ the mean intensity of the Spectralon panel at waveband λ. Once the hyperspectral images are converted to reflectance, it is necessary to segment the canopy from its background. The Normalized Difference Vegetation Index (*NDVI*) has widely been used to detect vegetation canopy [21]:

$$NDVI = \frac{\rho_{NIR} - \rho_{red}}{\rho_{NIR} + \rho_{red}} \tag{2}$$

where ρ_{NIR}, ρ_{red} are the reflectances at the *NIR* and red wavelengths, respectively. However, the *NDVI* is affected by several factors including shadows [21] that could lead to 0/0 undefined values. To avoid this, we used the Soil-Adjusted Vegetation Index (*SAVI*) that overcomes the issues of the *NDVI* [21] and selected those values where *SAVI* > 0.3 (Figure 3):

$$SAVI = 1.5 \frac{\rho_{NIR} - \rho_{red}}{(0.5 + \rho_{NIR} + \rho_{red})} \tag{3}$$

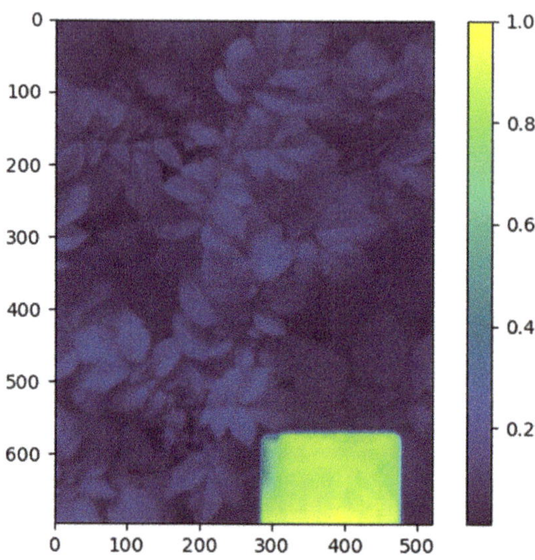

Figure 2. Normalized sum of red, green, blue, *NIR* bands.

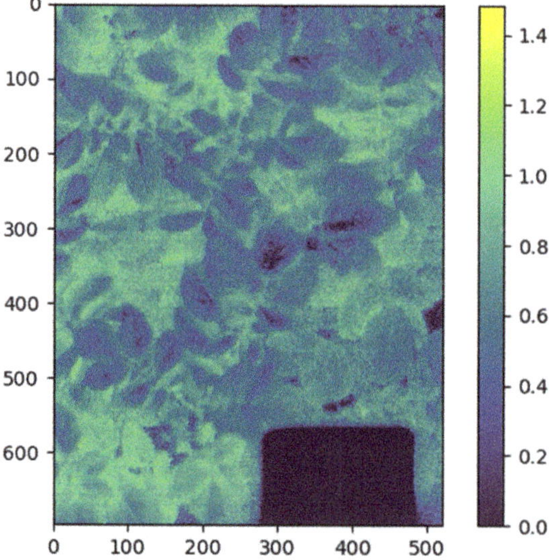

Figure 3. *SAVI* greater than 0.3 to detect leaves.

From the image campaign at the tubers differentiation phenological stage, 64 images were acquired to be used for the machine learning algorithms (stressed and control plants) with water stresses that range from 3 to 20 days. From the image campaign at the maximum tuberization phenological stage, 52 images were acquired to be used for the machine learning algorithms (stressed and control plants) with water stresses that range from zero to nine days. The reading and preprocessing of the hyperspectral images were done using Python 3.8.5 that comes with Anaconda [22]. The Python spectral library [23] was used to read the hyperspectral images.

There are control plants that provide images for the control class and there are several images for each stress condition, taken at different days after the application of each stress level.

2.3. Machine Learning Algorithms

Two supervised classification tasks for the two phenological stages of the potato crops were carried out: detection of water stress i.e., the plant is water-stressed or not (two classes) and the estimation of the water level of stress i.e., the plant is not water-stressed, is lightly water-stressed, is moderately stressed or severely stressed (four classes). To perform these classification tasks six well-known machine learning algorithms were used:

- Random decision forest (RF) [24] using 100 trees, with a balanced class weight. RF are an ensemble of decision trees, the class predicted corresponds to the class most voted for the decision trees.
- Multi-layer perceptron (MLP) [25] with an input layer having equal nodes as the number of bands (128) and an output layer having equal nodes as the number of classes (2 or 4). Each layer is followed by a batch normalization layer [26], a dropout layer [27] with a probability of 0.2, a rectified linear activation function (RELU, a function that will output the same input if it is positive, zero otherwise) [28] on the input layer, and a Softmax activation function [28] on the output layer for the case of four classes or a Sigmoid activation function [28] for the case of two classes classification (see Figure 5). An MLP neural network consists of layers of nodes: an input layer, hidden layers and an output layer. Except for the input nodes, each node is a neuron that uses a nonlinear activation function. Each node on a layer connects with each node of the following layer by a weight function. The neural network learns the weights from the training data.
- Convolutional neural networks (CNN) [29] with two convolutional layers using a kernel size of 3 and 20 filters each one. The two convolutional layers are followed by a batch normalization layer, a dropout (0.2) layer, and a RELU layer. After the two convolutional layers, a flatten layer follows to flatten out the last convolutional layer into MLP nodes. After the flatten layer, an input MLP layer of size equal to the half of nodes of the flatten layer follows, then a middle MLP layer with half the nodes of the previous layer and an output layer with equal nodes as the number of classes. Each MLP layer is followed by a batch normalization layer, a dropout (0.2) layer, and a RELU layer for the case of the first MLP layer and the second MLP layer. The last MLP layer is followed by a Softmax activation function in the case of four classes or a Sigmoid activation function, in the case of two classes (Figure 4). Convolutional neural networks are a kind deep learning neural network specialized on images, with convolutional layers applying different kinds of filters on patches of the images and then on previous convolutional layers, to capture variabilities at higher scales.
- Support vector machine (SVM) [30] using linear SVM with default parameters. SVM maps training examples to points in space so as to maximize the width of the gap between the classes.
- Extreme gradient boost (XGBoost) [31] using tree classifiers (gbtree) as weak learners and 100 estimators. Gradient boosting produces an ensemble of weak predictions (usually trees) models and generalizes them by the optimization of a differentiable loss function. XGBoost in an implementation of gradient boosting that uses a more regularized model formalization to control overfitting.
- AdaBoost (AB) [32] with 100 estimators. An AdaBoost classifier works by fitting a classifier that first fits the dataset and then fits additional copies of the classifier, but

giving more weight to the incorrectly classified instances, so subsequent classifiers focus on harder cases.

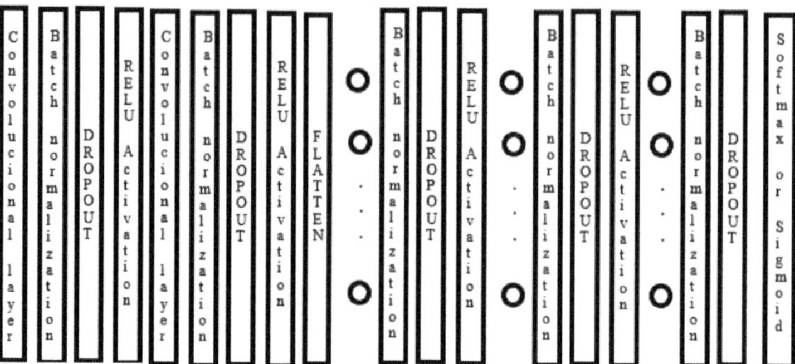

Figure 4. Convolutional neural network layout.

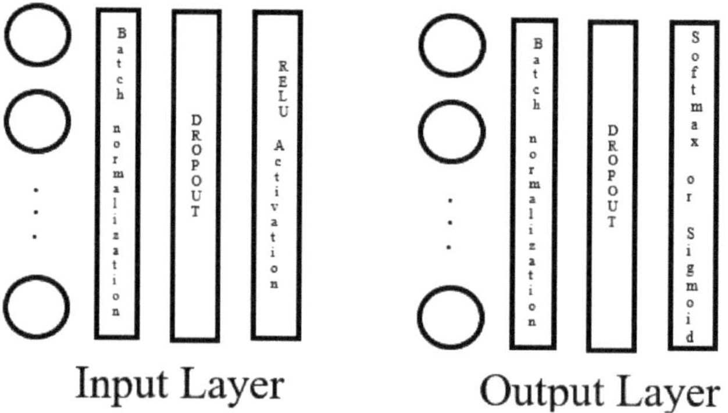

Figure 5. Multi-layer perceptron layout.

The RF, SVM, AB classifiers were implemented in Python 3.8 using the sklearn library. The MLP and CNN were implemented in Python 3.5 using the keras library with tensorflow under the hood in the High Performance Computing servers of Agrosavia, given the memory required by CNN. The XGBoost classifier was implemented using xgboost python library in Python 3.8.

Given the size of the images ($520 \times 696 \times 128$) and equipment memory constraints and processing times, only 10000 pixels were selected at random from the canopy (identified using $SAVI > 0.3$) on each image to train the classifiers forming a training dataset. In the case of CNN, a window of size $5 \times 5 \times 128$ was selected centered on each one of the 10,000 pixels selected at random in the canopy to form the CNN dataset. To evaluate the classifiers, five-fold cross-validation was employed to measure the probability of classfication overfitting, due to the tendency of classifiers to overfit the training dataset. Here, 80% of the dataset is used for training and 20% for testing the classifiers on each one of the five-fold cross-validation runs. In the case of MLP and CNN, 20% of the 80% available data for training is used for validation in such a way that the MLP or CNN models are saved only if the computed loss improves for the validation data, as an extra measure to avoid overfitting the dataset. Furthermore, the classifiers were trained with the full training dataset and then used to classify the whole canopy on each image (containing many more pixels unseen by the classifiers) using majority voting, i.e., selecting the class most pixels are classified with.

3. Results

Figure 6 shows the classification performance using two classes (water stress or control) for the phenological stage tubers differentiation using overall accuracy, sensitivity, and specificity (see confusion matrices in the Appendix B), where the standard deviation of the mean is indicated for accuracy, sensitivity, and specificity, as error bars. As can be seen from these results RF and XGBoost achieve the best classification performance, being XGBoost the best.

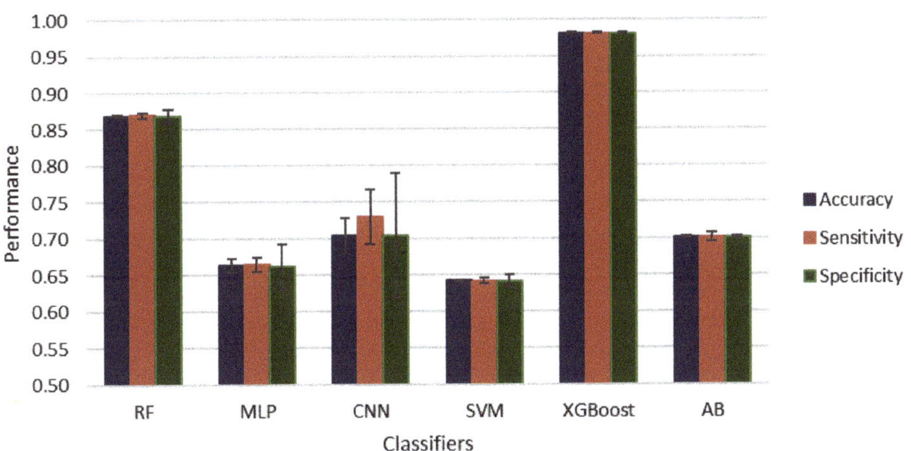

Figure 6. Classification performance, tubers differentiation phenological stage using two classes.

Table 1 compares the classification performance using the best three classifiers found: RF, XGBoost, and CNN alone and using Majority Voting (MV). This table shows that both RF and XGBoost correctly classify all the images using majority voting, followed by CNN.

Table 1. Comparison of classification performance of RF, XGBoost, and CNN alone and using MV for tubers differentiation phenological stage using two classes.

	RF	RF + MV	XGBoost	XGBoost + MV	CNN	CNN + MV
Accuracy	0.8691875	1	0.98120469	1	0.69395156	0.875
Sensitivity	0.86965626	1	0.98114434	1	0.71711594	0.875855327
Specificity	0.86857087	1	0.98123905	1	0.69385401	0.875855327

Figure 7 shows the classification performance for the tubers differentiation phenological stage and four classes: control and three levels of water stress: light, moderate, and severe (see confusion matrices in the Appendix B), where the standard deviation of the mean is indicated for accuracy, sensitivity, and specificity, as error bars. In this case, XGBoost performs best, followed by RF and MLP. Table 2 compares the classification performance of the three best classifiers: RF, XGBoost, and CNN alone and using MV. In this case, XGBoost performs best, followed by RF and CNN.

Table 2. Comparison of classification performance of RF, MLP, and CNN alone and using MV for tubers differentiation phenological stage using two classes.

	RF	RF + MV	XGBoost	XGBoost + MV	CNN	CNN + MV
Accuracy	0.811439063	0.90625	0.985457813	1	0.63530625	0.703125
Sensitivity	0.879199269	0.961538462	0.991209678	1	0.591587693	0.66889881
Specificity	0.707431992	0.829861111	0.978625726	1	0.495725174	0.517834596

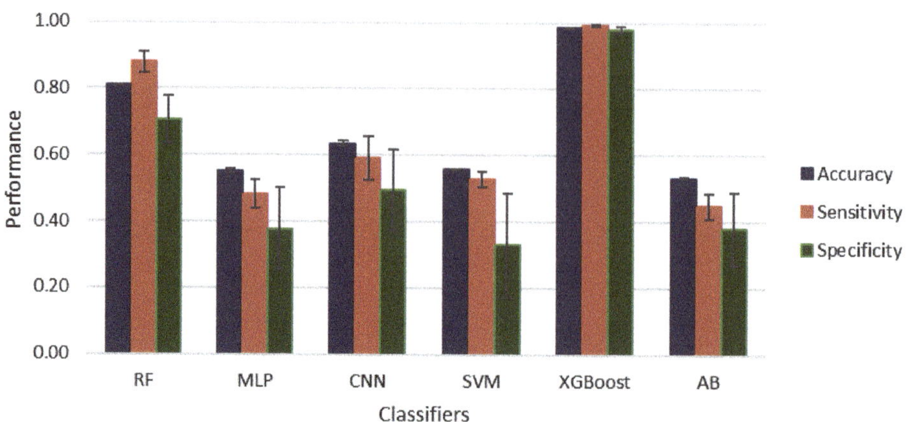

Figure 7. Classification performance, tubers differentiation phenological stage using four classes.

Figure 8 shows the classification performance at the maximum tuberization phenological stage using two classes: control and water stress (see confusion matrices in the Appendix B), where the standard deviation of the mean is indicated for accuracy, sensitivity, and specificity, as error bars. The best classifiers are XGBoost followed by RF and CNN. Table 3 compares the classification performance of RF, XGBoost, and CNN alone and using MV over all the images. This table shows RF and XGBoost both achieve perfect classification using MV of all the images taken at this phenological stage.

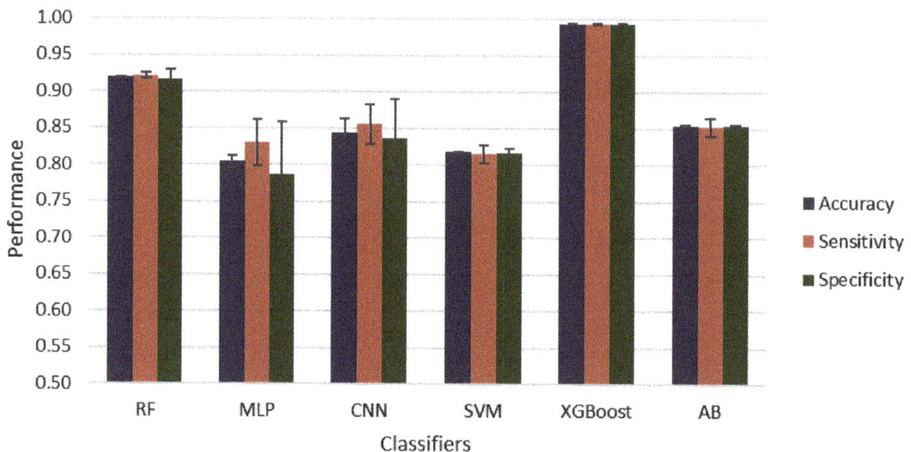

Figure 8. Classification performance, maximum tuberization phenological stage using two classes.

Table 3. Comparison of classification performance of RF, XGBoost, and CNN alone and using MV for tubers differentiation phenological stage using two classes.

	RF	RF + MV	XGboost	XGBoost + MV	CNN	CNN + MV
Accuracy	0.92025577	1	0.99373077	1	0.84420769	0.980769231
Sensitivity	0.92156596	1	0.99353142	1	0.8555923	0.979166667
Specificity	0.91676559	1	0.99376627	1	0.83643118	0.982758621

Figure 9 shows the classification performance at the maximum rate of tubers phenological stage using four classes: control, light, moderate, and severe water stress (see confusion matrices in the Appendix B), where the standard deviation of the mean is indicated for accuracy, sensitivity, and specificity, as error bars. Here, XGBoost obtains the best performance, followed by RF and CNN. As in the case of the two classes, the classification accuracies are good and allow estimation of the water stress from the first day. Table 4 compares the classification performance of RF, XGBoost, and CNN alone and using MV, where it can be noticed that XGBoost in combination with MV achieves perfect classification, followed by RF and CNN.

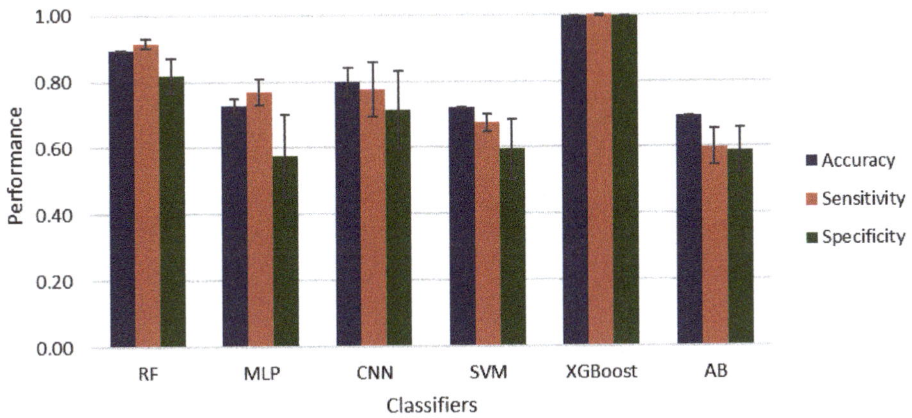

Figure 9. Classification performance, maximum tuberization phenological stage using four classes.

Table 4. Comparison of classification performance of RF, XGBoost, and CNN alone and using MV for the maximum tuberization phenological stage using four classes.

	RF	RF + MV	XGBoost	XGBoost + MV	CNN	CNN + MV
Accuracy	0.894794231	0.980769231	0.997425	1	0.797767308	0.961538462
Sensitivity	0.914904761	0.991666667	0.997991496	1	0.775300134	0.964285714
Specificity	0.818412336	0.964285714	0.996019078	1	0.713138567	0.982758621

Figure 10 shows XGBoost classification results on some images of the tubers differentiation phenological stage using four classes. The color code here is green for no water stress, blue for light stress, yellow for moderate stress, and red for severe stress. Figure 10a shows the classification for a control plant (no water stress). Figure 10b shows a plant that suffered light stress. Figure 10c shows a plant that suffered moderate stress. Figure 10d shows a plant that suffered severe stress.

Figure 11 shows some XGBoost classification results for the maximum tuberization phenological stage using the same color code as in Figure 10.

Figure 12 shows the band importance for RF classification in the detection (two classes) and estimation (four classes) of water stress at the phenological stage of tubers differentiation. Figure 13 shows the same band importance for RF classification of two and four classes at the phenological stage of the maximum tuberization. As indicated in Figures 12 and 13 the most important bands for classification in RF are the violet, the red edge, and a few wavelengths in the NIR.

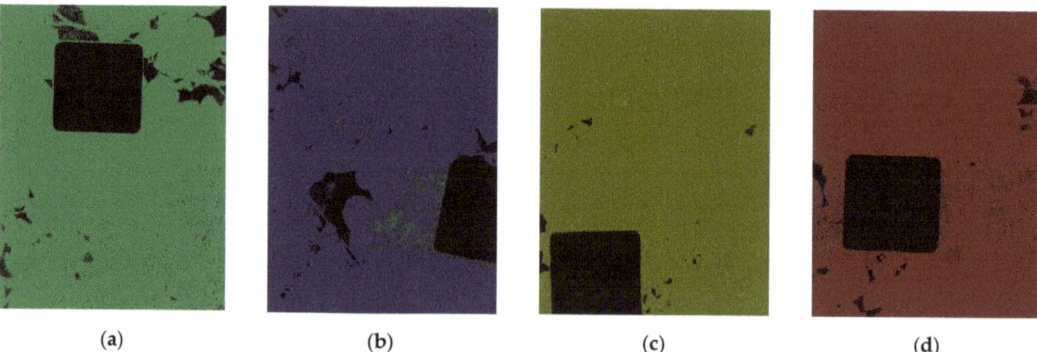

(a)   (b)   (c)   (d)

Figure 10. XGBoost classification of (**a**) image of a control plant, (**b**) image of a plant with light stress, (**c**) image of a plant with moderate stress, (**d**) image of a plant with severe stress.

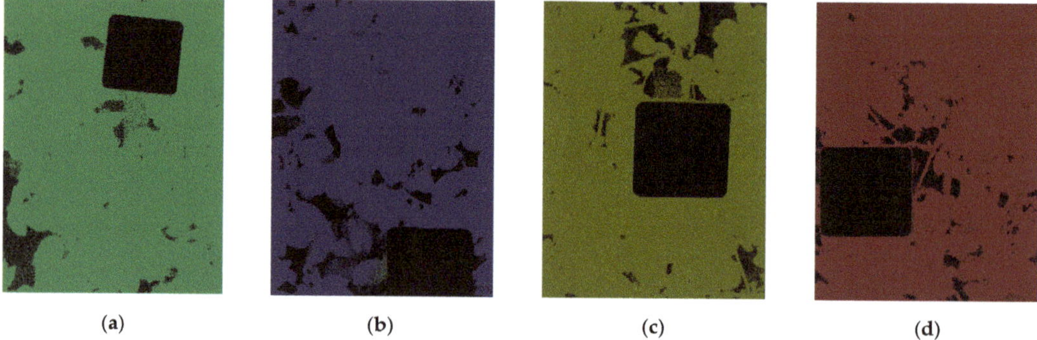

(a)   (b)   (c)   (d)

Figure 11. XGBoost classification of (**a**) control plant, (**b**) plant with light stress, (**c**) plant with moderate stress, (**d**) plant with severe stress.

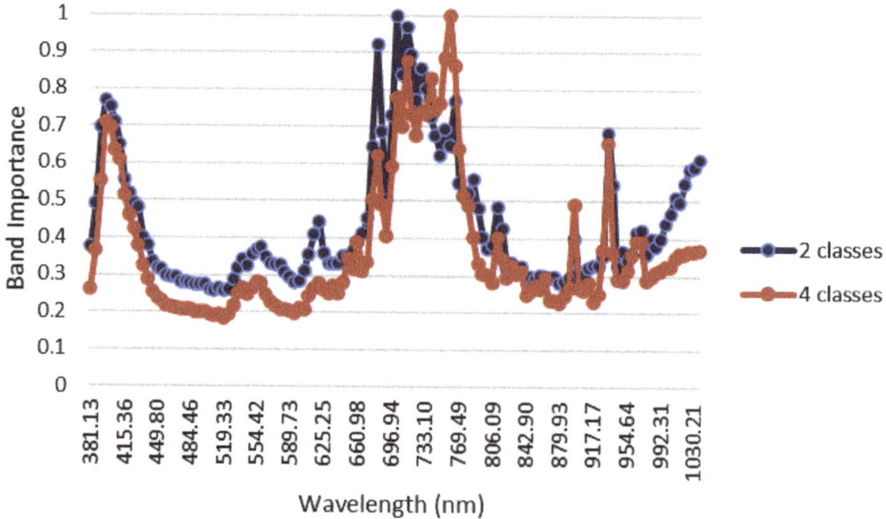

Figure 12. Band importance determined by RF on the tubers differentiation phenological stage.

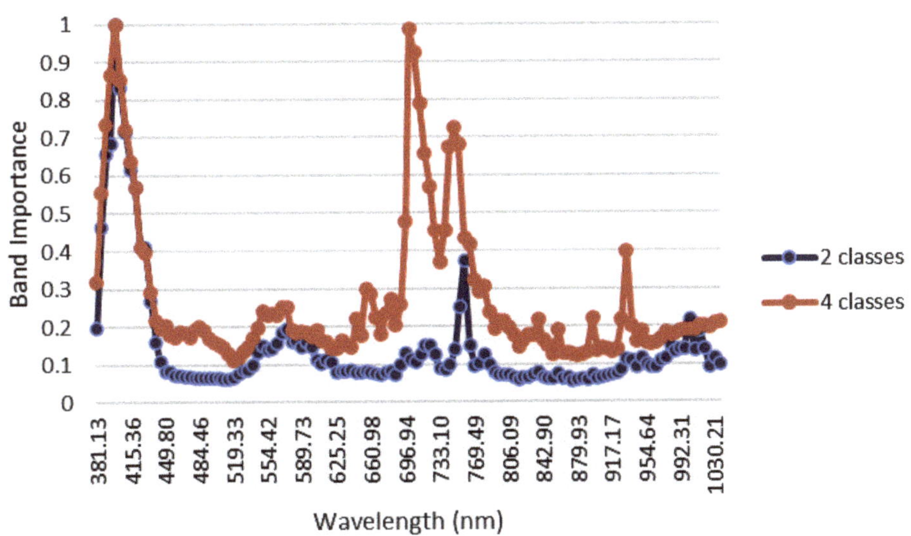

Figure 13. Band importance determined by RF on the maximum tuberization phenological stage.

Figure 14 shows the band importance for XGBoost classification in the detection (two classes) and estimation (four classes) of water stress at the phenological stage of tubers differentiation. Figure 15 shows the same band importance for XGBoost classification of two and four classes at the phenological stage of the maximum tuberization. From these figures, XGBoost considers important more bands than RF, i.e., it exploits better the spectral signature of the hyperspectral images. Band importance could help us identify which bands are better suited to detect water stress from multispectral imagery or to define water stress indices specially designed for potato crops.

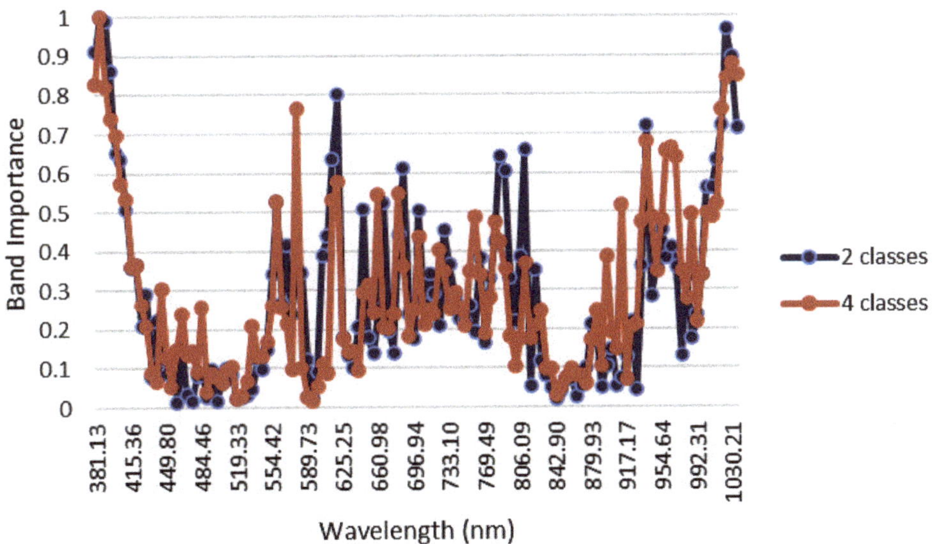

Figure 14. Band importance determined by XGBoost on the tubers differentiation phenological stage.

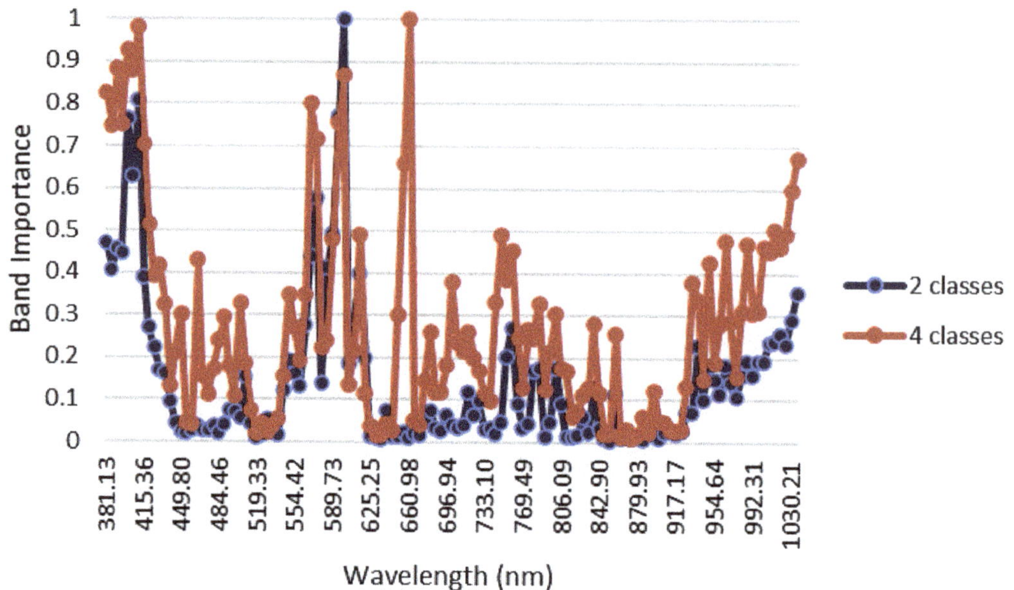

Figure 15. Band importance determined by XGBoost on the maximum tuberization phenological stage.

4. Discussion

The results indicate that even using a small subset of pixels, taken at random from the hyperspectral images, it is possible to obtain good classification accuracies for detecting and estimating water stress in potato crops. The results also indicate that as early as one day after the onset of the stress in the tubers differentiation phenological stage and on the same day of the onset of the stress in the maximum tuberization water stress can be detected and measured. Other researchers like [33] also found that hyperspectral imaging could be useful to detect water supply conditions of leafy vegetables growing under greenhouse, using modified partial least square regression algorithm, trained to classify different levels of leaf water potential, obtaining a correlation coefficient of 0.826. In this sense, hyperspectral imaging could become a useful tool for the design of precision irrigation systems that allow optimizing the use of water in crops such as potatoes, although it is necessary to develop more studies in real conditions of commercial cultivation.

It was evident that over all classification tasks and phenological stages XGBoost provides excellent classification accuracies alone or in combination with majority voting, followed closely by random forest. Random forest and XGBoost also provide a direct measure of band importance to detect and estimate water stress. In this case, XGBoost seems to better use the whole spectral signature of the canopy, while RF uses a reduced subset of bands. Although the SVM algorithm did not show the best results in this study, the authors of [34] reported promising results when using this algorithm (R = 0.7684) in combination with the Kullback–Leibler divergence (KLD) dimensionality reduction method to select the most relevant bands of hyperspectral images, in the detection of moisture content in maize leaves at the seedling stage. For future experiments, it may be useful to evaluate some combinations of algorithms that have proven to be efficient in the detection of relative water content in leaves, from remote hyperspectral sensing, as reported by [35] who used artificial neural networks (ANN) after selecting the most important bands through partial least squares regression (PLSR), improving the performance of ANN alone.

CNN is a deep learning neural network algorithm that extracts features from images. However, despite being the deep learning neural network most used to analyze images [10], its classification performance was lower than RF and XGBoost, and only by using majority voting, it was possible to improve its performance to classify all image pixels. This is probably because CNN exploits the spatial structure of the images (such as edges) and not the spectral signature of the images. In this case, the canopy consists of mostly leaves with no spatial clues related to water stress.

Our results indicate that using machine learning and spectral images constitute a phenotyping tool useful to detect and estimate water stress in potato plants, which can also be used in processes of genetic improvement, by choosing those phenotypes that better resist water stress. The reflectance images obtained may be sensitive to the physiological and biochemical changes of the substances and pigments that are degraded and mobilized due to water stress.

5. Conclusions

This work shows that detection of water stress, as well as estimation of the water stress level, is possible with good accuracy incremented on the whole canopy, using majority voting at the tubers differentiation and maximum rate of tuberization phenological stages. In particular, the classification results are more accurate and available from the first day of stress for both the tubers differentiation and maximum rate of tuberization phenological stages. Extreme gradient boost performed best overall phenological stages and classification tasks, followed by random decision forests. XGBoost and RF also provide a measure of the importance of each band to detect or estimate water stress in potato crops. In the case of RF, these bands are the violet, red edge, and some specific NIR bands, while in the case of XGBoost it includes some additional bands in the visible (green, yellow, red) and NIR, exploiting better the spectral signature.

These results could lead to the use of more specific normalized water indexes for water stress detection and estimation in potato crops using these machine learning algorithms. However, they are not intended to be used by producers, since this research work was conducted under greenhouse conditions. In this sense, these results are an important basis for further research considering actual potato crop field conditions and cultural practices. It will allow to design advanced tools for early detection of water stress, increasing the efficiency in the application of irrigation.

Author Contributions: Conceptualization, E.A.S.-A. and G.A.G.-V.; Methodology, E.A.S.-A.; Software, J.M.D.-C.; Validation, E.A.S.-A.; Formal analysis, L.M.T.-D. and O.D.O.-P.; Investigation, J.M.D.-C. and E.A.S.-A.; Resources, A.M.C.-M.; Data curation, L.M.T.-D. and O.D.O.-P.; Writing—original draft, J.M.D.-C.; Writing—review & editing, E.A.S.-A., G.A.G.-V., L.M.T.-D., O.D.O.-P. and A.M.C.-M. Visualization, L.M.T.-D.; Supervision, A.M.C.-M.; Project administration, A.M.C.-M.; Funding acquisition, A.M.C.-M.; Resources, G.A.G.-V. and A.M.C.-M. All authors have read and agreed to the published version of the manuscript.

Funding: This research was funded by the Science, Technology, and Innovation Fund of the General Royalty System, administered by the National Financing Fund for Science, Technology and Innovation Francisco José de Caldas, the Colombia BIO Program, the government of Cundinamarca—Colombia and the Ministry of Science, Technology, and Innovation (MINCIENCIAS), and Corporación Colombiana de Investigación Agropecuaria (AGROSAVIA).

Institutional Review Board Statement: Not applicable.

Acknowledgments: This work is part of a larger project in Agrosavia called Agroclimatic Information System for potato (*Solanum tuberosum* L.) crops within productive regions in Cundinamarca (SIAP in Spanish). We thank Jose Alfredo Molina Varón for his contribution with the experimental setup and the adaptation of measurement equipment, and Jhon Mauricio Estupiñán Casallas for his collaboration in the assembly of the irrigation systems.

Conflicts of Interest: The authors declare no conflict of interest.

Appendix A

Figure A1. Tubers differentiation stage: left and right photographs show the development of stolons: at the apex, "hook" and "matchstick" forms, these are morphological changes in the stolons in the tuber differentiation process.

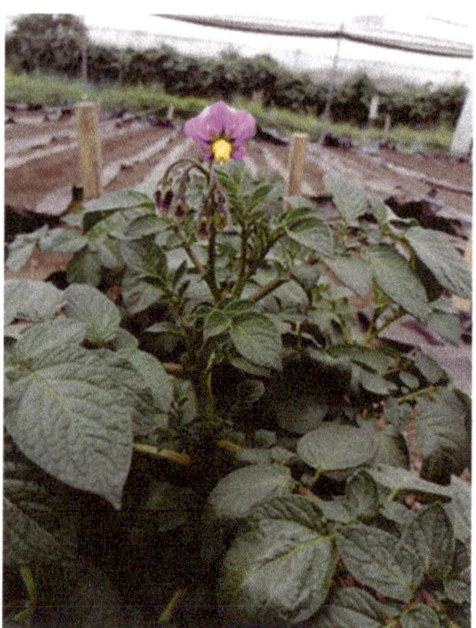

Figure A2. Maximum tuberization stage: left and right photographs show the development stage of flowering. Gómez et al. report that the stage of maximum tuberization and beginning of filling coincides with flowering.

Appendix B

Table A1. Four Classes, Tubers Differentiation.

RF			
63,977.80	246.00	340.00	1436.20
6514.60	10,013.60	38.60	1433.20
4497.20	2.40	7114.20	386.20
8758.40	393.20	89.80	22,758.60
SVM			
59,562.00	538.80	166.20	5733.00
13,379.40	1381.00	37.80	3201.80
9828.60	58.00	223.20	1890.20
21,264.40	388.00	47.60	10,300.00
CNN			
52,467.80	4374.40	1091.40	8066.40
7442.60	7230.20	79.40	3247.80
4198.00	3643.40	2123.40	2035.20
9497.20	2412.60	592.40	19,497.80
MLP			
53,304.20	3697.40	668.80	8329.60
10,808.20	4085.00	80.00	3026.80
5405.00	2299.80	1266.80	3028.40
17,965.20	1482.40	587.60	11,964.80
XGBoost			
329,137.00	90.00	69.00	704.00
4314.00	85,073.00	14.00	599.00
416.00	5.00	59,409.00	170.00
2869.00	32.00	25.00	157,074.00
Ada Boost			
51,866.40	2866.40	3189.60	8077.60
12,534.60	2287.20	498.00	2680.20
7042.40	114.80	3204.60	1638.20
18,669.00	1172.80	1264.00	10,894.20

Table A2. Four Classes, Maximum Tuberization.

RF			
56,866.20	51.20	995.00	87.60
1690.80	11,658.00	618.00	33.20
3996.60	539.40	19,275.80	188.20
1678.60	2.40	1060.40	5258.60
SVM			
52,848.20	1664.60	2907.80	579.40
3598.00	8849.20	1474.20	78.60
10,426.40	2783.60	9784.40	1005.60
2296.80	165.60	2138.20	3399.40
CNN			
50,919.00	2418.20	3942.80	720.00
1010.40	8773.20	3995.00	221.40
3454.80	299.40	18,737.60	1508.20
613.00	188.00	2661.00	4538.00

Table A2. *Cont.*

MLP			
56,069.80	626.20	690.80	613.20
5170.00	7803.20	942.60	84.20
14,354.60	992.60	7957.60	695.20
3486.80	20.60	895.00	3597.60
XGBoost			
289,738.00	150.00	91.00	21.00
520.00	69,382.00	95.00	3.00
291.00	26.00	119,683.00	0.00
131.00	3.00	8.00	39,858.00
Ada Boost			
48,421.80	3685.40	4595.60	1297.20
2715.60	8295.00	2920.80	68.60
6624.40	3534.20	11,884.60	1956.80
1909.00	104.80	2486.40	3499.80

Table A3. Two Classes, Tubers Differentiation.

RF		CNN		XGBoost	
58,628.0	7372.0	47,915.2	18,084.8	323,446.0	6554.0
9372.0	52,628.0	19,694.0	42,306.0	5475.0	304,525.0
SVM		MLP		Ada Boost	
43,555.4	22,444.6	47,570.6	18,429.4	46,091.2	19,908.8
23,343.4	38,656.6	24,596.0	37,404.0	18,315.6	43,684.4

Table A4. Two Classes, Maximum Tuberization.

RF		CNN		XGBoost	
54,926.80	3073.20	52,422.00	5578.00	288,103.00	1897.00
5220.20	40,779.80	10,624.40	35,375.60	1363.00	228,637.00
SVM		MLP		Ada Boost	
48,065.40	9934.60	54,502.60	3497.40	49,767.60	8232.40
8949.60	37,050.40	16,840.80	29,159.20	6870.40	39,129.60

References

1. Center, I.P. Potato Facts and Figures. Available online: https://cipotato.org/potato/potato-facts-and-figures/ (accessed on 25 June 2021).
2. van Loon, C.D. The effect of water stress on potato growth, development, and yield. *Am. Potato J.* **1981**, *58*, 51–69. [CrossRef]
3. Trenberth, K.E.; Dai, A.; van der Schrier, G.; Jones, P.D.; Barichivich, J.; Briffa, K.R.; Sheffield, J. Global warming and changes in drought. *Nat. Clim. Chang.* **2014**, *4*, 17–22. [CrossRef]
4. Romero, A.P.; Alarcón, A.; Valbuena, R.I.; Galeano, C.H. Physiological assessment of water stress in potato using spectral information. *Front. Plant. Sci.* **2017**, *8*. [CrossRef]
5. Caturegli, L.; Matteoli, S.; Gaetani, M.; Grossi, N.; Magni, S.; Minelli, A.; Corsini, G.; Remorini, D.; Volterrani, M. Effects of water stress on spectral reflectance of bermudagrass. *Sci. Rep.* **2020**, *10*, 1–12. [CrossRef]
6. Hoffmann, H.; Jensen, R.; Thomsen, A.; Nieto, H.; Rasmussen, J.; Friborg, T. Crop water stress maps for entire growing seasons from visible and thermal UAV imagery. *Biogeosciences Discuss.* **2016**, 1–30. [CrossRef]
7. Gago, J.; Douthe, C.; Coopman, R.E.; Gallego, P.P.; Ribas-Carbo, M.; Flexas, J.; Escalona, J.; Medrano, H. UAVs challenge to assess water stress for sustainable agriculture. *Agric. Water Manag.* **2015**, *153*, 9–19. [CrossRef]
8. Labbé, S.; Lebourgeois, V.; Jolivot, A.; Marti, R. Thermal infra-red remote sensing for water stress estimation in agriculture. In *Options Méditerranéennes*; CIHEAM: Zaragosa, Spain, 2012; Volume 67, pp. 175–184.
9. Gerhards, M.; Rock, G.; Schlerf, M.; Udelhoven, T. Water stress detection in potato plants using leaf temperature, emissivity, and reflectance. *Int. J. Appl. Earth Obs. Geoinf.* **2016**, *53*, 27–39. [CrossRef]

10. Saha, D.; Manickavasagan, A. Machine learning techniques for analysis of hyperspectral images to determine quality of food products: A review. *Curr. Res. Food Sci.* **2021**, *4*, 28–44. [CrossRef] [PubMed]

11. Amatya, S.; Karkee, M.; Alva, A.K.; Larbi, P.; Adhikari, B. Hyperspectral imaging for detecting water stress in potatoes. In Proceedings of the American Society of Agricultural and Biological Engineers Annual International Meeting 2012, ASABE 2012, Dallas, TX, USA, 29 July–1 August 2012; Volume 7, pp. 6134–6148. [CrossRef]

12. Susič, N.; Žibrat, U.; Širca, S.; Strajnar, P.; Razinger, J.; Knapič, M.; Vončina, A.; Urek, G.; Gerič Stare, B. Discrimination between abiotic and biotic drought stress in tomatoes using hyperspectral imaging. *Sens. Actuators B Chem.* **2018**, *273*, 842–852. [CrossRef]

13. Loggenberg, K.; Strever, A.; Greyling, B.; Poona, N. Modelling water stress in a Shiraz vineyard using hyperspectral imaging and machine learning. *Remote Sens.* **2018**, *10*, 202. [CrossRef]

14. El-Shirbeny, M.A.; Abutaleb, K. Sentinel-1 Radar Data Assessment to Estimate Crop Water Stress. *World J. Eng. Technol.* **2017**, *5*, 47–55. [CrossRef]

15. Van Emmerik, T.; Steele-Dunne, S.C.; Judge, J.; Van De Giesen, N. Dielectric Response of Corn Leaves to Water Stress. *IEEE Geosci. Remote Sens. Lett.* **2017**, *14*, 8–12. [CrossRef]

16. Fariñas, M.D.; Jimenez-Carretero, D.; Sancho-Knapik, D.; Peguero-Pina, J.J.; Gil-Pelegrín, E.; Gómez Álvarez-Arenas, T. Instantaneous and non-destructive relative water content estimation from deep learning applied to resonant ultrasonic spectra of plant leaves. *Plant. Methods* **2019**, *15*, 1–10. [CrossRef]

17. Gómez, M.I.; Magnitskiy, S.; Rodríguez, L.E. Normalized difference vegetation index, and K+ in stem sap of potato plants (Group Andigenum) as affected by fertilization. *Exp. Agric.* **2019**, *55*, 945–955. [CrossRef]

18. Hsiao, T.C. Plant responses to sawdust. *Proc. Indiana Acad. Sci.* **1931**, *41*, 125–126.

19. Tschaplinski, T.J.; Abraham, P.E.; Jawdy, S.S.; Gunter, L.E.; Martin, M.Z.; Engle, N.L.; Yang, X.; Tuskan, G.A. The nature of the progression of drought stress drives differential metabolomic responses in Populus deltoides. *Ann. Bot.* **2019**, *124*, 617–626. [CrossRef] [PubMed]

20. Purdue, Multispec. Available online: https://engineering.purdue.edu/~{}biehl/MultiSpec/ (accessed on 25 June 2021).

21. Xue, J.; Su, B. Significant remote sensing vegetation indices: A review of developments and applications. *J. Sens.* **2017**, *2017*. [CrossRef]

22. Anaconda. Anaconda Individual Edition. Available online: https://www.anaconda.com/products/individual (accessed on 25 June 2021).

23. Boggs, T. Spectral Python. Available online: http://www.spectralpython.net/ (accessed on 25 June 2021).

24. Ho, T.K. Random decision forests. In Proceedings of the 3rd International Conference on Document Analysis and Recognition, Montreal, QC, Canada, 14–16 August 1995; Volume 1, pp. 278–282. [CrossRef]

25. Wasserman, P.D.; Schwartz, T. Neural networks. II. What are they and why is everybody so interested in them now. *IEEE Expert* **1988**, *3*, 10–15. [CrossRef]

26. Ioffe, S.; Szegedy, C. Batch normalization: Accelerating deep network training by reducing internal covariate shift. In Proceedings of the 32nd International Conference on Machine Learning (ICML 2015), Lile, France, 6–11 July 2015; Volume 1, pp. 448–456.

27. Srivastava, M.; Hinton, G.; Krizhevsky, A.; Sutskever, I.; Salakhutdinov, R. Dropout: A Simple Way to Prevent Neural Networks from Overfitting. *Mach. Learn. Res.* **2014**, *15*, 1929–1958.

28. Nwankpa, C.E.; Ijomah, W.; Gachagan, A.; Marshall, S. Activation functions: Comparison of trends in practice and research for deep learning. *arXiv* **2018**, arXiv:1811.03378.

29. Krizhevsky, A.; Hinton, G. ImageNet Classification with Deep Convolutional Neural Networks. In Proceedings of the ImageNet Large Scale Visual Recognition Challenge, Florence, Italy, 7 May 2012; p. 27.

30. Cortez, C.; Vapnik, V. Support-Vector Networks. *Mach. Learn.* **1995**, *20*, 273–297. [CrossRef]

31. Cheng, T.; Guestrin, C. XGBoost: A Scalable Tree Boosting System. In Proceedings of the The Association for Computing Machinery's Special Interest Group on Knowledge Discovery and Data Mining, San Francisco, CA, USA, 13–17 August 2016; pp. 785–794.

32. Shen, Y.; Jiang, Y.; Liu, W.; Liu, Y. *Multi-Class AdaBoost ELM*; Springer Nature: Singapore, 2015; Volume 2, pp. 179–188. [CrossRef]

33. Tung, K.C.; Tsai, C.Y.; Hsu, H.C.; Chang, Y.H.; Chang, C.H.; Chen, S. Evaluation of Water Potentials of Leafy Vegetables Using Hyperspectral Imaging. *IFAC-PapersOnLine* **2018**, *51*, 5–9. [CrossRef]

34. Gao, Y.; Qiu, J.; Miao, Y.; Qiu, R.; Li, H.; Zhang, M. Prediction of Leaf Water Content in Maize Seedlings Based on Hyperspectral Information. *IFAC-PapersOnLine* **2019**, *52*, 263–269. [CrossRef]

35. Krishna, G.; Sahoo, R.N.; Singh, P.; Bajpai, V.; Patra, H.; Kumar, S.; Dandapani, R.; Gupta, V.K.; Viswanathan, C.; Ahmad, T.; et al. Comparison of various modelling approaches for water deficit stress monitoring in rice crop through hyperspectral remote sensing. *Agric. Water Manag.* **2019**, *213*, 231–244. [CrossRef]

 *horticulturae*

Article

Cowpea Ecophysiological Responses to Accumulated Water Deficiency during the Reproductive Phase in Northeastern Pará, Brazil

Denilson P. Ferreira [1], Denis P. Sousa [1], Hildo G. G. C. Nunes [2], João Vitor N. Pinto [1], Vivian D. S. Farias [3], Deborah L. P. Costa [1], Vandeilson B. Moura [4], Erika Teixeira [1], Adriano M. L. Sousa [2], Hugo A. Pinheiro [1,2] and Paulo Jorge de O. P. Souza [1,2,*]

[1] Programa de Pós-Graduação em Agronomia (PGAGRO), Universidade Federal Rural da Amazônia (UFRA), Avenida Presidente Tancredo Neves, No. 2501 Terra Firme, Belém 66077-830, PA, Brazil; pontes.agro@gmail.com (D.P.F.); denisdepinho@agronomo.eng.br (D.P.S.); jvitorpinto@gmail.com (J.V.N.P.); deborahpires.agro@gmail.com (D.L.P.C.); eriikateixeira@hotmail.com (E.T.); hugo.pinheiro@ufra.edu.br (H.A.P.)

[2] Instituto Socioambiental e dos Recursos Hídricos (ISARH), Universidade Federal Rural da Amazônia (UFRA), Avenida Presidente Tancredo Neves, No. 2501 Terra Firme, Belém 66077-830, PA, Brazil; garibalde13@gmail.com (H.G.G.C.N.); adriano.souza@ufra.edu.br (A.M.L.S.)

[3] Faculdade de Agronomia, Universidade Federal do Pará (UFPA), Altamira 68372-040, PA, Brazil; viviandielly19@yahoo.com.br

[4] Agência de Defesa Agropecuária do Pará, Monte Alegre 68220-000, PA, Brazil; vandeilsonbelfort@hotmail.com

* Correspondence: paulo.jorge@ufra.edu.br; Tel.: +55-91-3254-9729

Citation: Ferreira, D.P.; Sousa, D.P.; Nunes, H.G.G.C.; Pinto, J.V.N.; Farias, V.D.S.; Costa, D.L.P.; Moura, V.B.; Teixeira, E.; Sousa, A.M.L.; Pinheiro, H.A.; et al. Cowpea Ecophysiological Responses to Accumulated Water Deficiency during the Reproductive Phase in Northeastern Pará, Brazil. *Horticulturae* **2021**, 7, 116. https://doi.org/10.3390/horticulturae7050116

Academic Editors: Stefania Toscano, Giulia Franzoni and Sara Álvarez

Received: 24 March 2021
Accepted: 11 May 2021
Published: 18 May 2021

Publisher's Note: MDPI stays neutral with regard to jurisdictional claims in published maps and institutional affiliations.

Abstract: Cowpea (*Vigna unguiculata* (L.) Walp.) is a leguminous species widely cultivated in northern and northeastern Brazil. In the state of Pará, this crop still has low productivity due to several factors, such as low soil fertility and climatic adversity, especially the water deficiency. Therefore, the present study aimed at evaluating the physiological parameters and the productivity of cowpea plants under different water depths. The experiment was conducted in Castanhal/Pará between 2015 and 2016. A randomized block design was applied with six replications and four treatments, represented by the replacement of 100%, 50%, 25% and 0% of the water lost during crop evapotranspiration (ETc), starting from the reproductive stage. The rates of net photosynthesis (A), stomatal conductance (gs), leaf transpiration (E_{leaf}), substomatal CO_2 concentration (Ci), leaf temperature (T_{leaf}) and leaf water potential (Ψ_w) were determined in four measurements at the R5, R7, R8 and R9 phenological stages. Cowpea was sensitive to the water availability in the soil, showing a significant difference between treatments for physiological variables and productivity. Upon reaching a Ψ_w equal to -0.88 MPa, the studied variables showed important changes, which allows establishing this value as a threshold for the crop regarding water stress under such experimental conditions. The different water levels in the soil directly influenced productivity for both years, indicating that the proper water supply leads to better crop growth and development, increasing productivity.

Keywords: *Vigna unguiculata* (L.) Walp; water deficiency; physiological parameters; productivity

1. Introduction

The cowpea production in Brazil is concentrated in northeastern and northern regions and it was introduced in the state of Pará by migrants from northeastern Brazil, with the Vigna genus responsible for 80% of the state production, generating more than 70 thousand direct jobs [1]. However, this crop still has low productivity in the Pará state, reaching approximately 821 kg ha^{-1} [2] as a function of several factors such as improper seed management, low soil fertility and climatic adversity, mainly the water deficiency [3,4].

Hayatu and Mukhtar [5] found that the effect of water stress on cowpea genotypes leads to a reduction in several components (chlorophyll fluorescence, chlorophyll content, specific leaf area and shoot biomass) and it is more severe when it occurs during reproductive stages and in treatments of severe water stress. Therefore, there is no physiological variable that indicates tolerance to water deficiency in isolation [6]. According to Martínez-Vilalta and Garcia-Forner [7], it is advisable to evaluate more than one variable, including water potential, stomatal conductance, and the temperature and transpiration of leaves, considered important to evaluate the responses of plant species to water stress.

As a manner of increasing crop productivity, reducing production costs, enhancing income for rural producers, reducing environmental damage and maximizing the natural resource use, it is essential to adopt technologies, as well as the suitable irrigation management [8]. The proper irrigation depth must be considered to ensure a good water supply, avoiding crop stress and favoring plant growth [2]. Thus, the more detailed knowledge on cowpea development in response to water depths is an important factor to generate low-cost technology with increased local production.

In northern Brazil, there is a demand for studies on the physiological and productive behavior of cowpea according to water limitation for purposes of management and increasing water use efficiency and grain productivity. Thus, the present study aimed to evaluate the ecophysiological parameters and the productivity of cowpea in response to water deficiency during the reproductive phase.

2. Material and Methods

The experiment was carried out in the municipality of Castanhal, located in northeastern state of Pará, Brazil, from 2015 to 2016, in area of approximately 0.5 hectares, situated at the experimental farm belonging to the Federal Rural University of the Amazon (UFRA) ($1°19'24''$ S Latitude, $47°57'38''$ W Longitude, 41 m Altitude). The climate of the experimental area is defined as Am according to Köppen's climatic classification, a tropical climate, showing a moderate dry season with average annual rainfall from 2000 to 2500 mm. The driest season occurs between June and November, while the rainiest season is from December to May.

A randomized block design was used, with six replications and four treatments, evaluating different levels of water availability in the soil, starting from the reproductive phase of cowpea. The experimental units consisted of plots with 22×24 m, separated by a 1 m border area, with a spacing of 0.5 m between planting lines and 0.1 m between plants, composing a density of 200,000 plants per hectare.

To identify the physical and chemical attributes of the soil in the experimental area, two collections were carried out at a depth of 0–20 cm, which corresponded to a large part of the effective depth explored by the cowpea root system [4]. The analyses were performed at the Soil Laboratory of the Brazilian Agricultural Research Corporation–Eastern Amazon (Belém, Pará, Brazil), and the results are expressed in Table 1.

Table 1. Physical and chemical attributes of the soil from the experimental area.

Attributes	2015	2016
pH (H_2O)	4.9	3.7
N (%)	0.05	0.0
P (mg dm^{-3})	2	20
K^+ (mg dm^{-3})	26	30
Na^{2+} (mg dm^{-3})	9	2
Ca^{2+} ($cmol_c$ dm^{-3})	0.5	1.0
$Ca^{2+}+Mg^{2+}$ ($cmol_c$ dm^{-3})	0.8	1.2
Al^{3+} ($cmol_c$ dm^{-3})	0.8	0.6
Sand (g kg^{-1})	835	835
Silt (g kg^{-1})	125	125
Clay (g kg^{-1})	40	40
Soil density (g cm^{-3})	1.56	1.56
Field capacity (m^3 m^{-3})	0.20	0.20
Permanent wilting point (m^3 m^{-3})	0.11	0.11

Sowing was carried out manually on 23 September 2015 and on 17 September 2016. The cultivar used was BR3-Tracateua, which is recommended for the region [1]. Fertilizations were performed according to the soil chemical analysis, applying 350 kg ha^{-1} of chemical fertilizer with NPK formulation 10-20-20 for the 2015 experiment, and 195 kg ha^{-1} of chemical fertilizer with NPK formulation 6-18-15 for the 2016 experiment. Fertilization and other management practices were carried out following technical recommendations for the crop in the region [3].

Four treatments were tested: T1—the replacement of 100% of the water (irrigation + rainfall) lost by the crop evapotranspiration (ETc); T2—replacement of 50% of the water lost by ETc; T3—replacement of 25% of the water lost by ETc; and T4—without replacement of the water lost by ETc. For the T4 treatment of each block, mobile covers of 100-micron transparent polypropylene were built, with 1.5 m height, aiming at preventing the entry of water through rainfall, starting from the reproductive stages and installed only in the rainfall period, during the day. The plastic cover was not installed during the night to avoid possible heating by the retention of long-wave radiation at night. Coincidentally, no rain events occurred during the night.

A drip irrigation system was used. To determine the water depth, the reference evapotranspiration (ET$_0$) was calculated using the Penman–Monteith equation [9] using data obtained from the meteorological station of the National Institute of Meteorology (INMET), installed 2 km from the experimental area. The ET$_0$ was multiplied by the crop coefficient (Kc) of each cowpea phase available in the literature [2] to obtain the maximum crop evapotranspiration.

During the vegetative phase, all treatments were kept close to the field capacity—that is, with replacement of 100% of the ETc. The differentiation of water depths in T2 and T3 treatments, as well as the interruption of irrigation in T4, occurred 36 days after sowing (DAS) for 2015 and 2016, when the crop reached the reproductive phase. Irrigation was interrupted 58 DAS in 2015 and 61 DAS in 2016, when the grain ripening phase (R9) was reached.

It is important to mention that, during the experimental period, the irrigation was interrupted when precipitation exceeded the daily ETc, aiming not to raise the soil moisture above the field capacity, controlling the entry of water into the soil and restarting the irrigation when soil moisture reached the value before the rain event. It was monitored using Time Domain Reflectometer sensors installed in each treatment.

In the center of the experimental area, an automatic meteorological station was installed for meteorological data collection, including air temperature, relative air moisture, volumetric water content in the soil and rainfall. All sensors were connected to a CR10× datalogger (Campbell Scientific, Inc., Logan, UT, USA), configured for reading every ten seconds, recording total and average values every ten minutes. To quantify the deficiencies caused by treatments subjected to water deficit, the sequential water balance was carried out according to Carvalho et al.; for more details, see in Nunes et al. [10]. Accumulated soil water deficiency was obtained as a cumulative difference between daily ETc and daily actual evapotranspiration.

The phenological stages of cowpea were monitored daily. In order to achieve this, the Geptz and Fernández scale was used. For each treatment, in all blocks, 1 m-long lines containing 10 plants were selected, which were monitored from plant emergency. The change regarding the phenological phase was characterized when 50% + 1 of the plants from the line showed the characteristics described by Farias et al. [1].

Productivity was measured 65 DAS in 2015 and 68 DAS in 2016, when 90% of the plants reached the R9 phenological stage. In both years, the productivity was determined considering two planting lines previously separated in each treatment, from where three samples of 1 m^2 were collected, represented by 2 m-long lines. After collection, the grains were dried for 72 h, weighed, and the production was estimated for each treatment.

Determinations of net photosynthetic rate (A), stomatal conductance (gs), leaf transpiration (E_{leaf}), substomatal CO$_2$ concentration (Ci), and leaf temperature (T_{leaf}) were performed between 8 and 11 h am using a portable, open-system infrared gas analyzer (model LI-6400 XT,

LI-COR Biosci. Inc., Lincoln, Nebraska, USA), set up to work with a constant photosynthetic photon flux density of 1500 µmol photons $m^{-2} s^{-1}$, with a CO_2 flow of 400 µmol mol^{-1}. Measurements were carried out at 7, 14, 21, and 28 days after treatment onset (DAT), corresponding, respectively, to the phenological stages R5, R7, R8 and R9.

Air temperature (Tair) and relative humidity (RH) followed environmental conditions. The average Tair during leaf gas exchange measurement intervals in 2015 (2016) were 27.5 ± 0.03 °C (29.8 ± 0.27 °C); 27.6 ± 0.04 °C (29.8 ± 0.36 °C); 29.4 ± 0.03 °C (30.4 ± 0.26 °C) and 29.2 ± 0.07 °C (30.9 ± 0.34 °C) at 7, 14, 21 and 28 DAT, respectively. The average RH under same conditions were $67.1 \pm 2.47\%$ ($69.4 \pm 1.03\%$); $62.9 \pm 1.82\%$ ($70.6 \pm 1.91\%$); $63.4 \pm 1.84\%$ ($66.9 \pm 1.17\%$) and $66.0 \pm 1.61\%$ ($63.4 \pm 1.85\%$) at 7, 14, 21 and 28 DAT, in 2015 (2016), respectively.

Two healthy and fully expanded leaflets from the medium portion of the third or fourth leaf from the apex were sampled for leaf gas exchange measurements. After that, the same leaflets were excised and immediately placed in a Scholander-type pressure chamber (model 3115, Soilmoisture equipement Copr., Santa Barbara, California, USA) for leaf water potential (Ψ_w) determination. Ecophysiological data collections followed the same randomized block design, with six replications and four treatments, evaluating different levels of water availability in the soil, starting from the reproductive phase of cowpea, including two samples per treatment in the six blocks, composed of 48 plants per collection.

Results were submitted to regression analysis, and the significance of the generated equations was verified according to the F test [4], considering them valid as long as they were greater than 95% of probability. Productivity and ecophysiological variables data were subjected to analysis of variance and the means were compared by Tukey's test at 5% probability, using ORIGIN PRO 8.0v software (OriginLab Corp., Northampton, MA, USA) [11].

3. Results

The experiment of 2015 was carried out under effect of an El Niño phenomenon [12]. However, the daily averages of meteorological data observed between September and November 2015 and 2016 showed similar patterns in air temperature (T_{air}), reference evapotranspiration (ET_0), global solar radiation (Sin), and vapor pressure deficit (VPD) variables (Figure 1). The average Tair values were 28.0 and 27.2 °C for 2015 and 2016, respectively. The average ET_0 values were 5.0 mm day^{-1} in 2015 and 4.9 mm day^{-1} in 2016. The Sin showed average values equal to 20.6 and 19.5 MJ m^{-2} day^{-1} for 2015 and 2016, respectively, while the VPD valus were equal to 0.96 kPa for 2015 and 0.93 kPa for 2016.

The total water blade applied for all treatments in 2016 was higher than those applied in 2015, which can be explained by the rainfall of 141.2 mm during the cowpea vegetative phase in 2016, while there was no rainfall in the same period in 2015, directly influencing the total number of irrigations (Table 2). However, when comparing only the reproductive phase (differentiation of treatments), the water blade was higher in 2015 than in 2016, since the precipitation values in this interval were 30.5 and 12.2 mm, respectively.

Data of soil moisture and precipitation for 2015 and 2016 are shown in Figure 2. For both years, soil moisture was controlled during the vegetative period, in order that all treatments had the same water availability. Despite the treatments receiving the same blade, there are small differences in the volumetric water content between them, which may be associated with the differences generated by the installation locations of the sensors.

From the reproductive phase, the volumetric soil water content varied between treatments, exhibiting an expected pattern. In 2015, T1 showed the highest volumetric soil water content, equal to 0.21 m^3 m^{-3}, followed by T2 with 0.18 m^3 m^{-3}, T3 with 0.16 m^3 m^{-3}, and T4 with 0.14 m^3 m^{-3}, while in 2016 the results were 0.22, 0.18, 0.16, and 0.12 m^3 m^{-3}, for T1, T2, T3 and T4, respectively.

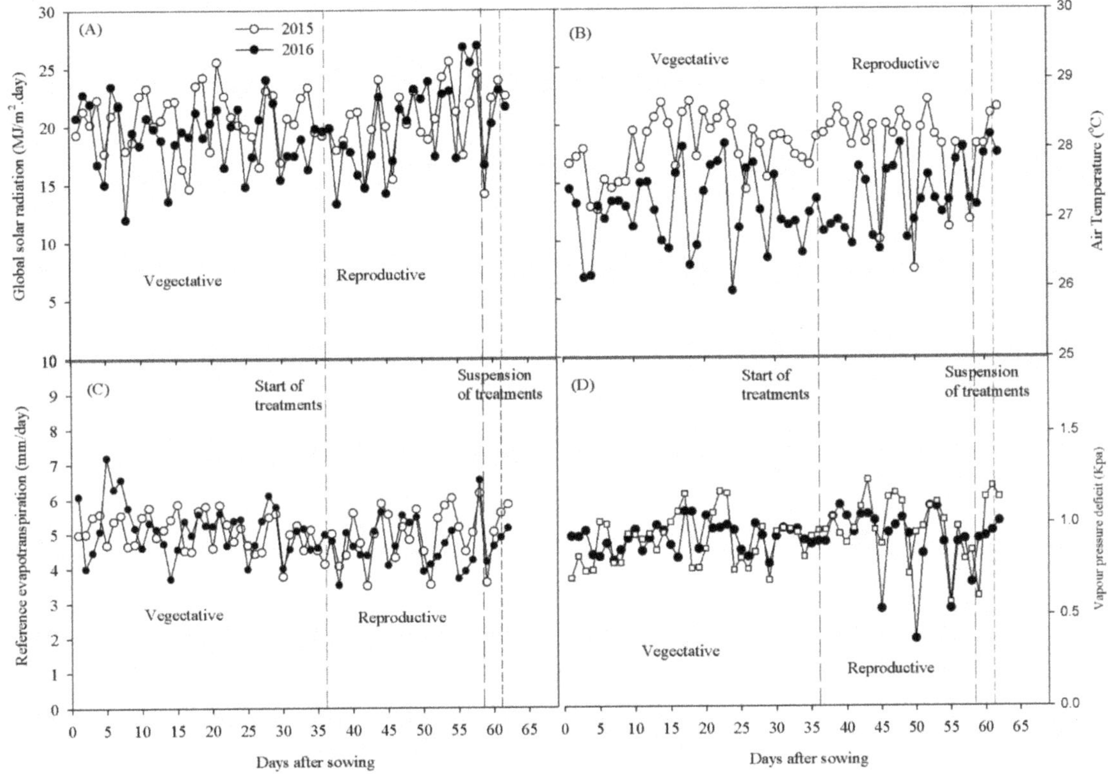

Figure 1. Meteorological conditions during the experimental period in 2015 and 2016. Global solar radiation (**A**), air temperature (**B**), reference evapotranspiration (**C**) and vapour pressure deficit (**D**).

Table 2. Water blade applied before the differentiation of treatments (irrigation + precipitation), water blade applied after the differentiation of treatments (irrigation + precipitation), total water blade during the experiment (total), number of irrigations (NI) and crop evapotranspiration (ETc) in 2015 and 2016.

Experiment	Treatments	Water Depth (mm)				Total	NI	Total ETc
		Vegetative Phase		Reproductive Phase				
		Irrigation	Rainfall	Irrigation	Rainfall			
2015	T1	173.83	0	113.45	30.47	317.75	58	308.23
	T2	173.83	0	56.73	30.47	261.03	58	241.03
	T3	173.83	0	28.36	30.47	232.66	58	207.45
	T4	173.83	0	0	0	173.83	35	173.83
2016	T1	87.64	141.18	113.81	12.19	354.82	40	304.98
	T2	87.64	141.18	56.09	12.19	297.10	40	241.05
	T3	87.64	141.18	28.45	12.19	269.46	40	209.09
	T4	87.64	141.18	0	0	228.82	17	177.12

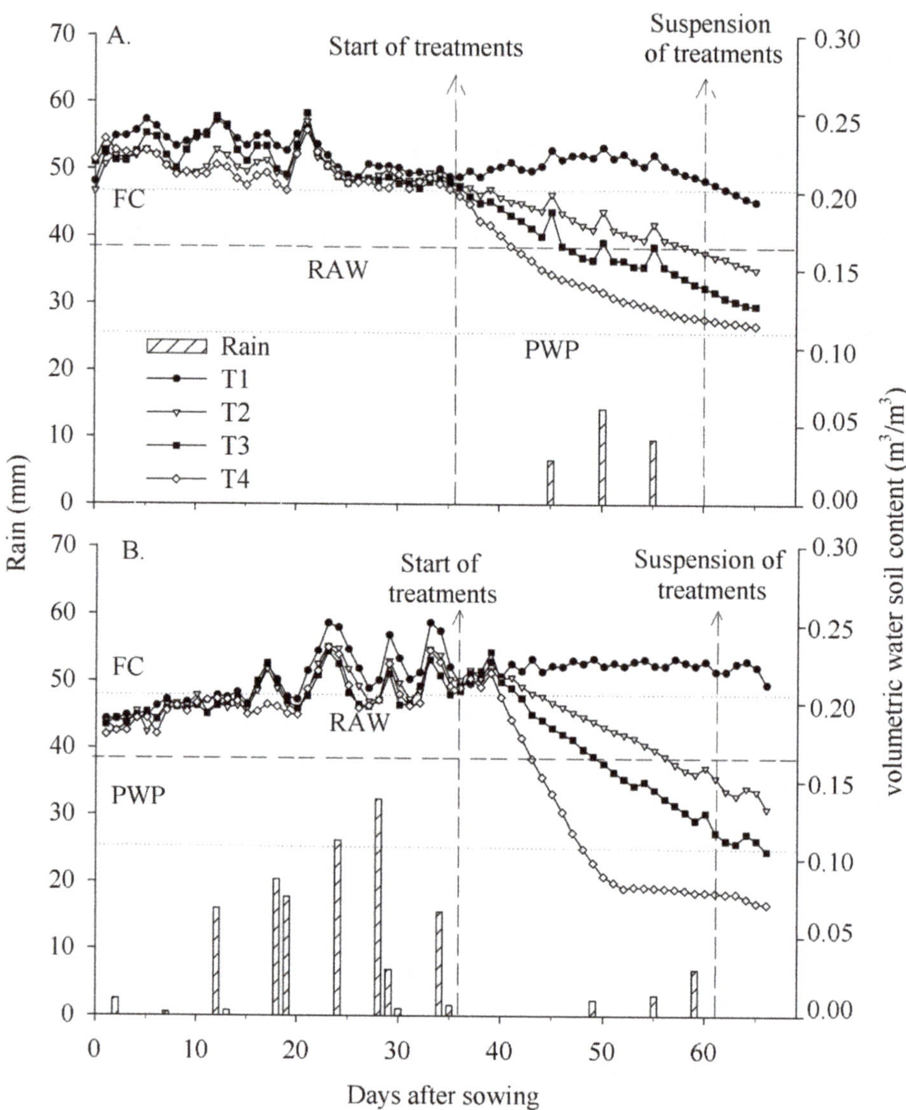

Figure 2. Volumetric soil water content and rain during the experimental period in 2015 (**A**) and 2016 (**B**). FC is the field capacity, PWP is the permanent wilting point, RAW is the readily available water.

When cowpea reached the R9 stage, irrigation was interrupted to reduce the contained grain moisture, accelerating the grain maturation process. At this time, the values of available soil water content were 108%, 54%, 33% and 3%, and 122%, 47%, 8% and 0% for T1, T2, T3, and T4 treatments in 2015 and 2016, respectively.

Using mobile covers, it was possible to control the entry of water into the soil in T4, which provided a greater change in volumetric soil water content for all treatments. In 2016, T4 reached 0.117 m^3 m^{-3} of volumetric soil water content at 47 DAS, close to the permanent wilting point (PWP), corresponding to 0% available soil water content. Ferreira et al. [13] report that each species differs regarding the response to soil moisture and that PWP in isolation is not a suitable criterion for establishing water availability to the plant.

Figure 3 shows the report of ANOVA (Tukey's test at 5% probability) applied for averages of Net photosynthetic rate (A), Leaf transpiration (E_{leaf}) Stomatal conductance (gs) and Substomatal CO_2 concentration (Ci) of cowpea in response to water deficiency under different reproductive phases during 2015 and 2016. It was noted that A, E_{leaf}, gs and Ci were higher ($p \leq 0.05$) in the absence of water deficiency regardless of the reproductive phase as well as the year evaluated. Considering that the T1 treatment represents a hypothetically ideal condition in terms of water availability, it is clear that throughout the reproductive phase the ecophysiological variables remained at close levels despite the phenological evolution, corroborating the hypothesis that the differences found over time were more related to the water deficiency factor than due to the natural aging of the plant.

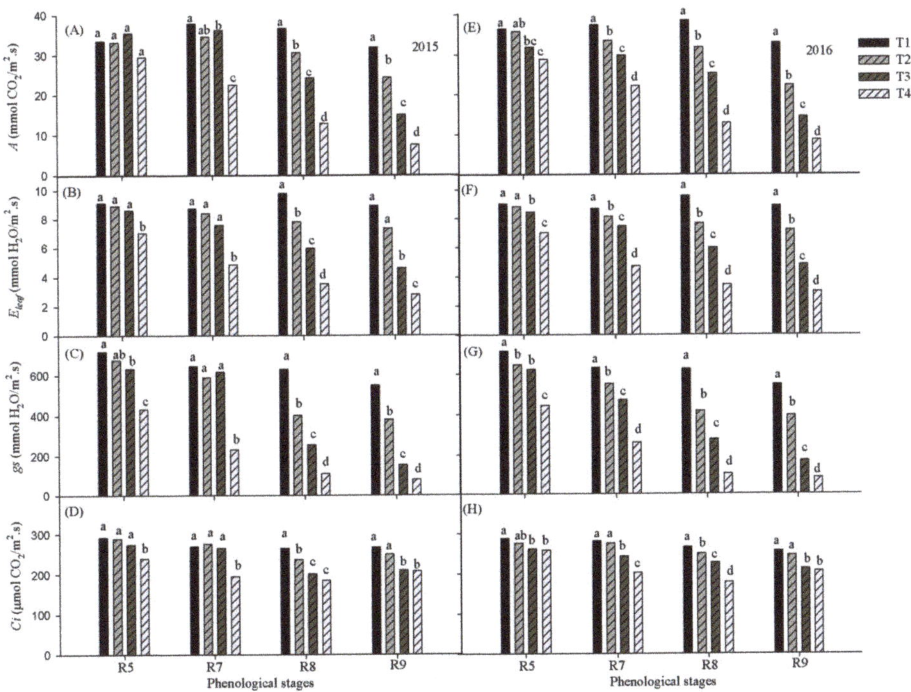

Figure 3. Net photosynthetic rate—A (**A**), Leaf transpiration—E_{leaf} (**B**), Stomatal conductance—gs (**C**) and Substomatal CO_2 concentration—Ci (**D**) of cowpea plants in differents reproductive phase in response to accumulated soil water deficiency in the 2015 and 2016 (**E–H**) experiments, respectively. Means followed by the same letter do not differ statistically between treatments (Tukey teste, $p \leq 0.05$).

The effects of accumulated water deficiency for the variables as a function of the imposed treatments were more significant ($p \leq 0.05$) after the reproductive phase R5, becoming more accentuated in phases R8 and R9, when all variables differed statistically between treatments ($p \leq 0.05$), corroborating that the water stress is more severe during the reproductive stage [5]. However, in the 2016 experiment, variables A, E_{leaf} and gs started to differ significantly ($p \leq 0.05$) in phase R7. Between phases R5 to R8, all ecophysiological variables showed, in treatments T2 and T3, respectively, values associated with leaf water potential levels (Ψ_w) greater and lesser than -0.8 MPa (data not shown).

The highest rate of A in response to water availability ($p \leq 0.05$) occurred in treatment T1 corresponding to 37.8 µmol CO_2 m^{-2} s^{-1} in phase R7 and 38.4 µmol CO_2 m^{-2} s^{-1} in phase R8, in 2015 and 2016, respectively. At the end of the experiment (phase R9), the rates of A, E_{leaf}, gs and Ci were reduced ($p \leq 0.05$) to 7.5 (8.5) µmol CO_2 m^{-2} s^{-1}; 2.8 (2.9) mmol

H_2O m^{-2} s^{-1}; 78.9 (82.8) mmol H_2O m^{-2} s^{-1}, 207.5 (206.9) mmol H_2O m^{-2} s^{-1} in the year 2015 (2016).

As there was no significant difference in the effect of water deficiency between the 2015 and 2016 experiments in the ecophysiological variables analyzed [14], data were grouped for regression analysis. Figure 4 shows leaf water potential (Ψ_w), Stomatal conductance (gs), Net photosynthetic rate (A), Substomatal CO_2 concentration (Ci), leaf transpiration (E_{leaf}) and leaf temperature (T_{leaf}) of cowpea over the reproductive phase under different water deficiency values. All variables' responses were fit to an exponential model (except leaf temperature) that could explain the effects of water deficiency on cowpea ecophysiology variables with high precision, especially E_{leaf}, A and gs ($R^2 > 0.9$).

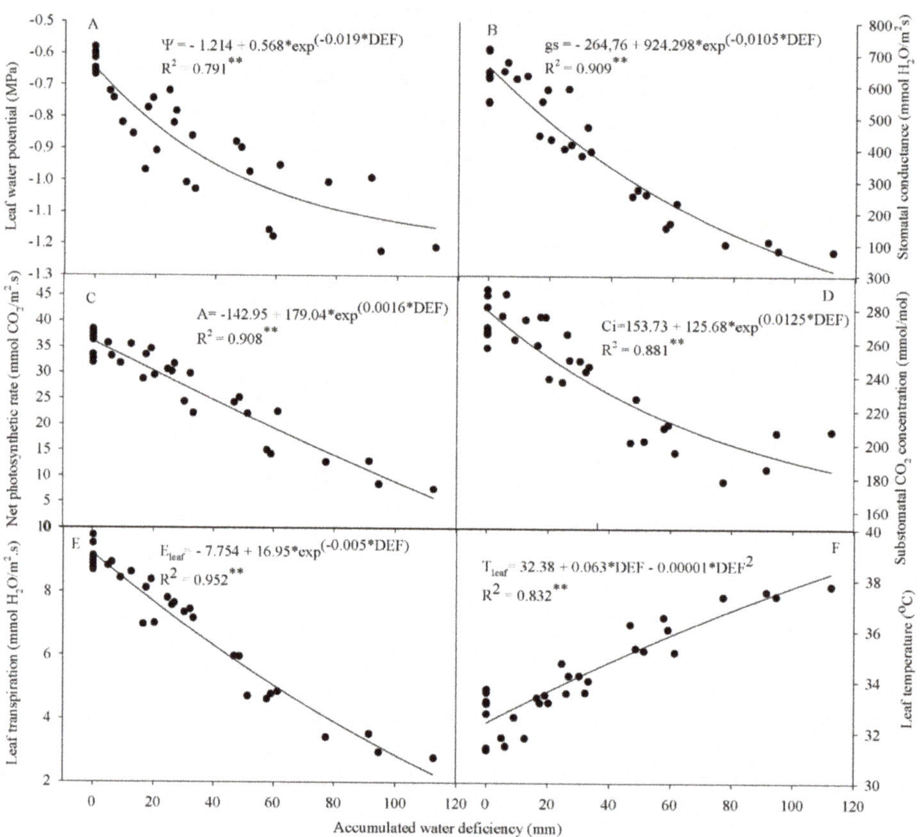

Figure 4. Leaf water potential—Ψ_w (**A**), stomatal conductance—gs (**B**), net photosynthetic rate—A (**C**), substomatal CO_2 concentration—Ci (**D**), leaf transpiration—E_{leaf} (**E**) and leaf temperature—T_{leaf} (**F**) of cowpea plants during the reproductive phase according to accumulated soil water deficiency in the 2015 and 2016 experiments, Castanhal, Pará. ** Significant by the F test ($p \leq 0.01$).

Ψ_w measurement results indicate that the highest values were found in T1 for both years, with averages close to −0.6 MPa (Figure 4A). The T4 treatment, which suffered the greatest influence from water deficiency among the treatments with interruption of irrigation, showed more negative values, reaching averages equal to −1.21 MPa in 2015 and −1.22 MPa in 2016 at the end of the experiment. The other treatments followed a natural trend for water availability in the soil, with averages of −1.01 and −1.03 MPa for T2 and −1.16 and −1.18 MPa for T3, in 2015 and 2016, respectively.

As expected, the Ψ_w was directly related to the soil moisture, decreasing according to increasing soil water deficiency (Figure 4A). The results are similar to those obtained by Dias and Bruggemann [15] that found leaf water potential to vary between -0.82 and -1.18 MPa, with a water deficit imposed in the reproductive phase, but differ from those obtained by Micheletto et al. [16], who observed Ψ_w values of -2.30 and -2.57 MPa, respectively, for common bean submitted to a water deficit. It is important to mention that the leaf water potential can vary according to the crop phenological stage, cultivar, water availability in the soil, vapor pressure deficit, and the time and place of recording [7].

For gs, as expected, lower average values were found in T4, showing 78.95 mmol H_2O m^{-2} s^{-1} in 2015 and 82.79 mmol H_2O m^{-2} s^{-1} in 2016, at the end of the cowpea crop cycle (Figure 4B). The T2 treatment presented 380.57 and 394.24 mmol H_2O m^{-2} s^{-1}, and T3 presented 153.74 and 168.31 mmol H_2O m^{-2} s^{-1}, in 2015 and 2016, respectively, at the last measurement date. These results indicate a direct relationship between gs and available water content—that is, the lower the water content in the plant, the lower the stomatal opening.

The T1 treatment, which was maintained close to the field capacity during the entire experimental period, showed a higher degree of stomatal opening than the other treatments, reaching average values of up to 716.17 mmol H_2O m^{-2} s^{-1} in 2015 and 712.20 mmol H_2O m^{-2} s^{-1} in 2016. Regarding the final average values of gs, there were reductions of 31% for T2, 72% for T3, and 86% for T4 in 2015, and 28% for T2, 69% for T3 and 85% for T4 in 2016.

In both years, there was a decline in the A of cowpea plants according to the advance of water deficiency (Figure 4C). T1 showed the highest rates during the experimental period, reaching averages of 31.91 and 32.75 µmol CO_2 m^{-2} s^{-1} in the last observation, while T4 reached only 7.57 and 8.49 µmol CO_2 m^{-2} s^{-1} at the end of the cycle in 2015 and 2016, respectively.

The A showed significant reductions throughout the experiment, following a exponential pattern (Figure 4C). Reductions of 24%, 53% and 76% for the T2, T3 and T4 treatments were observed at the end of the reproductive phase in 2015. In 2016, reductions of approximately 32%, 57% and 74% were found for T2, T3 and T4, compared to the T1 treatment, respectively.

In the two years of experimental conduction, for all evaluations, the values of Ci responded exponentially to the soil water supply, with higher values of Ci according to the greater water availability (Figure 4D). Thus, T4 always presented the lowest values, closely followed by T3, in relation to other treatments, showing 207.52 and 206.92 µmol CO_2 m^{-2} s^{-1} for T4, and 209.72 and 211.6 µmol CO_2 m^{-2} s^{-1} for T3, in 2015 and 2016, respectively, at the end of the reproductive phase. T1 and T2 also showed similar trends in 2015 and 2016, but with higher values, corresponding to 267.08 and 256.95 µmol CO_2 m^{-2} s^{-1}, and 249.32 and 245.98 µmol CO_2 m^{-2} s^{-1}, respectively.

The T1 treatment showed the highest average E_{leaf}, varying from 8.74 to 9.78 mmol H_2O m^{-2} s^{-1} in 2015 and from 8.68 to 9.53 mmol H_2O m^{-2} s^{-1} in 2016. The other treatments behaved as expected, with 7.36 and 7.17 mmol H_2O m^{-2} s^{-1} for T2, 4.63 and 4.78 mmol H_2O m^{-2} s^{-1} for T3, and 2.78 and 2.96 mmol H_2O m^{-2} s^{-1} for T4, in 2015 and 2016, respectively. With decreased water supply to the plant, E_{leaf} responded negatively, with reductions of 18%, 48% and 69% for T2, T3 and T4 in 2015, respectively, and 19% for T2, 46% for T3, and 67% for T4 in 2016, when compared to T1, in the last measurement.

With the stomatal closure to avoid water loss through transpiration, there was an increase in T_{leaf} (Figure 4B), which reduced the photosynthetic capacity due to the decrease in CO_2 inflow by reducing gs. In the present study, the increase in T_{leaf} between treatments reached a difference of up to 4 °C, with the highest T_{leaf} found for T4, equal to 37.77 °C in 2015 and 37.38 °C in 2016, while there was little variation ($\approx$2 °C) in T1 throughout the experimental period for the two years studied. The other treatments followed a natural trend for soil water availability, with final average values of 34.23 and 34.02 °C for T2 and 36.54 and 36.06 °C for T3, in 2015 and 2016, respectively.

When the Ψ_w reached -0.88 MPa (related to treatment T3), the analyzed variables significantly decreased (Figure 3), except leaf temperature, which increased (Figure 5A–D), corresponding to 60% for *gs*, 34% for *A*, and 39% for E_{leaf}, in addition to an increase of 3 °C in T_{leaf}. In both experimental years, the response of echopysiological variables was similar, with a reduction (increase) in such variables *A*, *gs*, E_{leaf} (T_{leaf}) according to the increase in water deficiency. Similar water deficiency effects were found by Rivas et al. [17], studying the tolerance of cowpea to water deficit, where by reducing the irrigation depth, there were reductions of 75% in Ψ_w, 83% in *gs* and an increase of 3.57% in T_{leaf}, also corroborating the results obtained by Medrano et al. [18], Dias and Bruggemann [15], Singh and Reddy [19] and Hayatu and Mukhtar [5]. Such results suggest this Ψ_w value as the threshold for water stress of cowpea in response to the water deficiency imposed.

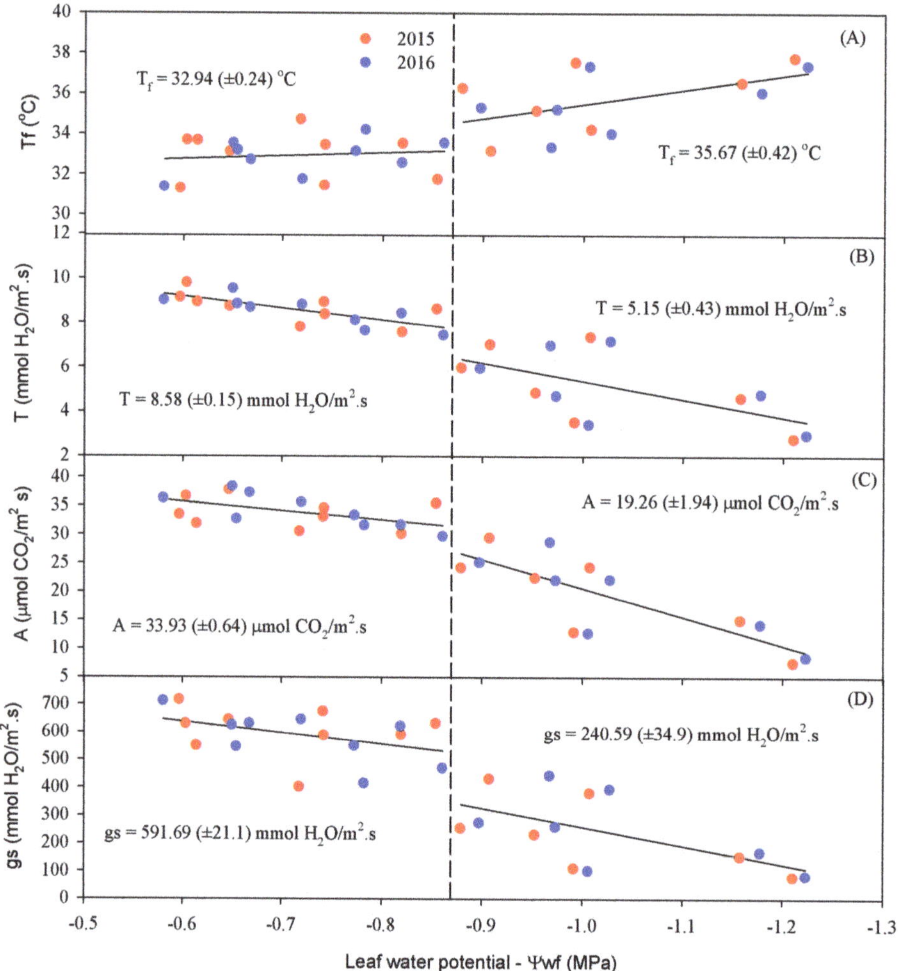

Figure 5. Leaf temperature—T_{leaf} (**A**), leaf transpiration—E_{leaf} (**B**), net photosynthetic rate—*A* (**C**) and stomatal conductance—*gs* (**D**) of cowpea plants according to the leaf water potential (Ψ_w) in the 2015 (red) and 2016 (blue) experiments, Castanhal, Pará.

In both years, the cowpea grain productivity proved that the differentiated soil water availability directly influenced production, as the greater the deficiencies caused by the treatments, the lower the final values of grain weight (Figure 6). The average values of productivity were 1474 (T1), 1098 (T2), 943 (T3), and 638 kg ha^{-1} (T4) in 2015, and 1597 (T1), 1295 (T2), 1069 (T3), and 684 kg ha^{-1} (T4) in 2016. Ainsworth and Rogers [20] report that plants submitted to water deficiency show less stomatal opening and a reduction in the interval that stomata remain open. In 2015, the decrease in productivity of the treatments was 25% for T2, 36% for T3 and 57% for T4. In 2016, these were 19%, 33% and 57%, for T2, T3 and T4, respectively.

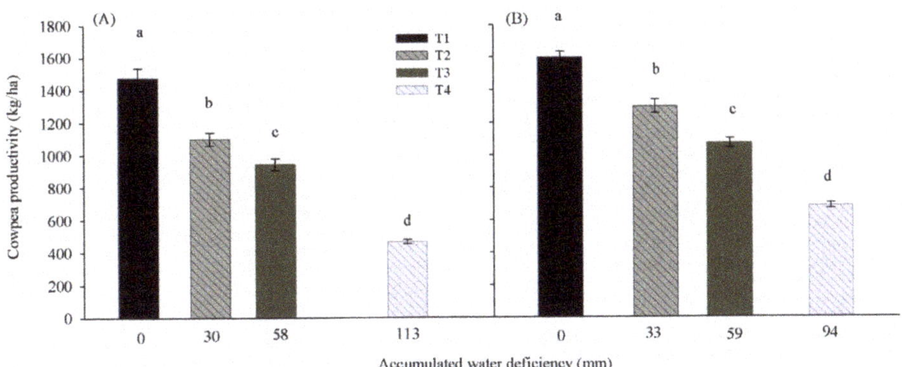

Figure 6. Productivity of cowpea plants in the 2015 (**A**) and 2016 (**B**) experiments, Castanhal, Pará. (Bars represent standard error.) Columns followed by the same letter do not differ statistically between treatments (Tukey teste, $p \leq 0.05$)

4. Discussion

The total water blade applied in T1 (100% of ETc replacement) was sufficient to supply the water demand of cowpea during both years (317.75 and 354.82 mm in 2015 and 2016, respectively), since according to Farias et al. [1] the total crop water consumption is approximately 267.73 ± 10.21 mm for the cultivar under the local conditions. Studying the response of different genotypes of cowpea for drought tolerance, Bastos et al. [21] found reduced plant production with 50% of available water, causing considerable changes in production components according to water availability.

In a study on cowpea plants submitted to water deficiency, reductions in Ψ_w, gs, and E_{leaf} were also observed, with increased diffusive resistance to water vapor by stomatal closure, reducing transpiration and CO_2 supply for photosynthesis [22]. Taiz and Zeiger [23] report that inhibitions of growth and photosynthesis occur in plants suffering from water deficiency during drought periods, indicating that the sensitivity of certain physiological processes to water deficiency is a consequence of the plant strategy to deal with the water availability variation.

The reduction observed in gs is explained by Ainsworth and Rogers [20], who report that plants keep their stomata closed to avoid water losses under conditions of water stress, with a certain turgor (higher water potential) maintained in some species, which is an important characteristic of drought tolerance as observed in studies with C_3 plants and on cowpea by Medrano et al. [18] and Singh and Reddy [19].

According to Taiz and Zeiger [23], gs is recognized as the variable that suffers the greatest influence, as it is controlled by the stomatal opening and closure, mainly according to environmental conditions, such as vapor pressure deficit, relative air moisture, and global solar irradiation. It is possible to observe that several factors can influence the stomata, but the effect of stress caused by water deficit on this parameter is evident, which can be used as an indicator of water deficiency [18].

The exponential decrease in the A occurred possibly as a result of the decline in gs, as the stomatal opening is the main responsible for the entry and exit of gases in the plant and affects the process of photosynthetic gain by controlling the CO_2 inflow [20]. The reduction in photosynthetic activity by reducing the CO_2 assimilation is an indicator of the water deficiency effect [24].

According to Ainsworth and Rogers [20], the greater the lack of water in the plant, the lower the stomatal opening degree and, consequently, the greater the resistance to the entry of atmospheric CO_2; therefore, the intrafoliar concentrations of CO_2 tend to decrease substantially. Souza et al. [22], evaluating the water relations and gas exchange of cowpea submitted to different irrigation managements, found that the reduction in irrigation leads to linear reductions in the plant photosynthetic rate.

The response pattern of the Ci as a function of accumulated soil water deficiency demonstrates that the Ci reduction occurs with decreased availability of water to the plant. According to Matthews et al. [24], the increase in Ci values is usually followed by increases in gs. Under such conditions, there is an increase in the activity of ribulose-1.5-bisphosphate carboxylase-oxygenase (rubisco), increasing CO_2 consumption [25]. The increase in the Ci under low water availability, as seen in Figure 4D, may be related to a decrease in the enzymatic activity involved in the CO_2 fixation process. Zhao et al. [26] suggest that an increase in the Ci associated with a decrease in gs, in conditions of water deficiency, may indicate a decrease in carboxylation efficiency. Therefore, when there is an increase in the Ci, with a decrease in gs, the decrease in A indicates that this limitation is not only due to the increase in stomatal resistance, but also to the effect of water stress on photosynthesis.

As observed in previous variables in the present study, with increased soil water deficiency, the transpiration levels decreased, a mechanism used by the plant to prevent dehydration through transpiration [23]. The E_{leaf} control by stomata is a mechanism used by many species to restrict water loss and overcome drought periods [7]. According to Medrano et al. [18], the maintenance of T_{leaf} equal or slightly lower than the environment indicates the cooling capacity of plants, through the maintenance of the gs, aiming at keeping the plant protected from very high thermal ranges. According to Lin et al. [27], E_{leaf} has important effects in tropical plants, such as cooling the leaf, since in order to evaporate the water from the leaf, the water removes its thermal energy, reducing the T_{leaf} from 2 to 3 °C compared to the air temperature. Singh and Reddy [19] found significant reductions in T_{leaf} in cowpea plants submitted to water deficiency during the vegetative and reproductive phases.

The increase in T_{leaf} showed a direct relationship with the water availability in the soil, with higher T_{leaf} values according to the greater water deficiency. The increase in T_{leaf} as a function of water stress can be explained by the reduction in loss of latent heat through transpiration, which usually reduces under these conditions [27]. Data from previous studies show that T_{leaf} is usually higher than the air temperature under dry conditions, resulting in an increase in the leaf/environmental temperature ratio [19].

The lower productivity was due to the reduced soil water supply. Bastos et al. [21] observed a reduction of 60% in the productivity levels (grain yields) studying cowpea genotypes under water deficiency. Similar results were found by Hayatu and Mukhtar [5] and Nunes et al. [10] that found reductions of 21.38%, 36.12% and 49.50% in treatments under water restriction. Souza et al. [4] found significant reductions in the productivity of this cultivar when it was submitted to rainfed conditions, corresponding to a reduction of 41% under water deficiency of 26 mm, and 72% under a deficiency equal to 76 mm.

According to Hetherington and Woodward [28], since gs has the function of regulating gas exchange, it also has great affinity for the photosynthetic process, participating directly in plant growth and development. In this sense, the plant suffers a reduction in gs and E_{leaf} and, as a consequence, there is an increase in T_{leaf} and a reduction in final productivity [23]. Under such conditions, the reduction in leaf area of plants as well limits productivity due to the decreased light interception and CO_2 absorption [29].

5. Conclusions

The reduction in soil water content reduces the water potential and gas exchange of the plant, reducing crop productivity.

The irrigation depth of T1 showed the best result for all the studied variables. Therefore, it is the most indicated for cowpea (BR3-Tracuateua) under the climatic conditions of northeast Pará.

The value of -0.88 MPa was established as threshold water potential, from which the water deficiency causes negative effects for cowpea (BR3-Tracuateua) grown under the climatic conditions of northeastern Pará.

Author Contributions: Cowpea experiment: P.J.d.O.P.S., D.P.F., D.P.S., H.G.G.C.N., J.V.N.P., V.D.S.F., and V.B.M.; data organization: V.D.S.F., E.T., D.L.P.C., and V.B.M.; data quality control: J.V.N.P., H.G.G.C.N., and D.L.P.C.; data analysis: D.P.F., D.P.S., H.G.G.C.N., and V.D.S.F., writing—original draft preparation: D.P.F. and P.J.d.O.P.S.; writing—review and editing: P.J.d.O.P.S., V.D.S.F., H.G.G.C.N., A.M.L.S., and H.A.P.; supervision: P.J.d.O.P.S.; project administration, P.J.d.O.P.S. and A.M.L.S.; funding acquisition, P.J.d.O.P.S. and A.M.L.S. All authors have read and agreed to the published version of the manuscript.

Funding: This research was funded by the National Council for Scientific and Technological Development (CNPq) through the Universal project (process n° 483402/2012-5).

Institutional Review Board Statement: Not applicable.

Informed Consent Statement: Not applicable.

Acknowledgments: The Federal Rural University of Amazon are acknowledged for the structural support during the experiment, as is Hugo Pinheiro (UFRA) for the availability of the IRGA and the Scholander pump, as well as the graduate program in agronomy (PGAGRO) for the support offered to students.

Conflicts of Interest: The authors declare no conflict of interest. The funders had no role in the design of the study; in the collection, analyses, or interpretation of data; in the writing of the manuscript, or in the decision to publish the results.

References

1. Farias, V.D.S.; Lima, M.J.A.; Nunes, H.G.G.C.; Sousa, D.P.; Souza, P.J.O.P. Water demand, crop coefficient and uncoupling factor of cowpea in the eastern amazon. *Rev. Caatinga* **2017**, *30*, 190–200. [CrossRef]
2. Nunes, H.G.G.C.; Farias, V.D.S.; Sousa, D.P.; Costa, D.L.P.; Pinto, J.V.N.; Moura, V.B.; Teixeira, E.O.; Lima, M.J.A.; Ortega-Farias, S.; Souza, P.J.O.P. Parameterization of the AquaCrop model for Cowpea and assessing the impact of sowing dates normally used on yield. *Agric. Water Manag.* **2021**, *252*, 106880–106914. [CrossRef]
3. Lima, J.V.; Lobato, A.K.S. Brassinosteroids improve photosystem II efficiency, gas exchange, antioxidant enzymes and growth of cowpea plants exposed to water deficit. *Physiol. Mol. Biol. Plants* **2017**, *23*, 59–72. [CrossRef]
4. Souza, P.J.O.P.; das Farias, V.D.; de Lima, M.J.A.; Ramos, T.F.; de Sousa, A.M.L. Cowpea leaf area, biomass production and productivity under different water regimes in Castanhal, Pará, Brazil. *Rev. Caatinga* **2017**, *30*, 748–759. [CrossRef]
5. Hayatu, M.; Mukhtar, F.B. Physiological responses of some drought resistance cowpea genotypes (*Vigna unguiculata* (L.) Walp) to water stress. *BAJOPAS* **2010**, *3*, 69–75.
6. Blanco-Cipollone, F.; Lourenço, S.; Silvestre, J.; Conceição, N.; Moñino, M.J.; Vivas, A.; Ferreira, M.I. Plant Water Status Indicators for Irrigation Scheduling Associated with Iso- and Anisohydric Behavior: Vine and Plum Trees. *Horticulturae* **2017**, *3*, 47. [CrossRef]
7. Martínez-Vilalta, J.; Garcia-Forner, N. Water potential, stomatal behaviour and hydraulic transport under drought: Deconstructing the iso/anisohydric concept. *Plant Cell Environ.* **2017**, *40*, 962–976. [CrossRef] [PubMed]
8. Alvino, A.; Ferreira, M.I.F.R. Refining Irrigation Strategies in Horticultural Production. *Horticulturae* **2021**, *7*, 29. [CrossRef]
9. Allen, R.G.; Pereira, L.S.; Howell, T.A.; Jensen, M.E. Evapotranspiration information reporting: I. Factors governing measurement accuracy. *Agric. Water Manag.* **2011**, *98*, 899–920. [CrossRef]
10. Nunes, H.G.G.C.; Sousa, D.P.; Moura, V.B.; Ferreira, D.P.; Pinto, J.V.N.; Vieira, I.C.O.; Silva, V.D.S.; Oliveira, E.C.; Souza, P.J.O.P. Perfomance of the AquaCrop model in the climate risk analysis and yield prediction of cowpea (*Vigna unguiculatta* L. Walp). *Aust. J. Crop Sci.* **2019**, *13*, 1105–1112. [CrossRef]
11. Souza, P.J.O.P.; Farias, V.D.; Pinto, J.V.N.; Nunes, H.G.G.C.; de Souza, E.B.; Fraisse, C.W. Yield gap in cowpea plants as function of water deficits during Reproductive stage. *Rev. Bras. Eng. Agríc. Ambiental* **2020**, *24*, 372–378. [CrossRef]
12. Pascolini-Campbell, M.; Zanchettin, O.; Bothe, O.; Timmreck, C.; Matei, D.; Jungclaus, J.H.; Graf, H.F. Toward a record of central Pacific El Niño Events since 1880. *Theor. Appl. Climatol.* **2015**, *119*, 379–389. [CrossRef]
13. Ferreira, R.O.; Souza, L.d.S.; do Nascimento, M.N.; Silveira, F.G. Permanent wilt point from two methods for different combinations of citrus rootstock. *Cienc. Rural* **2020**, *50*, e20190074. [CrossRef]

14. Souza, P.J.O.P.; Ferreira, D.P.; Sousa, D.P.; Nunes, H.G.G.C.; Barbosa, A.V.C. Gas exchange of cowpea cultivated in Norteast of Pará in response to imposed water deficit during Reproductive phase. *Rev. Bras. Meteorol.* **2020**, *35*, 13–22. [CrossRef]

15. Dias, M.C.; Bruggemann, W. Limitations of photosynthesis in Phaseolus vulgaris under drought stress: Gas exchange, chlorophyll fluorescence and Calvin cycle enzymes. *Photosynthetica* **2010**, *48*, 96–102. [CrossRef]

16. Micheletto, S.; Rodriguez-Uribe, L.; Hernandez, R.; Richins, R.D.; Cury, J.; O'Connell, M.A. Comparative transcript profiting in roots of Phaseolus acutifolus and P. vulgaris under water deficit stress. *Plant Sci.* **2007**, *173*, 510–520. [CrossRef]

17. Rivas, R.; Falcão, H.M.; Ribeiro, R.V.; Machado, E.C.; Pimentel, C.; Santos, M.G. Drought tolerance in cowpea species is driven by less sensitivity of leaf gas exchange to water deficit and rapid recovery of photosynthesis after rehydration. *S. Afr. J. Bot.* **2016**, *103*, 101–107. [CrossRef]

18. Medrano, H.; Escalona, J.M.; Bota, J.; Gulías, J.; Flexas, J. Regulation of photosynthesis of C3 plants in response to progressive drought: Stomatal conductance as a reference parameter. *Ann. Bot.* **2002**, *89*, 895–905. [CrossRef] [PubMed]

19. Singh, S.K.; Reddy, K.R. Regulation of photosynthesis, fluorescence, stomatal conductance and water-use efficiency of cowpea (*Vigna unguiculata* [L.] Walp.) under drought. *J. Photochem. Photobiol. C* **2011**, *105*, 40–50. [CrossRef]

20. Ainsworth, E.A.; Rogers, A. The response of photosynthesis and stomatal conductance to rising [CO_2]: Mechanisms and environmental interactions. *Plant Cell Environ.* **2007**, *30*, 258–270. [CrossRef]

21. Bastos, E.A.; Nascimento, S.P.; Silva, E.M.; Freire Filho, F.R.; Gomide, R.L. Identification of cowpea genotypes for drought tolerance. *Rev. Ciênc. Agron.* **2011**, *42*, 100–177. [CrossRef]

22. Souza, R.P.; Machado, E.C.; Silva, J.A.B.; Lagôa, A.M.M.A.; Silveira, J.A.G. Photosynthetic gas exchange, chlorophyll fluorescence and some associated metabolic changes in cowpea (*Vigna unguiculata*) during water stress and recovery. *Environ. Exper. Bot.* **2004**, *51*, 45–56. [CrossRef]

23. Taiz, L.; Zeiger, E. *Plant Physiology*, 3rd ed.; Sinauer Associates: Sunderland, UK, 2002.

24. Matthews, J.S.A.; Vialet-Chabrand, S.R.M.; Lawson, T. Diurnal Variation in Gas Exchange: The Balance between Carbon Fixation and Water Loss. *Plant Physiol.* **2017**, *174*, 614–623. [CrossRef]

25. Parry, M.A.J.; Keys, A.J.; Madgwick, P.J.; Carmo-Silva, A.E.; Andralojc, P.J. Rubisco regulation: A role for inhibitors. *J. Exp. Bot.* **2008**, *59*, 1569–1580. [CrossRef] [PubMed]

26. Zhao, W.; Liu, L.; Shen, Q.; Yang, J.; Han, X.; Tian, F.; Wu, J. Effects of Water Stress on Photosynthesis, Yield, and Water Use Efficiency in Winter Wheat. *Water* **2020**, *12*, 2127. [CrossRef]

27. Lin, H.; Chen, Y.; Zhang, H.; Fu, P.; Fan, Z. Stronger cooling effects of transpiration and leaf physical traits of plants from a hot dry habitat than from a hot wet habitat. *Funct. Ecol.* **2017**, *31*, 2202–2211. [CrossRef]

28. Hetherington, A.M.; Woodward, F.I. The role of stomata in sensing and driving environmental change. *Nature* **2003**, *424*, 901–908. [CrossRef]

29. Slaterry, R.A.; Ort, D.R. Perspectives on improving light distribution and light use efficiency in crop canopies. *Plant Physiol.* **2021**, *185*, 34–48. [CrossRef]

Review

Biochemical, Physiological, and Molecular Aspects of Ornamental Plants Adaptation to Deficit Irrigation

Maria Giordano [1], Spyridon A. Petropoulos [2], Chiara Cirillo [1] and Youssef Rouphael [1,*

[1] Department of Agricultural Sciences, University of Naples Federico II, Via Università 100, 80055 Portici, Italy; maria.giordano@unina.it (M.G.); chiara.cirillo@unina.it (C.C.)
[2] Department of Agriculture Crop Production and Rural Environment, University of Thessaly, Fytokou Street, N. Ionia, 38446 Magnissia, Greece; spetropoulos@uth.gr
* Correspondence: youssef.rouphael@unina.it

Abstract: There is increasing concern regarding global warming and its severe impact on the farming sector and food security. Incidences of extreme weather conditions are becoming more and more frequent, posing plants to stressful conditions, such as flooding, drought, heat, or frost etc. Especially for arid lands, there is a tug-of-war between keeping high crop yields and increasing water use efficiency of limited water resources. This difficult task can be achieved through the selection of tolerant water stress species or by increasing the tolerance of sensitive species. In this scenario, it is important to understand the response of plants to water stress. So far, the response of staple foods and vegetable crops to deficit irrigation is well studied. However, there is lack of literature regarding the responses of ornamental plants to water stress conditions. Considering the importance of this ever-growing sector for the agricultural sector, this review aims to reveal the defense mechanisms and the involved morpho-physiological, biochemical, and molecular changes in ornamental plant's responses to deficit irrigation.

Keywords: ornamental species; water deficit; water stress; defense mechanisms; climate change; stress responsive genes; stress adaptation

Citation: Giordano, M.; Petropoulos, S.A.; Cirillo, C.; Rouphael, Y. Biochemical, Physiological, and Molecular Aspects of Ornamental Plants Adaptation to Deficit Irrigation. *Horticulturae* **2021**, *7*, 107. https://doi.org/10.3390/horticulturae 7050107

Academic Editor: Alessandra Francini

Received: 26 April 2021
Accepted: 6 May 2021
Published: 10 May 2021

Publisher's Note: MDPI stays neutral with regard to jurisdictional claims in published maps and institutional affiliations.

1. Introduction

Climate change refers to anomalous atmospheric conditions, as well as sudden unexpected climatic events, such as floods, hurricanes, intense and/or prolonged drought, extreme temperatures, etc. Drought is among the environmental stressors that has the most severe impact on crops throughout the world [1–3]. One-third of arable lands are already defined as arid or semi-arid ones [4], and the severity of drought shows increasing trends [5] since a 5 °C increase in mean air temperature is expected in the following years [6–10]. According to experts, the drylands on Earth will increase by 30% and the drier summers and reduced rainfall are expected to affect mostly Asian mid-continental regions, southern Europe, Northern and South Africa [11]. The reduction of usable water sources and the continuous demographic growth make it necessary to improve water use efficiency in the farming sector in order to ensure food security for the years to come. A big step towards this goal has been made by the introduction of soilless cropping systems, where the use of irrigation water is under continuous control [12]. However, the appropriate supply of water to crops, even in soilless conditions, requires the monitoring of various parameters, such as the growth substrate humidity, the climatic and microclimatic conditions, and most importantly, the water status of plants [13], which is more complex to quantify than climatic and growth substrate related parameters [14]. Furthermore, there may be differences between species or even cultivars of the same species in terms of water stress, especially under deficit irrigation conditions where a genotype dependent response is observed. Scientists are looking for mechanisms that regulate the response of plants to water stress, aiming to either identify the most tolerant species or increase tolerance in the sensitive ones. For this purpose, genetic studies are based on breeding and genetic engineering of model plants, such as *Arabidopsis thaliana* [1,15,16], so that the obtained responses could be extrapolated to other crops such as staple food, medicinal, aromatic, and fiber plants. The efficient use of water is a crucial point in cultivating ornamental plants which have to respond to different needs, e.g., moderate use of natural resources, climate change, environmental pollution, increasing production costs, and maximizing profits [17,18]. Unfortunately, there is still no standard protocol for the irrigation of ornamental species, and water requirements of plants are covered based on growers' personal experience [14,19].

Knowing the response of different species to water stress conditions would allow the identification of morphological indices and biochemical markers useful for distinguishing sensitive and tolerant species to water deficit stress [20–22]. Therefore, in this review, the morphological, biochemical, physiological, and molecular responses of the main ornamental plants cultivated throughout the world have been studied. Moreover, a literature update regarding the genes involved in ornamental plants' response to water stress is also presented and discussed.

2. The Effect of Deficit Irrigation on Morphology, Growth, and Quality of Ornamental Plants

The growth and morphology of ornamental plants have an aesthetic value and are very important parameters which guide the consumer's choice. The effects of deficit irrigation on the leaf are related to orientation changes, to reduction of leaf area and leaves number, to reduction of trichomes and canopy area, and to increase in leaf thickness as plant responses to avoid water losses [23–26] (Figure 1, Table 1).

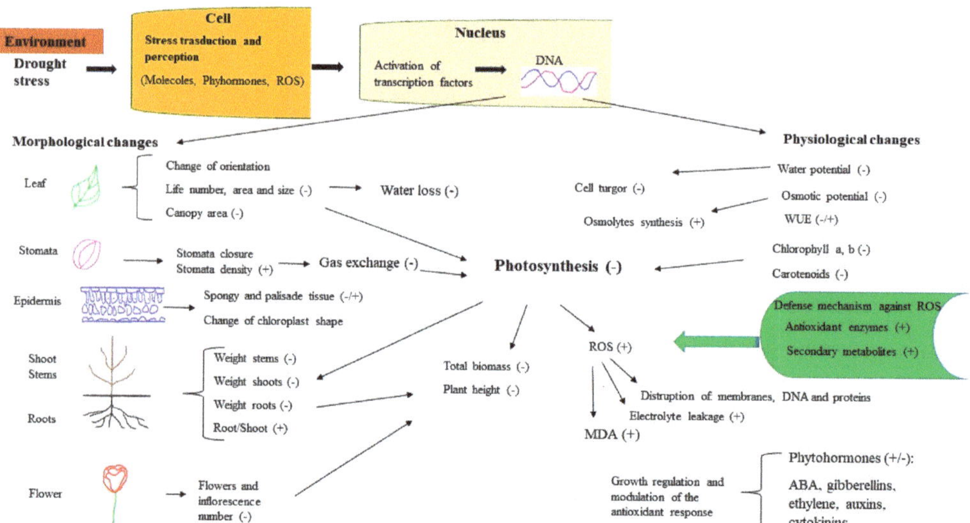

Figure 1. Water stress-induced morphological and physiological changes. (−) reduction due to water stress; (+) increase due to water stress. abscisic acid (ABA); malondialdehyde (MDA); reactive oxygen species (ROS); water use efficiency (WUE).

Lantana and *Ligustrum*, two important ornamental plants of the Mediterranean area, showed an increase in spongy and palisade tissue, following severe water stress [24]. The change in the leaf anatomy serves to increased diffusion of CO_2 from the external atmosphere to the spaces between cells [25,27], while thicker leaves presented higher chlorophyll content and photosynthetic activity [27]. Therefore, these responses related to leaf anatomy constitute an avoidance mechanism to reduce water losses.

Water stress has an impact on the morphology of *Chrysanthemum morifolium* Ramat cv. Hj inflorescences, an ornamental plant characterized by ray and disc florets [28]. The reduction of soil moisture reduces the number and shape of ray florets, while the number of disc florets increases. In *Callistemon citrinus*, the number of inflorescences did not change under moderate stress but reduced when severe stress was implemented [14]. Avoidance mechanisms are also evident in *Viburnum opulus* L. and *Photinia × fraseri*, two Mediterranean species which show alterations of leaf parameters under both moderate (60% evapotranspiration [ET]) and severe (30% ET) water stress conditions [27]. The changes in leaf parameters depend on the intensity of water deficit as well as on the genotype.

Reduction in leaf thickness in terms of epidermal thickness, palisade, and spongy tissue, and higher stomatal density have been associated with greater water stress sensitivity in *Passiflora alata* plants [29], whereas *Passiflora setacea* has shown fewer leaf anatomical and is considered more tolerant to deficit irrigation. Moreover, deficit irrigation may change the shape of chloroplasts in *Paeonia ostii* plants, e.g., from an oval shape in control plants to a more rounded shape in stressed plants [30]. All the above-mentioned examples reveal the diversity in plants' responses to water stress related to leaf parameters and highlight the complexity of the defense mechanisms against water stress.

Growth reduction is one of the first manifestations that plants are subjected to with water stress. For example, the application of water stress for one, two, or three weeks decreased the growth of poinsettia (*Euphorbia pulcherrima*) in terms of plant height (67.4, 57.0, and 49.0 cm, respectively) and leaf area (2.91, 1.22, and 0.93 cm^2, respectively) [18]. In addition, *Rosa damascena* Mill., a rose from Damascus which is widespread all over the world for its perfume and use in cosmetics and medicine, was subjected for 90 days to 100% of field capacity (FC), moderate water stress (50% FC), and severe water stress (25% FC) [10]. On the other hand, the number of leaves was not reduced by stress, so the reduction in aerial biomass was mainly attributed to a reduction in leaf area [31].

Antirhinum majus cv. Butterfly is an ornamental plant widely used to beautify urban areas and gardens, which also responds to water stress with a reduction of plant growth parameters (leaves, shoots, flowers), as well as with changes in plant nutritional status (the content of N, P, K, Mg, and Ca) [32]. Similarly, two cultivars of *Matthiola incana* L., an ornamental plant of the *Brassicaceae* family widely appreciated for its beautiful and colorful flowers, was subjected to 5 levels of water stress, namely 90%, 80%, 70%, 60% of field capacity [33].

Adonis amurensis and *Adonis pseudoamurensis*, two species belonging to the *Ranunculaceae* family [7] (Table 1), exhibited reduced growth only in the last days of deficit irrigation treatment, indicating that they can tolerate water deficit conditions. Moreover, water stress reduced shoot dry mass in purple coneflower plants (*Echinacea purpurea* L.) by 51.5% [34], while five species of *Passiflora* spp. (*P. edulis*, *P. gibertii*, *P. cincinnata*, *P. alata*, *P. setacea*) showed a reduction in growth within the range of 50–75%, following water deficit conditions [29].

Water stress may also increase the root-to-shoot ratio. This is an adaptive response to deficit irrigation as a result of the increase in the root system growth and the concomitant reduction in the aerial part of the plant [14]. In this way, the plant tries to cope with reduced water availability by increasing water absorption though roots and reducing water loss from leaves at the same time [25,35,36]. Water stress may also cause changes in roots architecture. For example, in *Callistemon citrinus* plants subjected to water stress, the main roots were longer, whereas the growth of small roots, lateral and thinner ones, was eliminated [37]. Similar results were reported for *Nerium oleander* L., *Pittosporum tobira* Thunb., and *Ligustrum japonicum* Thunb. 'Texanum' (Mediterranean ornamental shrubs) plants [12], subjected to four levels of water stress (90%, 80%, 70%, and 60% of container capacity).

Rafi et al. [26] examined the morphological response to water stress in two native, and therefore already adapted to the local climate conditions, ornamental species, namely *Althea rosea* and *Malva sylvestris*, and two exotic ones, namely *Rudbeckia hirta* and *Callistephus chinensis*. The results showed that, concerning roots length, volume, and density, a decreasing trend was observed with increasing water stress severity in the case of *C. chinensis* and *M. sylvestris*. In contrast, in *A. rosea*, the length of the roots increased as the deficit irrigation levels increased, while roots density decreased in *R. hirta* plants when water stress was more severe.

Three potted Bougainvillea genotypes (*B. glabra* var. Sanderiana, *B.* × *buttiana* 'Rosenka', *B.* '*Lindleyana*' (=*B.* '*Aurantiaca*') were grown on three irrigation levels (100%, 50%, and 25% of substrate moisture) and two canopy shapes (globe and pyramid), aiming to identify the most tolerant genotype and the most useful shape [38]. Moreover, the results showed that total dry biomass was reduced as water stress increased, with the *B.* '*Lindleyana*' genotype recording the highest reduction (33%), followed by *B. glabra* var. Sanderiana (20%) and *B.* × *buttiana* 'Rosenka' (5.5%). The effect of water stress on leaves number was the highest in the case of *B.* '*Lindleyana*' plants (reduced by 43%), followed by *B. glabra* var. Sanderiana (reduced by 33%) and *B.* × *buttiana* 'Rosenka' (reduced by 19%). The authors also suggested that the leaf area was reduced (by 43%) by water stress when canopy shape was pyramidal compared to the global one, while water deficit also reduced the content of N, P, and K in the three genotypes examined [38]. Moreover, according to Rouphael et al. [39], water stress is responsible for the reduction in leaf macronutrient contents in plants, probably because of the lower solubilization due to the water deficit, and therefore the lower absorption and translocation of nutrients [40].

Tolerance mechanisms have also been recorded in *Nerium oleander* L., an evergreen shrub belonging to the *Apocynaceae* family which is widespread in dry and semi-arid regions, such as the Mediterranean ones. In the work of Kumar et al. [1], 1-year-old *Oleander* plants were pot grown in a greenhouse and were normally irrigated until acclimatized. Subsequently, they were subjected to water stress and plants were analyzed after 15 and 30 days of stress initiation. The results showed that there were no effects on stem elongation (cm) and fresh weight of leaves (g) after 15 days of stress, whereas the effects became significant after 30 days of stress.

Four species belonging to the genus *Sedum* L. (*Crassulaceae* family), namely *Sedum spurium*, *S. ochroleucum*, *S. album*, and *S. sediforme*, also called "Green roofs" and being used to adorn the urban area and mitigate area pollution, showed different tolerance to water stress implemented with interruption of irrigation for 4 weeks [22]. All species showed a reduction in plant growth, and changes in morphological parameters (stem length, fresh weight) which allowed to establish a gradual tolerance to deficit irrigation.

Table 1. The effect of water stress on ornamental plants growth and morphology. (−) reduction due to water stress and compared to the control (C); (+) increase due to water stress and compared to the control.

Species	Plant Habit	Deficit Irrigation Treatment	Plant Growth Stage at the Beginning of Treatment	Modulation of Growth and Morphology by Water Stress	References
Lantana camara, *Ligustrum lucidum*	Shrub	C = 100% of water container capacity; Stress: 75%, 50%, and 25% of C	Two month old rooted cuttings	Dry weight (−) Leaf number (−) Leaf area (−) Leaf thickness (−) Thickness of the spongy and palisade tissue (+)	[24]
Polygala and *Viburnum*		10%, 20%, 30%, 40% of water content of the pot volume		Thickness of the spongy and palisade tissue (+)	[25]
Malva sylvestris, Althea rosea, Callistephus chinensis and *Rudbeckia hirta*	Herbaceous plants	C = 100% of ET_0 (local reference evapotranspiration) Stress: 25%, 50%, 75% of ET_0	1-month-old seedlings grown in the field and acclimatized for one month before treatment begun	Root length, root volume, root density: (−) in *C. chinensis*, and *M. sylvestris*) Root length: (+) in *Althea rosea*) Root density: (+) in *R. hirta*	[26]
Bougainvillea glabra var. Sanderiana, *Bougainvillea buttiana* 'Rosenka', Bougainvillea 'Lindleyana' (=*B.* 'Aurantiaca')	Rooted cuttings	C = 100% of substrate moisture Stress: 50% and 25% of control	Plants grown in greenhouse into pots filled with 3 L of peat-moss, irrigated with water and nutrient solution	Total dry biomass (−) Leaves number and leaf area (−) Number and flower index (no. dm^{-2} leaf area) (+) N, P, K (−)	[38]
Geranium macrorrhizum L. (Bevan variety from UK, and wild type from Hungary)	Cuttings from rhizome division	Interruption of irrigation for six weeks	Plants grown in greenhouse for 5 months and then a lath house for 7 months, into pots filled with 90% turf, 10% clay, irrigated manually with water	Different leaf area ratio (ratio between the leaf area and total weight of the plant, LAR m^2 kg^{-1}). Different leaf mass fraction (LMF, leaf biomass/total biomass; kg kg^{-1}) and root mass fraction (RMF, root biomass/total biomass; kg kg^{-1})	[41]
Nerium oleander L.	Seeds sampled in the wild	Interruption of irrigation for 15 and 30 days	One-year-old seedlings grown in greenhouse, into pots filled with peat-perlite-vermiculite (50%, 25%, 25%), irrigated with nutrient solution for a week before treatment begun	Stem elongation (−), Leaf fresh weight (−), Leaf water content percentage (−) K^+/Na^+ in roots (−)	[1]
Chrysanthemum morifolium Ramat. cv. Hj	Germplasm	35–40%, 65–70%, 95–100% of soil water holding capacity (WHC), for 62 days	Four-month-old seedlings grown in greenhouse, into plastic pots	Ray florets (−) Disc floret (+)	[28]

Table 1. *Cont.*

Species	Plant Habit	Deficit Irrigation Treatment	Plant Growth Stage at the Beginning of Treatment	Modulation of Growth and Morphology by Water Stress	References
Viburnum opulus L. and *Photinia* × *fraseri* 'Red robin'	Shrubs	C = 100% ET (Evapotraspiration) Moderate water deficit = 60% ET Severe water deficit = 30% ET, for 5 months	Plants grown in open air and grenhouse, into pots filled with peat, pumice, and osmocote.	Stem elongation (−) Leaf area (−) Number of leaves (−) Foliar biomass (−) Spongy tissue thickness (+) Shoot/root (+)	[27]
Sedum spurium, S. ochroleucum, S. album, and *S. sediforme*	Herbs, and sub-shrubs	C = irrigation twice a week Stress: interruption of irrigation per 4 weeks	Two-month-old seedlings grown in growth chamber, into pots filled with peat, perlite, and vermiculite, irrigated with nutrient solution	Total stem length (−) Leaves fresh weight (−)	[22]
Antirhinum majus cv. butterfly	Seeds	C = 80% of soil water content Stress = 60%, 40%, 20% of soil water content, for 10 weeks	Seedlings grown in greenhouse, into pots filled with sandy loamy soil, irrigated with tap water for three weeks before treatment begun	Shoot height and diameter (−) Number and leaf area (−) Fresh and dry weight of flowers (−) N, P, K, Mg and Ca content (−)	[32]
Passiflora spp. (*P. edulis, P. gibertii, P. cincinnata, P. alata, P. setacea*)	Germplasm	C = 100% of field capacity Stress = interuption of irrigation until apparent wilting (about 96 days)	Seedling grown in greenhouse, into pots.	Plant height (−) Plants dry weight (−) Leaf area (−) Leaves number (−) Different variation of leaf anatomy Stomatal density (+)	[29]
Paeonia ostii (*Paeonia* section Moutan DC)		C = plants watered daily Stress = interruption of irrigation for 4, 8, 12 days	3-year-old plants grown into pots and watered daily	Change of chloroplasts shape	[30]

3. Effect of Water Stress on Physiological Parameters, Hormonal Activity, and Biochemical Changes

3.1. Gaseous Exchange

The complete or partial closure of stomata to reduce water losses in the instance of water stress involves variations in gaseous exchange in leaves (Figure 1, Table 2). Several parameters are considered to measure the changes in gaseous exchange, e.g., stomatal conductance (gs), transpiration rate (E), and leaf relative water content (RWC) [26]. In Damask rose, the stomatal conductance was reduced by 19% in mild stress (50% of field capacity) and by 36% in severe water stress (25% of field capacity) compared to the control treatment (100% of field capacity) [10]. The transpiration rate increased twofold in mild stress (0.88 mmol H_2O m^{-2} s^{-1}) and remained unchanged under severe stress conditions (0.43 mmol H_2O m^{-2} s^{-1}), compared to the control (0.44 mmol m^{-2} s^{-1}). In the same context, stomatal conductance was reduced with increasing water stress in *Nerium oleander* L., *Pittosporum tobira* Thunb., and *Ligustrum japonicum* Thunb. *'Texanum'*, while the values for the same parameter were higher in *N. oleander* than in *P. tobira* and *L. japonicum*. [12]. In addition, *N. oleander* had a larger leaf area than the other two species. These results showed that *N. oleander* was more tolerant to water stress than the other two Mediterranean shrubs. In another study, stomatal conductance was reduced in all five species of *Passiflora* spp. which were subjected to water stress until stomatal closure and rehydrated when plants exhibited wilting symptoms [29]. Moreover, at the time of rehydration, the five species exhibited different conductance recovery rates, demonstrating different adaptation to deficit irrigation as well as different adaptation strategies [29].

In tolerant plants, leaf RWC decreases as soil moisture is reduced [7]. In four species examined by Rafi et al. [26] (*Althea rosea*, *Malva sylvestris*, and two exotic *Rudbeckia hirta* and *Callistephus chinensis*), there was a reduction trend for the RWC parameter as water stress increased, while the most sensitive species were *C. chinensis* and *M. sylvestris*, recording lower relative water content by 59.0% and 52.5% compared to untreated plants, respectively. A reduction in relative leaf water content relative water content was also observed in *Adonis amurensis* and *Adonis pseudoaumernsis* [7] (Table 1), while for both species, the relative water content decreased slowly at the onset of stress, and then decreased rapidly.

Leaf water potential (Ψw) and osmotic potential ($\Psi\pi$) are two physiological parameters related to leaf water content and cell turgor. They reduce with increasing stress, as shown in *Bougainvillea* plants subjected to water stress [38]. Moreover, water deficit may reduce evapotranspiration values, stomatal conductance, and water potential, as shown in the case of *Callistemon citrinus* plants [14].

Navarro-Rocha et al. [41] compared the morphological and physiological responses to deficit irrigation in *Geranium macrorrhizum*, a plant widely used for its ornamental characteristics (in particular, for its pink and white flowers), and the presence of germacron sesquiterpene, an important essential oil constituent. The authors examined two varieties of two different origins, namely a variety selected in England (Bevans' (BV)), and a wild Hungarian geranium (GH) [41]. Cuttings of both varieties were grown in greenhouses within pots for 5 months, and after that, some pots were selected and subjected to stress with water holding for six weeks. In both genotypes, water potential did not increase excessively during the deficit irrigation period, and the authors attributed resistance to water stress to the closure of stomata which allowed to regulate water losses. The water potential remained constant for 20 days and then increased, resulting in accelerated water losses from the plants. The greater foliar growth and the better water status of leaves in GH variety were at the expense of root biomass, which was greater in the BV genotype (root mass fraction = root biomass/total biomass = 0.87 kg kg^{-1}). Moreover, both genotypes had similar root water contents which also indicates that GH plants might have a higher transpiration rate. In effect, under adequate water availability conditions, the larger leaf area means higher growth rate, while under water shortage, it results in rapid water losses through increased transpiration. The authors concluded that *G. macrorrhizum* can tolerate water stress for at least one month. Although belonging to the same species, the two varieties had different morphological and physiological responses to water stress, suggesting that Bevan variety is more suitable for ornamental purposes under water stress conditions [41].

In another experiment, *Viburnum opulus* L. and *Photinia* × *fraseri* 'Red robin' were grown both in open air and greenhouse conditions and subjected to moderate and severe water deficit. In both species, the water potential of leaves decreased as the water deficit increased, with more negative values being observed in the greenhouse experiment, while the response of *P.* × *fraseri* plants was delayed compared to *V. opulus*. On the other hand, in the field experiment, severe stress reduced stomatal conductance in *V. opulus* and photosynthetic activity in *P.* × *fraseri* plants, while under greenhouse conditions, the reduction of stomatal conductance, transpiration, and photosynthesis already occurred even with moderate and severe stress in the case of *V. opulus* and *P.* × *fraseri*, respectively. The various physiological changes observed under moderate stress suggested that the decidual *V. opulus* was more sensitive to water stress, compared to the evergreen *P.* × *fraseri* [27].

The closure of the stomata and the reduction of gaseous exchanges imply a reduction in photosynthetic activity. Moreover, water use efficiency defines the relationship between photosynthesis and transpiration (P_n/E). According to the literature, an increase in WUE under water stress conditions is associated with an adaptation to deficit irrigation, while WUE reduction is associated with sensitive species [42–44]. However, plants with low WUE were more competitive in arid environments because they consumed more resources more rapidly thus suppressing competitors. On the other hand, plants with high WUE show a better performance in the absence of competition and regardless of water availability, probably because they had better water and nitrogen reserves [45]. The WUE can increase, decrease, or remain unchanged under water deficit conditions, depending on the genotype and the water stress level [46].

In *Callistemon citrinus*, the water deficit increased the ratio between photosynthesis and stomatal conductance (P_n/gs) [14]. Thus, photosynthesis increased as stomatal conductance decreased up to a stomatal conductance of approximately 100 mmol m^{-2} s^{-1}, whereas for stomatal conductance values less than 100 mmol m^{-2} s^{-1}, photosynthesis was rapidly reduced, suggesting that other parameters (biochemical limitations) may influence photosynthesis. The effect of water deficit on P_n/gs may vary based on many factors, such as the species, variety, and stress intensity [14]. For example, in *Callistemon* plants, photosynthesis remained at acceptable values when stomatal conductance had values between 100 and 200 mmol m^{-2} s^{-1}, which correspond to moderate water stress [14]. Moreover, the moderate water stress in *Callistemon* determined higher P_n/gs and root/shoot ratios, indicating the formation of small plants but of good quality with reduced losses of water and inflorescences similar to the control.

3.2. Chlorophyll Content and Photosynthesis

The physiological status of plants can be assessed via the integrity of the photosynthetic apparatus, and therefore the efficiency of the photosystems [25]. Adverse environmental conditions, such as water stress, can damage the photosystems [25]. For example, in Damask rose, the photosynthetic activity was reduced by 31% with moderate water stress (4.5 μmol CO_2 m^{-2} s^{-1}) and by 55% with severe water stress (2.9 μmol CO_2 m^{-2} s^{-1}), compared to the control (7.5 μmol CO_2 m^{-2} s^{-1}) [10]. An indirect measurement to evaluate this damage is the fluorescence of chlorophyll *a*. In particular, the values of this parameter increase when photosystem II does not work efficiently due to an imbalance between the number of electrons present in the photosystem and their use [47]. The F_v/F_m ratio records the maximum quantum yield of PSII reaction centers and it is used to measure the degree of plant stress [25] and an F_v/F_m ratio between 0.78–0.85 indicates the absence of stress [25]. Ornamental plants of the Mediterranean area, such as *Callistemon* [48], were considered tolerant to water stress since, during the treatment with different levels of deficit irrigation, they kept constant optimum values of F_v/F_m (0.8), showing that they have adopted particular strategies to dissipate the reducing power created during the stress conditions [49]. The tolerance of the species is observed in practice with the recovery of plant when the stress is over or lessened [25]. In contrast, maximum quantum yield of PSII (F_v/F_m) and net photosynthesis were reduced in *Paeonia ostii* plants when subjected to water stress [30].

In other species such as *Althea rosea, Malva sylvestris, Rudbeckia hirta*, and *Callistephus chinensis*, water stress significantly affected chlorophyll *a* and *b* content in all four species, while total chlorophyll content was reduced by 16%, 18%, 31%, and 55% in *A. rosea, R. hirta, C. chinensis*, and *M. sylvestris*, respectively [26]. In *Nerium oleander* L. plants, chlorophyll *a* did not show a reduction after 15 days of stress but it was reduced by more than 50% after 30 days of stress. On the other hand, chlorophyll *b* increased in the first 15 days of stress and decreased similarly to chlorophyll *a* at prolonged stress conditions. In contrast, the carotenoids content was reduced even after 15 days of stress.

Oleander appears to be resistant to water stress because the symptoms related to plant growth, water loss, and reduction of chlorophyll *a* and *b* content are visible only after a month of stress [1]. The reduction of photosynthetic pigments in conditions of water deficit is also shown for *Antirhinum majus* cv. Butterfly [32], *Sedum* sp. L. [22], *Matthiola incana* L. [33], and *Paeonia ostii* [30], indicating sensitivity to water stress conditions.

In purple coneflower (*Echinacea purpurea* L.) plants subjected to water deficit conditions, the chlorophyll content was reduced by up to 37.3%, and that of carotenoids increased by up to 83%, compared to control plants. The increase in carotenoids attenuates the oxidative stress caused by deficit irrigation, as carotenoids prevent the production of singlet oxygen, thus mitigating the damage experienced by this radical [34].

In *Rhododendron delavayi*, the application of 9-days of water stress resulted in reduced photosynthetic activity and damage to chloroplasts, along with a reduction in stomatal conductance and transpiration [50]. Moreover, chloroplasts had an oval shape in control plants, whereas under stress, the chloroplasts became swollen. However, when plants were re-watered, the photosynthetic activity and other parameters were recovered, demonstrating a strong tolerance capacity of this species [50].

3.3. Oxidative Stress: ROS Production and Adaptive Responses

Water stress causes an excess of excitation energy due to the slowdown of photosynthetic activity. This energy causes the formation of oxygen free radicals or ROS in chloroplasts, mitochondria, and peroxisomes [25]. ROS include superoxide anion (O_2^-), hydrogen peroxide (H_2O_2), hydroxyl radical (OH^-), singlet oxygen ($^1O_2$), and ozone (O_3). These molecules are very reactive due to the presence of single electrons at their outer orbitals and may convert to other forms either spontaneously or enzymatically, e.g., O_3 decomposes into H_2O_2, O_2^- and $^1O_2$; O_2^- can be transformed into H_2O_2, and H_2O_2 can react with Fe^{2+} to form OH [25]. ROS are produced by plants not only under stress conditions since they are by-products of aerobic metabolism and are also used as signal molecules, while at normal conditions, their level is kept low by antioxidant enzymes activity [22]. Abiotic or biotic stress may raise the content of ROS, including water stress [25]. An excess of ROS indicates a condition of oxidative stress because, being radical, these molecules are very reactive and may damage or cause cell death [51]. Oxygen radicals affect membranes, proteins, and the genome, therefore cellular structures and metabolism are severely altered [52,53].

Various molecules can be used as an index of oxidative stress, such as H_2O_2 and MDA (malondialdehyde), and electrolyte leakage. H_2O_2 at low concentrations is a signal molecule for the development of tolerance to various biotic and abiotic stresses, while when its concentration increases, it may contribute to oxidative stress as it can oxidize the thiol groups of enzymes by inactivating them [7]. For example, a high increase in H_2O_2 and O_2^- with increasing water stress was observed in *Paeonia* section Moutan DC plants subjected to 12 days of water stress [30].

On the other hand, malondialdehyde (MDA) is a marker molecule of lipid peroxidation and it is formed by the oxidation of polyunsaturated fatty acids caused by ROS. In the case of purple coneflower plants (*Echinacea purpurea* L.), water stress increased the MDA content by up to 75.8% compared to non-stressed plants, highlighting the important information that can be revealed regarding the susceptibility of various species to stressors [34]. Moreover, an increase in H_2O_2 and MDA was recorded and shown for *Adonis amurensis* and *A. pseudoamurensis* plants subjected to water stress [7]. In particular, in the case of water-stressed plants of *A. amurensis*, H_2O_2 increased from 2.07 µmol g^{-1} FW to a maximum of 4.56 µmol g^{-1} FW, while in *A. pseudomurensis*, the increase was greater and up to 9.13 µmol g^{-1} FW in the first 20 days of water stress and then decreased. Concerning MDA content, *A. pseudomurensis* contained higher amounts, demonstrating that it was more susceptible to water stress than *A. amurensis*. Similarly, Koźmińska et al. [22] examined the response to water stress in four species of *Sedum* L. and suggested that the MDA presence may confirm the sensitivity of the species to this stressor. In the same context, the lack of changes in MDA content detected in other species may indicate the presence of effective defense mechanisms against oxidative stress. Finally, electrolyte leakage is another index for stress evaluation which indicates membrane stability under stress conditions. Therefore, water deficit tolerant plants are expected to present low electrolyte leakage values [26].

However, plants have an "innate" defense mechanism which can either block the formation of ROS or block their oxidative activity when they are formed. This innate immunity refers to secondary metabolites and antioxidant enzymes that plants synthesize to protect themselves against stressors [51,54]. Among the detoxifying enzymes, the most commonly measured are superoxide dismutase (SOD), catalase (CAT), peroxidase (POD), glutathione peroxidase (GPX), and ascorbate peroxidase (APX). Catalase is found in peroxisomes, while the rest of the enzymes are found in different organelles [55]. The quantity and presence of antioxidant molecules or enzymes can reveal the plant's response to stress.

For example, in *Adonis amurensis* plants, the CAT and POD enzymes reduce their activity within the first 10 days of stress initiation (2.08 and 521.15 U $g^{-1}min^{-1}$, respectively), while after 30 days of stress, both enzymes increase their activity (3.42 and 695.39 U $g^{-1}min^{-1}$, respectively) compared to the control at the same day (2.62 and 554.31 U $g^{-1}min^{-1}$, respectively) [7]. The POD enzyme also showed a similar trend in *A. pseudoamurensis*, examined by the same authors. In both species (*A. amurensis* and *A. pseudoamurensis*), SOD enzyme reached the maximum of its activity in 10 days after stress (7.76 × 106 and 7.02 × 106 U $g^{-1} h^{-1}$ FW, respectively), and then it reduced as stress retained (2.49 × 106 and 4.12 × 106 U $g^{-1} h^{-1}$ FW, respectively). Similarly, APX reaches its maximum activity at 30 days of stress in both species [7]. Moreover, in both species, H_2O_2 and MDA were detected at low concentration at the beginning of deficit irrigation implementation, probably due to the concomitant accumulation of antioxidant molecules and enzymes. Then, the concentration of H_2O_2 and MDA increased with the persistence of stress, a finding which indicates that in conditions of severe stress, both species were unable to reduce oxidative stress, despite the increase of antioxidant enzymes content, probably due to the disruption of the antioxidant defense mechanism [56].

In the case of *Nerium Oleander* L. [1], water stress induced a 6-fold increase in APX (ascorbate peroxidase) content compared to the control treatment after 15 days of stress and 4.5 times after 30 days of stress, while GR (glutathione reductase) increased its activity by 1.6 times after 30 days of stress. The activation of other antioxidant enzymes tested, such as SOD and CAT, was not observed in *Oleander* plants, indicating they were not involved in the plant defense mechanism.

An increase in all enzymes tested, namely (catalase (CAT), peroxidase (POX), ascorbate peroxidase (APX), and superoxide dismutase (SOD), especially the activity of CAT, was observed in Purple coneflower plants (*Echinacea purpurea* L.) subjected to water deficit [34]. Moreover, after 12 days of deficit irrigation application in *Paeonia ostii* plants, a significant increase in the activity of peroxidase (POD) and ascorbate peroxidase (APX) was observed [30]. On the other hand, the SOD enzyme activity was increased in the first four days of stress, and then it was reduced in the 8 following days compared to control plants.

Apart from enzymes, secondary metabolites are responsible for plant's tolerance to stressors. The main antioxidant secondary metabolites are tocopherol, ascorbate, glutathione, phenols, alkaloids, flavonoids, and proline [25,47,51,57–60]. Phenolic compounds, including flavonoids, were found to be increased in response to water stress, indicating their important role in the overall defense mechanism of plants [34,61,62].

In *Nerium Oleander* L., the total phenols content was slightly increased within the first 15 days of stress and further increased after 30 days of stress. Flavonoids behave as an inducible defense mechanism and their concentration increases only in conditions of severe stress (e.g., after 30 days of deficit irrigation) [1]. In PanAmerican and Cinderella, two cultivars of *Matthiola incana* L. subjected to water deficiency, the anthocyanin content increased from 0.92 to 1.31 (g FW) and from 0.90 to 1.44 (g FW), respectively, while the phenolic compounds content increased from 0.22 to 0.43 (mg GAE g^{-1} FW) and from 0.27 to 0.38 (mg GAE g^{-1} FW) [33]. Moreover, water stress increased the total phenols content by 17%, 29%, and 38% in plants of *C. chinensis*, *A. rosea*, and *R. hirta* respectively, compared to control plants [26], while an increase in the content of secondary metabolites such as chlorogenic acid, luteoloside, and 3,5-dicaffeoylquinic acid was also observed in the flowers of *Chrysanthemum morifolium* under water stress conditions [28]. Finally, the increase of phenols and flavonoids are an index of sensitivity to deficit irrigation in the case of *S. album* and *S. sediforme*, compared to more tolerant *S. spurium* and *S. ochroleucum* in which phenols or flavonoids are not formed after stress. However, phenolic compounds alone are not a safe index for stress tolerance and other molecules and enzymes have to be measured to evaluate plants response to water stress.

Ascorbic acid is another important antioxidant molecule which regulates the concentration of pro-oxidants, such as H_2O_2 and the closure of stomata and photosynthetic activity. Its action is reflected in leaf growth, flowering, and senescence [63,64]. The content of ascorbic acid, and other antioxidant compounds, such as phenols and flavonoids, is highly affected by various abiotic stressors, such as salinity, high temperature, and water stress [63]. For example, in *Conocarpus lancifolius* Engl., an ornamental species belonging to the *Combretaceae* family and considered as tolerant to semi-arid environments [63], the increase in phenols and flavonoids content in response to water stress was not accompanied by an equal increase of ascorbic acid content. According to the authors, this response to deficit irrigation is the result of a balance between various antioxidant molecules trying to cope with the oxidative stress, or of the faster synthesis of phenols and flavonoids compared to ascorbic acid. The same authors also suggested that phenols, such as caffeic acid and quercetin, have greater antioxidant power than ascorbic acid, hence the higher content detected [63].

3.4. Biochemical Changes

Water stress affects the osmotic balance due to changes in plant water status [65,66]. The main physiological responses of plants trying to adapt to the osmotic stress caused by deficit irrigation are the osmotic adjustment (OA) [3,67], or the accumulation of solutes in cells at levels that allow water uptake [31,68,69]. These solutes are proline, amino acids, glycine betaine, sugars [67,70,71]. However, the energy used and committed for the synthesis of these molecules cannot be used for growth and is called "fitness cost".

Proline has been found to accumulate in plants following numerous abiotic stressors [72,73]. In addition to its osmoprotective activity, proline is also an antioxidant and activator of antioxidant enzymes and is involved in the activation of genes activated by stress [74]. Its accumulation is considered an index of stress tolerance [26]. However, in some species, the higher proline content is associated with stress conditions rather than stress tolerance, meaning that plants with higher proline accumulation are considered sensitive to water stress [26]. This is confirmed by the negative correlation which is usually found between RWC and proline content [75]. In particular, water stress increased proline content by 363%, 115%, 103%, and 83%, in *M. sylvestris*, *C. chinensis*, *R. hirta*, and *A. rosea*, respectively, compared to control plants. However, the proline content was higher in

M. sylvestris and *C. Chinensis* which are considered sensitive to water stress, compared to the other two species (*R. hirta* and *A. rosea*) which are considered tolerant to water stress [75]. Moreover, in the Damask rose, the proline content increases from 14.5 mM (C) to 33.8 mM (50% FC), and 75.5 mM (25% FC), under water deficit [10]. An increase in proline content under severe water conditions (30 days of withholding water) was also found in *Adonis amurensis* and *Adonis pseudoamurensis* [7].

Osmolytes such as soluble sugars and proteins may increase at a certain level of stress and then reduce as stress progresses, denoting the fact that this mechanism is effective at first when the plant tries to defend itself and up to a point where stress interferes too much with plant physiological processes, seriously compromising the synthesis of soluble sugar and proteins. For example, under deficit irrigation conditions, *Oleander* plants accumulated much more sugar than proline and glycine betaine, which only slightly increased their concentration with stress (about 1.3 times, compared to the unstressed control treatment). It could be suggested that in *Oleander* plants, sugars assume a more important role as osmoregulatory compounds compared to proline and glycine betaine, thus demonstrating their importance in plant metabolism and in defense mechanism as well [1].

3.5. Hormonal Activity

Hormones hold a key position in plant defense mechanism against abiotic stresses [7]. Abscisic acid (ABA) plays an important role in resistance to water stress [76] since it regulates stomata closure to and reduces transpiration. Moreover, ABA is also involved in the increase of the antioxidant response of plants against ROS [77]. Some studies showed that adaptation of plants to arid environments is linked to the reduction in gibberellins (GA) and a concomitant increase in ABA content [78,79]. For example, an increase in ABA and a decrease in GA content was recorded in *Adonis amurensis* and *Adonis pseudoamurensis* plants subjected to deficit irrigation. Since GA is a growth-promoting hormone [80], its reduction may indicate a plant strategy of reducing water consumption needed for plant growth and biomass production, while increasing tolerance to stress at the same time.

Ethylene is also important in plants' response to stress and it has been found to induce leaves senescence under deficit irrigation conditions [25]. Moreover, in the work of Gadzinowska et al. [81], an attempt was made to study the biochemical mechanism which regulates the adaptation of sweet briar rose (*Rosa rubiginosa* L.) to arid lands, through analyzing auxin, cytokinin, and gibberellin synthesis. The authors reported that after 30 days of stress, a 3-fold increase (39 µg/g DW) in abscisic acid concentration was observed in stressed sweet briar seedlings compared to control plants (approximately 13 µg/g DW), demonstrating the significant role of abscisic acid in the species response to prolonged stress. Moreover, a series of gibberellins were detected, namely GA1, GA3, GA4, GA5, GA6, GA7, GA8, GA9, among which GA3, GA4, GA5, and GA6 increased with stress, especially GA3 which increased by 329.8% (3-fold compared to the control) [81]. On the other hand, GA9 content was reduced by 65.5% compared to the control. According to the authors, the tolerance of rose plants to water stress was due to the reduction of specific gibberellins (e.g., GA7, GA8, and GA9), since through gibberellins, deficiency plants may reduce their growth and use excessive energy towards the defense mechanisms against water stress, thus confirming the concept of "fitness cost" [81].

The same authors also showed that deficit irrigation resulted to the accumulation of specific auxins, such as indole-3-acetic acid (IAA), indole-3-acetic acid methyl ester (IAA-Met), indole-3-carboxylic acid (IAA-CarbA), indole-3-acetyl-l-aspartic acid (IAA-AsA), indole-3-acetyl-l-glutamic acid (IAA-GluA), and indole-3-butyric acid (IBA). In contrast, the content of other auxins, such as Oxo-IAA (oxindole-3-acetic acid), 4-Cl-IAA (4-chloroindole-3acetic acid), and 5-Cl-IAA (5-chloroindole-3-acetic acid), was reduced under water stress conditions [81]. The role of auxins against water stress consists in the increase of lateral roots and induction of stress genes which allow the synthesis of ABA and the modulation of antioxidant enzymes [82].

Concerning cytokinins, a varied response was observed and 8 cytokinins were increased, whereas 6 others were reduced. In particular, the cytokinin Kinetin riboside increased up to 136.2% compared to the control [81]. According to the authors, the reduction in cytokinins content due to an over-expression of the cytokinin oxidase/dehydrogenase (CKX) enzyme also resulted in reduced growth of roots and the entire plant, allowing the accumulation of bioactive molecules [81]. Besides, some cytokinins may activate transcription factors to increase tolerance to water stress through the stimulation of salicylic acid. Finally, the authors, after comparing the total amount of auxins, cytokinins, and gibberellins, highlighted that stress increased the total content of gibberellins at the expense of auxins and cytokinins [81]. This finding suggests that in the rose plants examined, the overall hormonal balance is more important for plants response to water stress than the changes in specific groups of hormones.

Table 2. Effect of water stress on physiological parameters of ornamentals plant.

Species	Plant Habit	Deficit Irrigation Treatment	Plant Growth Stage at the Beginning of Treatment	Modulation of Physiological Parameters by Water Stress	References
Sweet briar rose (*Rosa rubiginosa* L.)	Shrub	Reduced irrigation for 30 days: 11.2 L of water in control plants and 3.6 L in stressed plants (67.9% less). Plants did not receive water in the last three days of experiment	Seedlings, grown in a garden tunnel, into plastic boxes filled with Klasmann-Deilmann TS1 substrate and sand (v/v: 1:2) and irrigated with 11.2 L of water per box	ABA (+3-fold) Gibberellin (+/−). Auxine (+/−). Cytokinin (+/−).	[81]
Adonis amurensis and *Adonis pseudoamurensis*	Middle and lower part of the hillside grassland	C = 32% of soil moisture Stress: interruption of irrigation for 5, 10, 20, and 30 days	Seedlings grown in natural conditions, into polyethylene plastic pots filled with turf and sand, irrigated with water for 4 weeks before experiment begun	RLWC (−); H_2O_2 (+); MDA (+); Pro (+); Total phenols (+); flavonoids (+); CAT, POD, APX, SOD (+/−); ABA (+); GA (+/−)	[7]
Malva sylvestris, Althea rosea, Callistephus chinensis and *Rudbeckia hirta*	Herbaceous plants	C = 100% of ET_0 (local reference evapotranspiration) Stress: 25%, 50%, 75% of ET_0	1-month-old seedlings grown in the field and acclimated for one month before treatment begun	RLWC (−); Chl *a* and Chl *b* (−); Pro (+), Total phenolic compounds (+); EL (+)	[26]
Geranium macrorrhizum L. (Bevan variety from UK, and wild type from Hungary)	Cuttings from rhizome division	Interruption of irrigation for six weeks	Plants grown in greenhouse for 5 months and then a lath house for 7 months, into pots filled with 90% turf, 10% clay, irrigated manually with water	Water potential (Ψ) (+) Different amounts of water that the aerial parts (WSL) and roots (WSR) were able to store	[41]
Chrysanthemum morifolium Ramat. cv. Hj	Germplasm	35–40%, 65–70%, 95–100% of soil water holding capacity (WHC), for 62 days	Four-month-old seedlings grown in greenhouse, into plastic pots	Chlorogenic acid; luteoloside, and 3,5-dicaffeoylquinic acid (−)	[28]
Viburnum opulus L. and *Photinia* × *fraseri* 'Red robin'	Shrubs	C = 100% ET Moderate water deficit = 60% ET Severe water deficit = 30% ET, for 5 months	Plants grown in open air and grenhouse, into pots filled with peat, pumice, and osmocote.	Leaf water potential (−) gs (−); ET (−); P_n (−) WUE (+)	[27]
Sedum spurium, S. ochroleucum, S. album and *S. sediforme*	Herbs and sub-shrubs	C = irrigation twice a week Stress: interruption of irrigation per 4 weeks	Two-month-old seedlings grown in growth chamber, into pots filled with peat, perlite, and vermiculite, irrigated with nutrient solution	Chlorophyll *a, b* and carotenoids (−); MDA (+); Total phenols (+); Total flavonoids (+); Pro (+)	[22]

Table 2. *Cont.*

Species	Plant Habit	Deficit Irrigation Treatment	Plant Growth Stage at the Beginning of Treatment	Modulation of Physiological Parameters by Water Stress	References
Callistemon citrinus cv Firebrand' (*Crimson bottlebrush*)	Rooted cuttings of 2 year-old	C = 100% of container capacity Stress: moderate stress (50% of control) and severe stress (25% of control)	Two-year-old seedlings grown in greenhouse, into pots filled with coconut fiber, peat, and perlite, irrigated with water for three weeks before treatment begun	ET (−); RLWC (−); gs (−); WUE (+)	[14]
Rhododendron delavayi	Shrub	C = daily irrigation Stress = interruption of irrigation for 5 and 9 days	Five-year-old plants grown in greenhouse, into pots filled with peat and coconut coir	A (−); ROS (+); Damage to chloroplast ultrastructures	[50]
Matthiola incana L. (PanAmerican and Cinderella cultivar)	Seeds	C = 100% of field capacity Stress: 90%, 80%, 70% 60% of field capacity	Seedlings grown in greenhouse, into plastic pots filled with loam, decayed leaves, rotten manure, and river sand (50:25:12.5:12.5), irrigated with tap water, until plants reached the eighth true leaf	Chl (−); CAT (+); anthocyanin content (+); phenolic compounds (+) Pro (+)	[33]
Conocarpus lancifolius Engl. (Combretaceae)	Shrub	C = daily irrigation Stress = interruption of irrigation for 12 days	Shoots at the 13-15 leaf growth stage grown in greenhouse, into pots filled with sandy soil and peat, irrigated with distilled water	A (−); Electrons transport (−); Ascorbic acid (−); Flavonoids (+), Phenols (+)	[63]
Purple coneflower (*Echinacea purpurea* L.)	Seedlings	C = 100% of field capacity Stress = 20, 40, 60% of field capacity, until full flowering stage	Seedlings grown in a farm on soil, irrigated until four leaf stage	Chl *a* and *b* (−), Carotenoids (+); Pro (+); MDA (−) Enzymes antioxidant activity (+); Phenols (+); Flavonoids (+)	[34]
Paeonia ostii (*Paeonia* section Moutan DC)		C = plants watered daily Stress = interruption of irrigation for 4, 8, 12 days	3-year-old plants grown into pots and watered daily	H_2O_2 (+); O_2^- (+); RLWC (−); Pro (+); MDA (+); Chl (−); Carotenoids (−); POD, APX activity (+); SOD activity (+/−); F_v/F_m (−) A (−)	[30]

Evapotranspiration rate(ET) = mmol H_2O m^{-2} s^{-1}; gs = stomatal conductance (mmol m^{-2} s^{-1}); photosynthesis rate (A) (P_n, µmol m^{-2} s^{-1}); water use efficiency (WUE) (µmol CO_2/mmol H_2O); leaf water potential (Ψw) = MPa; leaf osmotic potential ($\Psi \pi$) = MPa; RLWC (relative leaf water content); malondialdehyde (MDA); electrolyte leakage (EL); abscisic acid (ABA); Chl = chlorophyll; (APX) ascorbate peroxidase; (SOD) superoxide oxidase; (POD) peroxidase; (CAT) catalase; (GR) glutathione reductase; glycine betaine (GB); total soluble sugars (TSS); proline (Pro); glycine betaine (GB); oxygen reactive species (ROS).

4. Stress Genes Involved in Plant Tolerance Mechanism against Water Stress

The first perception of water stress by plants is achieved through the root system. The plant responds to stress with physiological, biochemical, and molecular changes and this response depends on the activation of specific genes. Studies on *Arabidopsis thaliana* revealed the transcription products of these genes and identified transcription factors synthesized during the water stress response [25]. From these studies, it emerged that the intensity of stress activates specific genes involved in the response [83]. A target example of the response to water stress is the synthesis of dehydrin as well as the activation of ABA and ethylene pathways. Among the transcription factors involved in this response are ABRE, AREB, AREB/ABFs, DREB/CBF, ABF/AREB, NAC, WRKY, AP2, ethylene response elements [84], MYB2, and MYC2 [85].

Genes involved in the response to deficit irrigation also encode proteins, such as the late embryogenesis abundant (LEA) [86,87], and membrane proteins, such as aquaporins, i.e., the water channels [25].

Dendrobium catenatum is a species belonging to the *Orchidaceae* family, appreciated not only as an ornamental plant but also for its pharmacological properties [88]. The polysaccharides contained in the stems of the species possess anti-inflammatory and antioxidant properties. The content of these polysaccharides is very sensitive to the amount of light and water available to the plant. Huang et al. [88] performed a genetic analysis of superoxide dismutase (SOD) in *Dendrobium catenatum*. SOD enzymes could be found in different cellular compartments, and were distinguished according to the cofactor they were bound to, e.g., Cu, Fe, and Mn (Cu/ZNSOD, Fe/SOD, and Mn/SOD) [89]. Genetic screening led to the identification of 8 genes that code for the SOD enzyme, namely 4 genes for Cu/ZNSOD: *DcaCSD1, DcaCSD2, DcaCSD3, DcaCSD4*, with probable localization of the gene products being chloroplast and cytoplasm; 3 genes coding for FeSOD: *DcaFSD1, DcaFSD2, DcaFSD3*, with localization of the gene product being chloroplasts (excluding *DcaFSD3*); and 1 gene coding for MnSOD: *DcaMSD1*, which product was located in the mitochondrion (Table 3). Furthermore, *DcaCSD2, DcaCSD3, DcaCSD4*, and *DcaMSD1* genes were expressed more in flowers and leaves than in roots and stems. Through phylogenetic analysis, Huang et al. [88] also found that these genes were phylogenetically linked to gene sequences of *Arabidopsis, Oryza sativa, Phalaenopsis equestris*, and *Apostasia shenzhenica*. The authors then identified the gene regions in these genes involved in the synthesis of hormones (gibberellins, abscisic acid, salicylic acid), and the response to cold, light, and water stress, while they also revealed that all SOD genes were upregulated under severe deficit irrigation conditions [88]. *DcaCSD2* and DcaCSD1 genes were upregulated by up to 6 times and three times under water stress, respectively, compared to control [88]. Finally, the authors highlighted that *FeSOD* and *MnSOD* are usually found in algae and bryophytes, while *Cu/ZnSOD* is present only in higher plants, indicating that this form evolved later, and probably due to environmental stresses which became more complex over time [88].

Table 3. Water stress responsive genes.

Gene	Species	Cellular or Subcellular Localisation	Activity during Water Stress	References
DcaCSD1-2-3-4	*Dendrobium catenatum*	Chloroplast (DcaCSD1), citoplasm	Cu/ZnSOD synthesis	[88]
DcaFSD1-2-3	*Dendrobium catenatum*	Chloroplast	Fe/SOD synthesis	[88]
DcaMSD1	*Dendrobium catenatum*	Mitochondrion		[88]
HjCYC2c	*Chrysanthemum morifolium*	Young inflorescence	Adjusting of shape flowers of *Chrysanthemum morifolium*	[28]
FLS	*Chrysanthemum morifolium*	Young inflorescence	Adjusting of pathways of flavonoids during water stress	[28]
Lhca, Lhcb (18 genes), *Psa* (11 genes), *Psb* (15 genes) (all involved in photosynthetic apparatus synthesis), *F3H, DFR, ANS* (flavonoids biosynthesis) PP2C (abscisic acid synthesis), BAK1 and BRI1 (brassinosteroids synthesis)	*Rhododendron delavayi*	Leaves	Response to stimulus; Biosynthesis of secondary metabolites (flavonoids and brassinosteroids); Synthesis of photosystem I and II proteins, and electron transport chain proteins; Synthesis of ATP synthase	[50]
F3H, CCOAOMT, CYP98A CAD, GLU, ZEP, NCED, CCD, TKL, RPI, FBP, KCS, ECH, PPT, LOX, CYP, ORP	*Paeonia ostii*	Leaves	Increase of proline, flavonoids, stilbenoid, diarylheptanoid, and gingerol. Reduction of chlorophylls, carotenoids, phenylpropane and fatty acids.	[30]

In studies with *Rhododendron delavayi* plants, an evergreen ornamental species, subjected to deficit irrigation for 9 days, it was revealed through transcriptome sequencing analyses the expression of 22,728 differentially expressed genes (DEGs) [50]. DEGs encoding photosystem I and II proteins, electron transport chain proteins, and light-harvesting chlorophyll-protein complex (*Lhca, Lhcb, Psa, Psb* genes) were found to be downregulated in the presence of deficit irrigation treatment, whereas the same DEGs were upregulated in the absence of stress (control or re-watered plants), allowing the recovery of photosynthetic activity. Other DEGs involved in the antioxidant response system (synthesis of flavonoids, anthocyanins, and antioxidant enzymes SOD, CAT, POD, GSH, APX) and in the transduction of the hormonal signaling were also upregulated during stress (Table 3). According to the authors, the presence and expression of these genes allowed *Rhododendron delavayi* plants to protect their photosynthetic activity and to exhibit a strong tolerance to water stress [50]. In fact, *Rhododendron delavayi* was shown to have a high concentration of MDA, SOD activity, and proline, and soluble sugars content during stress, while the values of the same parameters were reduced with re-watering [50].

Zhao et al. [28] sequenced the *HjCYC2c* gene in *Chrysanthemum morifolium* Ramat. cv. Hj, which is downregulated in ray florets but upregulated in disc florets, after water stress. They also identified the *FLS* gene, which is involved in flavonoids biosynthesis and determines the symmetry of *Chrysanthemum* flowers. It was also observed that in the case of water stress, *FLS* was downregulated in ray florets and upregulated in disc florets. According to the authors, these two genes interacted with each other in both the synthesis of flavonoids and the regulation of flower symmetry in *Chrysanthemum morifolium* under water stress conditions [28]. Moreover, the gene expression analysis of water-stressed *Paeonia ostii* plants revealed 22,870 DEGs, of which 12,246 were up-regulated and 10,624 were downregulated. Those upregulated were mostly DEGs involved in the biosynthesis of proline, arginine, flavonoids and stilbenoids (*F3H, CCOAOMT, CYP98A*), where the down-regulated ones were mainly involved in the biosynthesis of pigments, phenylpropanoids, fatty acids, and in photosynthesis (*CAD, GLU, ZEP, NCED, CCD, TKL, RPI, FBP, KCS, ECH, PPT, LOX, CYP, ORP*) [30].

The response of sensitive and tolerant ornamental plants to water stress is shown in Figure 2, where in sensitive plants, morphological and physiological changes appear at low and middle levels of stress above which plants generally fail to survive, whereas in tolerant plants, morphological and physiological changes appear at levels between middle and high stress.

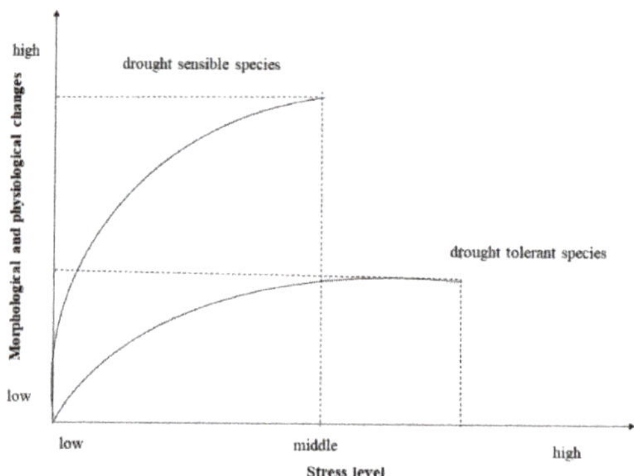

Figure 2. Graphic representation of morphological and physiological changes in drought-sensible and drought-tolerant ornamental plants as water stress level increases.

5. Agricultural Practices to Mitigate Water Stress in Ornamental Plants

A sustainable practice to cope with water stress, with wide spreading use in agriculture, is the application of biostimulants. They are substances of natural origin and microorganisms, such as fungi and bacteria, that are beneficial to plants. Recent studies have revealed that they can mitigate or eliminate the oxidative damage caused by biotic and abiotic stresses on vegetable crops. Furthermore, the use of biostimulants in agriculture can help reduce the excessive use of fertilizers and pesticides [90].

Mycorrhizal fungi have been found to increase the resistance of plants to water stress [25]. They absorb water through their hyphae, which they transfer to the plant. They can also regulate the stomatal opening through hormonal signals. Furthermore, they are involved in osmotic adjustment with greater accumulation of solutes, such as proline, in plants treated with mycorrhiza [25]. Besides, they improve the nutritional status of plants. In the work by Asrar et al. [32], *Glomus deserticola* (AMF) was used to inoculate seeds of *Antirhinum majus* cv. Butterfly. The seedlings were then subjected to various treatments, e.g., 80% (control treatment), 60%, 40%, and 20% of soil water content. The authors showed that *Glomus deserticola* increased tolerance to water stress in *A. majus* since mycorrhiza-treated plants showed increased leaf water potential and leaf water content, and reduced leaf electrolyte leakage, compared to non-mycorrhiza-treated stressed plants [32]. Furthermore, in the presence of fungi plants had a better growth and higher yield of flowers, a better nutritional status (in terms of macro elements content), and a greater accumulation of chlorophyll. The main effect of AMF seems to be the increased surface area, the improved architecture and the higher length of roots which allow the greater absorption of water from the soil. Furthermore, the lower proline accumulation in mycorrhiza-treated stressed plants indicated their higher tolerance compared to the non-mycorrhiza-treated stressed plants.

Another example of the beneficial effect of biostimulant was the better performance of *Petunia* spp., *Viola tricolor*, and *Cosmos* spp. plant grown under water deficit conditions which was achieved through the use of *Ascophyllum nodosum* extracts [91]. Biowaste soluble hydrolysates also increased the photosynthetic activity and gas exchange of *Hibiscus* spp. subjected to water deficit [92]. According to some authors, the positive effects of the various biostimulants is the higher accumulation of biomass, the increased number of flowers, and finally, the production of hormones, such as gibberellins and cytokinins, which stimulate growth under stressful conditions [93].

Darvizheh et al. [34] showed that the exogenous application of salicylic acid and polyamine spermine in purple coneflower plants (*Echinacea purpurea* L.), an ornamental plant also known in medicine for the anti-inflammatory and antioxidant properties of its extracts, increased the antioxidant defense, the pigment contents (chlorophyll and carotenoids), plant biomass, flavonoid, and proline content, whereas it reduced MDA content when plants were subjected to water stress.

Another way to increase resistance to water stress is to expose plants to irrigation cycles and water stress. In these cases, plants responded with better recovery, meaning they adapted to water stress by modulating their physiology for survival [23]. In fact, the plants reduced gas exchanges to reduce transpiration while maintaining good photosynthetic activity. Moreover, light water stress is used to reduce plant growth in pots, as shown in *Cornus alba*, *Lonicera periclymenum*, and *Forsythia* × *intermedia* plants in the work by Davies et al. [94].

6. Conclusions

Concerning the response of plants to environmental stress, such as drought, the species or even the cultivars within the same species, are divided into sensitive and tolerant. Sensitive genotypes generally cannot sustain their growth under prolonged or severe stress. On the other hand, tolerant genotypes manage to survive severe or prolonged stress, but up to specific limits which vary among the species and varieties. From the different species analyzed in this review, it emerged that both sensitive and tolerant plants have an innate defense mechanism which includes morphological changes, such increase of leaf thickness, and the reduction of stomata density and plant growth, as well as physiological changes, such as the restoration of osmotic balance, the closure of stomata, and the synthesis of antioxidant molecules and enzymes. The response to water stress also includes hormonal activity, transcription factors, and the activation of specific genes. Therefore, in tolerant species, the stress response is greater than in sensitive plants, in terms of the amount of molecules produced and enzymes activity. The better understanding of the defense mechanisms of plants against water stress is of major importance in order to apply targeted practices that will increase tolerance and allow the survival of crops under unfavorable conditions. In this context, the use biostimulants is a novel and eco-sustainable agricultural practice which may ensure not only improved water use efficiency in both sensitive and tolerant ornamental plants, but also high yields under deficit irrigation. Another practical application could be the irrigation management according to species or variety specific requirements that could allow revegetation and landscaping even in regions with limit water resources. Therefore, future studies are needed in order to better understand the synergistic effects of biostimulants and the innate defense system of plants under stress, as well as to establish specific agronomic protocols that allow sustainable cropping of ornamental plants under stressful conditions. Finally, considering the species- or variety-dependent response of plants to stressors and to biostimulant products application, further studies are needed to identify those combinations that allow better crop performance under water limitations.

Author Contributions: Conceptualization, M.G. and Y.R.; writing—original draft preparation, M.G., S.A.P. and Y.R.; writing—review and editing, M.G., S.A.P., C.C. and Y.R.; visualization, M.G., S.A.P., C.C. and Y.R.; supervision, M.G., S.A.P., C.C., and Y.R.; project administration, Y.R.; funding acquisition, M.G. and Y.R. All authors have read and agreed to the published version of the manuscript.

Funding: This research received no external funding.

Institutional Review Board Statement: Not applicable.

Informed Consent Statement: Not applicable.

Data Availability Statement: Not applicable.

Conflicts of Interest: The authors declare no conflict of interest.

References

1. Kumar, D.; Al Hassan, M.; Naranjo, M.A.; Agrawal, V.; Boscaiu, M.; Vicente, O. Effects of salinity and drought on growth, ionic relations, compatible solutes and activation of antioxidant systems in oleander (*Nerium oleander* L.). *PLoS ONE* **2017**, *12*, e0185017. [CrossRef] [PubMed]
2. Wang, Z.; Li, J.; Lai, C.; Wang, R.Y.; Chen, X.; Lian, Y. Drying tendency dominating the global grain production area. *Glob. Food Secur.* **2018**, *16*, 138–149. [CrossRef]
3. Marín-de la Rosa, N.; Lin, C.W.; Kang, Y.J.; Dhondt, S.; Gonzalez, N.; Inzé, D.; Falter-Braun, P. Drought resistance is mediated by divergent strategies in closely related *Brassicaceae*. *New Phytol.* **2019**, *223*, 783–797. [CrossRef] [PubMed]
4. Vurukonda, S.S.K.P.; Vardharajula, S.; Shrivastava, M.; SkZ, A. Enhancement of drought stress tolerance in crops by plant growth promoting rhizobacteria. *Microbiol. Res.* **2016**, *184*, 13–24. [CrossRef] [PubMed]
5. Okunlola, G.O.; Olatunji, O.A.; Akinwale, R.O.; Tariq, A.; Adelusi, A.A. Physiological response of the three most cultivated pepper species (*Capsicum* spp.) in Africa to drought stress imposed at three stages of growth and development. *Sci. Hortic.* **2017**, *224*, 198–205. [CrossRef]
6. Sherwood, S.C.; Alexander, M.J.; Brown, A.R.; McFarlane, N.A.; Gerber, E.P.; Feingold, G.; Scaife, A.A.; Grabowski, W.W. Climate processes: Clouds, aerosols and dynamics. In *Climate Science for Serving Society: Research, Modeling and Prediction Priorities*; Asrar, G.R., Hurrell, J.W., Eds.; Springer: Dordrecht, The Netherlands, 2013; pp. 73–103. [CrossRef]
7. Gao, S.; Wanga, Y.; Yua, S.; Huanga, Y.; Liua, H.; Chena, W.; He, X. Effects of drought stress on growth, physiology and secondary metabolites of two *Adonis* species in Northeast China. *Sci. Hortic.* **2020**, *259*, 108795. [CrossRef]
8. Hameed, M.; Moradkhani, H.; Ahmadalipour, A.; Moftakhari, H.; Abbaszadeh, P.; Alipour, A. A review of the 21st century challenges in the food-energy-water security in the Middle East. *Water* **2019**, *11*, 682. [CrossRef]
9. Lombardini, L.; Rossi, L. Ecophysiology of plants in dry environments. In *Dryland Ecohydrology*; Springer: Cham, Switzerland, 2019; pp. 71–100. [CrossRef]
10. Al-Yasi, H.; Attia, H.; Alamera, K.; Hassana, F.; Alia, E.; Elshazlya, S.; Siddiqued, K.H.M.; Hessini, K. Impact of drought on growth, photosynthesis, osmotic adjustment, and cell wall elasticity in Damask rose. *Plant Physiol. Biochem.* **2020**, *150*, 133–139. [CrossRef] [PubMed]
11. IPCC. Intergovernmental panel on climate change. In *Proceeding of the 5th Assessment Report, WGII, Climate Change 2014: Impacts, Adaptation, and Vulnerability*; Cambridge University Press: Cambridge, UK, 2014; Available online: http://www.ipcc.ch/report/ar5/wg2/ (accessed on 16 July 2018).
12. Zuccarini, P.; Galindo, A.; Torrecillas, A.; Pardossi, A.; Clothier, B. Hydraulic relations and water use of mediterranean ornamental shrubs in container. *J. Hortic. Res.* **2020**, *28*, 49–56. [CrossRef]
13. Gu, Z.; Qi, Z.; Burghate, R.; Yuan, S.; Jiao, X.; Xu, J. Irrigation scheduling approaches and applications: A review. *J. Irrig. Drain. Eng.* **2020**, *146*, 04020007. [CrossRef]
14. Alvarez, S.; Sanchez-Blanco, M.J. Changes in growth rate, root morphology and water use efficiency of potted *Callistemon citrinus* plants in response to different levels of water deficit. *Sci. Hortic.* **2013**, *156*, 54–62. [CrossRef]
15. Bita, C.; Gerats, T. Plant tolerance to high temperature in a changing environment: Scientific fundamentals and production of heat stress tolerant crops. *Front. Plant Sci.* **2013**, *4*, 273. [CrossRef]
16. Fita, A.; Rodrõ̂guez-Burruezo, A.; Boscaiu, M.; Prohens, J.; Vicente, O. Breeding and domesticating crops adapted to drought and salinity: A new paradigm for increasing food production. *Front. Plant Sci.* **2015**, *6*, 978. [CrossRef] [PubMed]
17. Flörke, M.; Schneider, C.; McDonald, R.I. Water competition between cities and agriculture driven by climate change and urban growth. *Nat. Sustain.* **2018**, *1*, 51–58. [CrossRef]
18. Nackley, L.L.; de Sousa, E.F.; Pitton, B.J.L.; Sisneroz, J.; Oki, L.R. Developing a water-stress index for potted Poinsettia production. *HortScience* **2020**, *55*, 1295–1302. [CrossRef]
19. Grant, O.M.; Davies, M.J.; Longbottom, H.; Harrison-Murray, R. Evapotranspiration of container ornamental shrubs: Modelling crop-specific factors for a diverse range of crops. *Irrig. Sci.* **2012**, *30*, 1–12. [CrossRef]
20. Ji, K.; Wang, Y.; Sun, W.; Lou, Q.; Mei, H.; Shen, S.; Chen, H. Drought-responsive mechanisms in rice genotypes with contrasting drought tolerance during reproductive stage. *J. Plant Physiol.* **2012**, *169*, 336–344. [CrossRef]
21. Nxele, X.; Klein, A.; Ndimba, B.K. Drought and salinity stress alters ROS accumulation, water retention, and osmolyte content in sorghum plants. *S. Afr. J. Bot.* **2017**, *108*, 261–266. [CrossRef]
22. Koźmińskaa, A.; Al Hassana, M.; Wiszniewskab, A.; Hanus-Fajerskab, E.; Boscaiuc, M.; Vicentea, O. Responses of succulents to drought: Comparative analysis of four *Sedum* (Crassulaceae) species. *Sci. Hortic.* **2019**, *243*, 235–242. [CrossRef]
23. Toscano, S.; Scuderi, D.; Giuffrida, F.; Romano, D. Responses of Mediterranean ornamental shrubs to drought stress and recovery. *Sci. Hortic.* **2014**, *178*, 145–153. [CrossRef]
24. Toscano, S.; Ferrante, A.; Tribulato, A.; Romano, D. Leaf physiological and anatomical responses of *Lantana* and *Ligustrum* species under different water availability. *Plant Physiol. Biochem.* **2018**, *127*, 380–392. [CrossRef] [PubMed]

25. Toscano, S.; Ferrante, A.; Tribulato, A.; Romano, D. Response of Mediterranean ornamental plants to drought stress. *Horticulturae* **2019**, *5*, 6. [CrossRef]

26. Rafi, Z.N.; Kazemi, F.; Tehranifar, A. Morpho-physiological and biochemical responses of four ornamental herbaceous species to water stress. *Acta Physiol. Plant.* **2019**, *41*, 7. [CrossRef]

27. Ugolini, F.; Bussotti, F.; Raschi, A.; Tognetti, R.; Roland Enno, A. Physiological performance and biomass production of two ornamental shrub species under deficit irrigation. *Trees* **2015**, *29*, 407–422. [CrossRef]

28. Zhang, W.; Wang, T.; Guo, Q.; Zou, Q.; Yang, F.; Lu, D.; Liu, J. Effect of soil moisture regimes in the early flowering stage on inflorescence morphology and medicinal ingredients of *Chrysanthemum morifolium* Ramat. Cv. 'Hangju'. *Sci. Hortic.* **2020**, *260*, 108849. [CrossRef]

29. Souza, P.U.; Kenneddy, L.; Lima, S.; Soares, T.L.; de Jesus, O.N.; Filho, M.A.C.; Girardi, E.A. Biometric, physiological and anatomical responses of *Passiflora* spp. to controlled water deficit. *Sci. Hortic.* **2018**, *229*, 77–90. [CrossRef]

30. Zhao, D.; Zhang, X.; Fang, Z.; Wu, Y.; Tao, J. Physiological and transcriptomic analysis of tree Peony (*Paeonia* section Moutan DC.) in response to drought stress. *Forests* **2019**, *10*, 135. [CrossRef]

31. Hessini, K.; Issaoui, K.; Ferchichi, S.; Saif, T.; Abdelly, C.; Siddique, K.H.; Cruz, C. Interactive effects of salinity and nitrogen forms on plant growth, photosynthesis and osmotic adjustment in maize. *Plant Physiol. Biochem.* **2019**, *139*, 171–178. [CrossRef] [PubMed]

32. Asrar, A.A.; Abdel-Fattah, G.M.; Elhindi, K.M. Improving growth, flower yield, and water relations of snapdragon (*Antirhinum majus* L.) plants grown under well-watered and water-stress conditions using arbuscular mycorrhizal fungi. *Photosynthetica* **2012**, *50*, 305–316. [CrossRef]

33. Jafari, S.; Garmdare, S.E.H.; Azadegan, B. Effects of drought stress on morphological, physiological, and biochemical characteristics of stock plant (*Matthiola incana* L.). *Sci. Hortic.* **2019**, *253*, 128–133. [CrossRef]

34. Darvizheh, H.; Zahedi, M.; Abbaszadeh, B.; Razmjoo, J. Changes in some antioxidant enzymes and physiological indices of purple coneflower (*Echinacea purpurea* L.) in response to water deficit and foliar application of salicylic acid and spermine under field condition. *Sci. Hortic.* **2019**, *247*, 390–399. [CrossRef]

35. Mejri, M.; Siddique, K.H.; Saif, T.; Abdelly, C.; Hessini, K. Comparative effect of drought duration on growth, photosynthesis, water relations, and solute accumulation in wild and cultivated barley species. *J. Plant Nutr. Soil Sci.* **2016**, *179*, 327–335. [CrossRef]

36. Farhat, N.; Belghith, I.; Senkler, J.; Hichri, S.; Abdelly, C.; Braun, H.P.; Debez, A. Recovery aptitude of the halophyte Cakile maritima upon water deficit stress release is sustained by extensive modulation of the leaf proteome. *Ecotoxicol. Environ. Saf.* **2019**, *179*, 198–211. [CrossRef]

37. Bañón, S.; Ochoa, J.; Franco, J.; Alarcón, J.; Sánchez-Blanco, M.J. Hardening of oleander seedlings by deficit irrigation and low air humidity. *Environ. Exp. Bot.* **2006**, *56*, 36–43. [CrossRef]

38. Cirillo, C.; Rouphael, Y.; Caputo, R.; Raimondi, G.; De Pascale, S. The influence of deficit irrigation on growth, ornamental quality, and water use efficiency of rhree potted *Bougainvillea* genotypes grown in two shapes. *HortScience* **2014**, *49*, 1284–1291. [CrossRef]

39. Rouphael, Y.; Cardarelli, M.; Schwarz, D.; Franken, P.; Colla, G. Effects of drought on nutrient uptake and assimilation in vegetable crops. In *Plant Responses to Drought Stress: From Morphological to Molecular Features*; Aroca, R., Ed.; Springer: Berlin/Heidelberg, Germany, 2012; pp. 171–195.

40. Garg, B.K. Nutrient uptake and management under drought: Nutrient-moisture interaction. *Curr. Agric.* **2003**, *27*, 1–8. [CrossRef]

41. Navarro Rocha, J.; Burillo-Alquézar, J.; Aibar-Lete, J.; González-Coloma, A. Adaptability of two accessions of *Geranium macrorrhizum* L. to drought stress conditions. *J. Appl. Res. Med. Aromat. Plants* **2017**, *7*, 149–152. [CrossRef]

42. Forner, A.; Valladares, F.; Bonal, D.; Granier, A.; Grossiord, C.; Aranda, I. Extreme droughts affecting Mediterranean tree species' growth and water-use efficiency: The importance of timing. *Tree Physiol.* **2018**, *38*, 1127–1137. [CrossRef] [PubMed]

43. Bai, T.; Li, Z.; Song, C.; Song, S.; Jiao, J.; Liu, Y.; Dong, Z.; Zheng, X. Contrasting drought tolerance in two apple cultivars associated with difference in leaf morphology and anatomy. *Am. J. Plant Sci.* **2019**, *10*, 709–722. [CrossRef]

44. Jin, N.; Ren, W.; Tao, B.; He, L.; Ren, Q.; Li, S.; Yu, Q. Effects of water stress on water use efficiency of irrigated and rainfed wheat in the Loess Plateau, China. *Sci. Total Environ.* **2018**, *642*, 1–11. [CrossRef]

45. Campitelli, B.E.; Des Marais, D.L.; Juenger, T.E. Ecological interactions and the fitness effect of water-use efficiency: Competition and drought alter the impact of natural MPK12 alleles in *Arabidopsis*. *Ecol. Lett.* **2016**, *19*, 424–434. [CrossRef]

46. Cameron, R.W.F.; Harrison-Murray, R.S.; Atkinson, C.J.; Judd, H.L. Regulated deficit irrigation: A means to control growth in woody ornamentals. *J. Hortic. Sci. Biotechnol.* **2006**, *81*, 435–443. [CrossRef]

47. Reddy, A.R.; Chiatanya, K.V.; Vivekanandan, M. Drought induced responses of photosynthesis and antioxidant metabolism in higher plants. *J. Plant Physiol.* **2004**, *161*, 1189–1202. [CrossRef]

48. Álvarez, S.; Navarro, A.; Nicolás, E.; Sánchez-Blanco, M.J. Transpiration, photosynthetic responses, tissue water relations and dry mass partitioning in *Callistemon* plants during drought conditions. *Sci. Hortic.* **2011**, *129*, 306–312. [CrossRef]

49. Flexas, J.; Medrano, H. Energy dissipation in C3 plants under drought. *Funct. Plant Boil.* **2002**, *29*, 1209–1215. [CrossRef] [PubMed]

50. Cai, Y.-F.; Wang, J.H.; Zhang, L.; Song, J.; Peng, L.C.; Zhang, S.B. Physiological and transcriptomic analysis highlight key metabolic pathways in relation to drought tolerance in *Rhododendron delavayi*. *Physiol. Mol. Biol. Plants* **2019**, *25*, 991–1008. [CrossRef]

51. Impa, S.M.; Nadaradjan, S.; Jagadish, S.V.K. Drought stress induced reactive oxygen species and anti-oxidants in plants. In *Abiotic Stress Responses in Plants*; Springer: New York, NY, USA, 2012; pp. 131–147. [CrossRef]

52. Anjum, S.; Xie, X.Y.; Wang, L.C.; Saleem, M.F.; Man, C.; Wang, L. Morphological, physiological and biochemical responses of plants to drought stress. *Afr. J. Agric. Res.* **2011**, *6*, 2026–2032.

53. Lawlor, D.W.; Tezara, W. Causes of decreased photosynthetic rate and metabolic capacity in water-deficient leaf cells: A critical evaluation of mechanisms and integration of processes. *Ann. Bot.* **2009**, *103*, 561–579. [CrossRef] [PubMed]

54. Foyer, C.H.; Noctor, G. Oxidant and antioxidant signalling in plants: A re-evaluation of the concept of oxidative stress in a physiological context. *Plant Cell Environ.* **2005**, *8*, 1056–1071. [CrossRef]

55. Cruz de Carvalho, R.; Catala, M.; Silva, J.M.D.; Branquinho, C.; Barreno, E. The impact of dehydration rate on the production and cellular location of reactive oxygen species in an aquatic moss. *Ann. Bot.* **2012**, *110*, 1007–1016. [CrossRef] [PubMed]

56. Pandey, V.; Ranjan, S.; Deeba, F.; Pandey, A.K.; Singh, R.; Shirke, P.A.; Pathre, U.V. Desiccation-induced physiological and biochemical changes in resurrection plant, *Selaginella bryopteris*. *J. Plant Physiol.* **2010**, *167*, 1351–1359. [CrossRef] [PubMed]

57. Chen, C.; Dickman, M.B. Proline suppresses apoptosis in the fungal pathogen *Colletotrichum trifolii*. *Proc. Natl. Acad. Sci. USA* **2005**, *102*, 3459–3464. [CrossRef] [PubMed]

58. Jaleel, C.A.; Riadh, K.; Gopi, R.; Manivannan, P.; Ines, J.; Al-Juburi, H.J.; Chang-Xing, Z.; Hong-Bo, S.; Panneerselvam, R. Antioxidant defense responses: Physiological plasticity in higher plants under abiotic constraints. *Acta Physiol. Plant.* **2009**, *31*, 427–436. [CrossRef]

59. Gill, S.S.; Tuteja, N. Reactive oxygen species and antioxidant machinery in abiotic stress tolerance in crop plants. *Plant Physiol. Biochem.* **2010**, *48*, 909–930. [CrossRef] [PubMed]

60. Ahmad, P.; Jaleel, C.A.; Salem, M.A.; Nabi, G.; Sharma, S. Roles of enzymatic and non-enzymatic antioxidants in plants during abiotic stress. *Crit. Rev. Biotechnol.* **2010**, *30*, 161–175. [CrossRef]

61. Sánchez-Rodríguez, E.; Moreno, D.A.; Ferreres, F.; del Mar Rubio-Wilhelmi, M.; Ruiz, J.M. Differential responses of five cherry tomato varieties to water stress: Changes on phenolic metabolites and related enzymes. *Phytochemistry* **2011**, *72*, 723–729. [CrossRef] [PubMed]

62. Bautista, I.; Boscaiu, M.; Lidon, A.; Llinares, J.V.; Lull, C.; Donat, M.P.; Mayoral, O.; Vicente, O. Environmentally induced changes in antioxidant phenolic compounds levels in wild plants. *Acta Physiol. Plant.* **2016**, *38*, 9. [CrossRef]

63. Redha, A.; Al-Mansor, N.; Suleman, P.; Al-Hasan, R.; Afzal, M. Modulation of antioxidant defenses in Conocarpus lancifolius under variable abiotic stress. *Biochem. Syst. Ecol.* **2012**, *43*, 80–86. [CrossRef]

64. Azzedine, F.; Gherroucha, H.; Baka, M. Improvement of salt tolerance in *Durum* wheat by ascorbic acid application. *J. Stress Physiol. Biochem.* **2011**, *7*, 27–37.

65. Hessini, K.; Martinez, J.P.; Gandour, M.; Albouchi, A.; Soltani, A.; Abdelly, C. Effect of water stress on growth, osmotic adjustment, cell wall elasticity and water use efficiency in *Spartina alterniflora*. *Environ. Exp. Bot.* **2009**, *67*, 312–319. [CrossRef]

66. Negrão, S.; Schmöckel, S.M.; Tester, M. Evaluating physiological responses of plants to salinity stress. *Ann. Bot.* **2017**, *119*, 1–11. [CrossRef] [PubMed]

67. Turner, N.C. Turgor maintenance by osmotic adjustment: 40 years of progress. *J. Exp. Bot.* **2018**, *69*, 3223–3233. [CrossRef]

68. Blum, A. Osmotic adjustment is a prime drought stress adaptive engine in support of plant production. *Plant Cell Environ.* **2017**, *40*, 4–10. [CrossRef] [PubMed]

69. Hessini, K.; Kronzucker, H.J.; Abdelly, C.; Cruz, C. Drought stress obliterates the preference for ammonium as an N source in the C4 plant *Spartina alterniflora*. *J. Plant Physiol.* **2017**, *213*, 98–107. [CrossRef] [PubMed]

70. Martínez, J.P.; Lutts, S.; Schanck, A.; Bajji, M.; Kinet, J.M. Is osmotic adjustment required for water stress resistance in the Mediterranean shrub (*Atriplex halimus* L.)? *J. Plant Physiol.* **2004**, *161*, 1041–1051. [CrossRef]

71. Ferchichi, S.; Hessini, K.; Dell'Aversana, E.; D'Amelia, L.; Woodrow, P.; Ciarmiello, L.F.; Carillo, P. *Hordeum vulgare* and *Hordeum maritimum* respond to extended salinity stress displaying different temporal accumulation pattern of metabolites. *Funct. Plant Biol.* **2018**, *45*, 1096–1109. [CrossRef] [PubMed]

72. Szabados, L.; Savoure, A. Proline: A multifunctional amino acid. *Trends Plant Sci.* **2010**, *15*, 89–97. [CrossRef] [PubMed]

73. Cicevan, R.; Al Hassan, M.; Sestras, A.F.; Prohens, J.; Vicente, O.; Sestras, R.E.; Boscaiu, M. Screening for drought tolerance in cultivars of the ornamental genus *Tagetes* (Asteraceae). *PeerJ* **2016**, *4*, e2133. [CrossRef]

74. Magdy, M.; Mansour, M.; Farouk, E. Evaluation of proline functions in saline conditions. *Phytochemistry* **2017**, *140*, 52–68. [CrossRef]

75. Pourghayoumi, M.; Rahemi, M.; Bakhshi, D.; Aalami, A.; Kamgar-Haghighi, A.A. Responses of pomegranate cultivars to severe water stress and recovery: Changes on antioxidant enzyme activities, gene expression patterns and water stress responsive metabolites. *Physiol. Mol. Biol. Plants* **2017**, *23*, 321–330. [CrossRef] [PubMed]

76. Hassan, M.S.; Elnemr, K.F. Plant response to drought stress simulated by ABA application: Changes in chemical composition of cuticular waxes. *Environ. Exp. Bot.* **2013**, *86*, 70–75. [CrossRef]

77. Ban, S.G.; Selak, G.V.; Leskovar, D.I. Short- and long-term responses of pepper seedlings to ABA exposure. *Sci. Hortic.* **2017**, *225*, 243–251. [CrossRef]

78. Kowitcharoen, L.; Wongs-Aree, C.; Setha, S.; Komkhuntod, R.; Srilaong, V.; Kondo, S. Changes in abscisic acid and antioxidant activity in sugar apples under drought conditions. *Sci. Hortic.* **2015**, *193*, 1–6. [CrossRef]

79. Zhang, S.H.; Xu, X.F.; Sun, Y.M.; Zhang, J.L.; Li, C.Z. Influence of drought hardening on the resistance physiology of potato seedlings under drought stress. *J. Integr. Agric.* **2018**, *17*, 336–347. [CrossRef]

80. Luo, H.H.; Han, H.Y.; Zhang, Y.L.; Zhang, W.F. Effects of drought and re-watering on endogenous hormone contents of cotton roots and leaves under drip irrigation with mulch. *J. Appl. Ecol.* **2013**, *24*, 1009–1016.

81. Gadzinowska, J.; Dziurka, M.; Ostrowskaa, A.; Hura, K.; Hura, T. Phytohormone synthesis pathways in sweet briar rose (*Rosa rubiginosa* L.) seedlings with high adaptation potential to soil drought. *Plant Physiol. Biochem.* **2020**, *154*, 745–750. [CrossRef] [PubMed]

82. Zhang, Q.; Li, J.J.; Zhang, W.J.; Yan, S.N.; Wang, R.; Zhao, J.F.; Li, Y.J.; Qi, Z.G.; Sun, Z.X.; Zhu, Z.G. The putative auxin efflux carrier OsPIN3t is involved in the drought stress response and drought tolerance. *Plant J.* **2012**, *72*, 805–816. [CrossRef] [PubMed]

83. Tommasini, L.; Svensson, J.T.; Rodriguez, E.M.; Wahid, A.; Malatrasi, M.; Kato, K.; Wanamaker, S.; Resnik, J.; Close, T.J. Dehydrin gene expression provides an indicator of low temperature and drought stress: Transcriptome-based analysis of barley (*Hordeum vulgare* L.). *Funct. Integr. Genom.* **2008**, *8*, 387–405. [CrossRef] [PubMed]

84. Klay, I.; Gouia, S.; Liu, M.; Mila, I.; Khoudi, H.; Bernadac, A.; Bouzayen, M.; Pirrello, J. Ethylene Response Factors (ERF) are differentially regulated by different abiotic stress types in tomato plants. *Plant Sci.* **2018**, *274*, 137–145. [CrossRef]

85. Abe, H.; Urao, T.; Ito, T.; Seki, M.; Shinozaki, K.; Yamaguchi-Shinozaki, K. Arabidopsis AtMYC2 (bHLH) and AtMYB2 (MYB) function as transcriptional activators in abscisic acid signaling. *Plant Cell* **2003**, *15*, 63–78. [CrossRef] [PubMed]

86. Hundertmark, M.; Hincha, D.K. LEA (late embryogenesis abundant) proteins and their encoding genes in *Arabidopsis thaliana*. *BMC Genom.* **2008**, *9*, 118. [CrossRef]

87. Magwanga, R.O.; Lu, P.; Kirungu, J.N.; Lu, H.; Wang, X.; Cai, X.; Zhou, Z.; Zhang, Z.; Salih, H.; Wang, K.; et al. Characterization of the late embryogenesis abundant (LEA) proteins family and their role in drought stress tolerance in upland cotton. *BMC Genet.* **2018**, *19*, 6. [CrossRef] [PubMed]

88. Huang, H.; Wang, H.; Tong, Y.; Wang, Y. Insights into the Superoxide Dismutase Gene Family and Its Roles in *Dendrobium catenatum* under Abiotic Stresses. *Plants* **2020**, *9*, 1452. [CrossRef] [PubMed]

89. Wang, W.; Xia, M.; Chen, J.; Yuan, R.; Deng, F.; Shen, F. Gene expression characteristics and regulation mechanisms of superoxide dismutase and its physiological roles in plants under stress. *Biochemistry* **2016**, *81*, 465–480. [CrossRef] [PubMed]

90. Rouphael, Y.; Colla, G. Synergistic biostimulatory action: Designing the next generation of plant biostimulants for sustainable agriculture. *Front. Plant Sci.* **2018**, *9*, 1655. [CrossRef] [PubMed]

91. Battacharyya, D.; Babgohari, M.Z.; Rathor, P.; Prithiviraj, B. Seaweed extracts as biostimulants in horticulture. *Sci. Hortic.* **2015**, *196*, 39–48. [CrossRef]

92. Massa, D.; Lenzi, A.; Montoneri, E.; Ginepro, M.; Prisa, D.; Burchi, G. Plant response to biowaste soluble hydrolysates in hibiscus grown under limiting nutrient availability. *J. Plant Nutr.* **2018**, *41*, 396–409. [CrossRef]

93. Calvo, P.; Nelson, L.; Kloepper, J.W. Agricultural uses of plant biostimulants. *Plant Soil* **2014**, *383*, 3–41. [CrossRef]

94. Davies, M.J.; Harrison-Murray, R.; Atkinson, C.J.; Grant, O.M. Application of deficit irrigation to container-grown hardy ornamental nursery stock via overhead irrigation, compared to drip irrigation. *Agric. Water Manag.* **2016**, *163*, 244–254. [CrossRef]

horticulturae

Review

Response Mechanism of Plants to Drought Stress

Xinyi Yang, Meiqi Lu, Yufei Wang, Yiran Wang, Zhijie Liu and Su Chen *

State Key Laboratory of Tree Genetics and Breeding, Northeast Forestry University, Harbin 150040, China; yangxinyi@nefu.edu.cn (X.Y.); lumeiqi0408@163.com (M.L.); wangyufei@nefu.edu.cn (Y.W.); wangyiran@nefu.edu.cn (Y.W.); hjdyyy69@163.com (Z.L.)
* Correspondence: chensu@nefu.edu.cn

Abstract: With the global climate anomalies and the destruction of ecological balance, the water shortage has become a serious ecological problem facing all mankind, and drought has become a key factor restricting the development of agricultural production. Therefore, it is essential to study the drought tolerance of crops. Based on previous studies, we reviewed the effects of drought stress on plant morphology and physiology, including the changes of external morphology and internal structure of root, stem, and leaf, the effects of drought stress on osmotic regulation substances, drought-induced proteins, and active oxygen metabolism of plants. In this paper, the main drought stress signals and signal transduction pathways in plants are described, and the functional genes and regulatory genes related to drought stress are listed, respectively. We summarize the above aspects to provide valuable background knowledge and theoretical basis for future agriculture, forestry breeding, and cultivation.

Keywords: drought stress; osmotic regulation; LEA protein; ROS; signaling; drought-responsive gene

Citation: Yang, X.; Lu, M.; Wang, Y.; Wang, Y.; Liu, Z.; Chen, S. Response Mechanism of Plants to Drought Stress. *Horticulturae* **2021**, *7*, 50. https://doi.org/10.3390/horticulturae7030050

Academic Editors: Stefania Toscano, Giulia Franzoni and Sara Álvarez

Received: 20 February 2021
Accepted: 11 March 2021
Published: 13 March 2021

Publisher's Note: MDPI stays neutral with regard to jurisdictional claims in published maps and institutional affiliations.

1. Introduction

Drought is one of the most important factors restricting agricultural production, which seriously affects crop yield [1,2]. Moreover, as one of the main restraining factors in the process of plant growth, drought can hinder plant respiration, photosynthesis, and stomatal movement; thus, affecting plant growth and physiological metabolism. In response to drought stress, plants activate their drought response mechanisms, such as morphological and structural changes, expression of drought-resistant genes, synthesis of hormones, and osmotic regulatory substances to alleviate drought stress. To better reveal the mechanism of drought resistance of plants, based on a lot of previous work, we summarized the status quo and progress of studies on the morphological structure, physiological and biochemical mechanism changes, internal signal transduction system, and molecular regulation mechanism of plants under drought stress in recent years. Under drought conditions, plants sense water stress signals and produce signal molecules, such as abscisic acid (ABA), Ca^{2+}, inositol-1, 4, 5-triphosphate (IP3), cyclic adenosine $5'$-diphosphate ribose (cADPR), NO, etc., and directly or indirectly lead to the morphological and physiological changes of plants through signal transduction. Indirectly, drought stress signals induce the expression of downstream genes. Functional gene products, such as proline (pro), glycine betaine (GB), soluble sugar (SS), late embryogenesis abundant (LEA) proteins, and aquaporin (AQP) can be involved in plant metabolism and, thus, affect plant state. Regulatory gene products, such as calcium-dependent protein kinases (CDPKs), mitogen-activated protein kinases (MAPKs), HD-zip/bZIP, AP2/ERF, NAC, MYB, and WRKY can cause changes in plant morphology or physiology by regulating signal transduction pathways or acting as transcription factors to regulate the expression of downstream genes, and further enable plants to successfully survive in the arid environment (Figure 1).

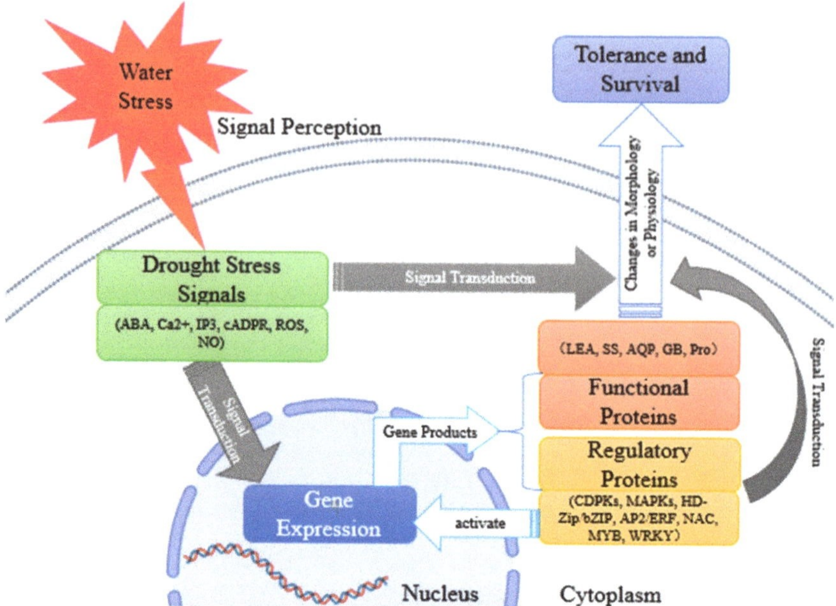

Figure 1. The process of plant drought-tolerance development.

We will elaborate from the following four parts. The first is the effect of drought stress on the external morphology and internal structure of plants. The second part elaborates the physiological and biochemical responses from the perspectives of osmotic regulation metabolism, drought-induced protein metabolism, and reactive oxygen metabolism. Here we summarize some important drought-regulating substances and we also briefly summarize the generation and scavenging process of reactive oxygen species (ROS). The third part is the signal transduction pathway in plants. We describe common signals in detail and elucidation of intracellular signal transduction pathways. The fourth part is about the molecular regulation mechanism of plants. From the perspective of genes, the anabolism and regulation mechanisms of osmotic regulation-related substances, drought-induced proteins, signaling path-related substances, and transcription factors are summarized respectively. All of the advances indicate that it is of great significance to study the effects of drought stress on plants and explore the mechanism of drought tolerance.

2. Effects of Drought Stress on Plant Morphological Characteristics

When plants are subjected to drought stress, they will first respond to changes in external form and internal structure. The most significant effect of water loss is that the plant grows slowly and even dies. Studies have shown that plants under abiotic stress can adapt to changing environmental factors through phenotypic plasticity. Therefore, under the influence of the environment, xerophytes have formed certain morphological characteristics in the process of evolution, and adapted themselves to drought in their ontogenetic development under these characteristics. The drought-resistant plants have morphological and structural characteristics that were adapted to the arid environment in terms of leaves, stems, roots, and so on.

2.1. Drought Stress and the External Form of Plants

The obvious symptoms of water deficit during the vegetative period are plant height decreased, leaf wilting, number and area of leaves changed. Plant height, severely affected by drought, is closely related to cell enlargement and leaf senescence. The decrease in plant height is mainly due to decreased cell expansion, increased leaf shedding, and impaired mitosis under drought conditions. Some studies have reported that plant height of lily [3], maize [4], cane [5], and rice [6] decreased significantly under drought stress. In addition to the changes in plant height, different organs of plants also differ significantly in morphology. As an indicator of the degree of water shortage in direct response, leaves are the main organs for plant assimilation and transpiration. Plant leaves generally adopt smaller leaf areas, larger leaf thickness, and higher leaf tissue density to adapt to drought [7]. The change of leaf area, which directly affects plant photosynthesis and yield, is one of the most easily observed features of plant leaves under drought stress. Previous studies have shown that the main reasons for the change of plant leaf area are the leaf turgor pressure, canopy temperature, and availability of photoassimilates [8]. Under the condition of drought, the leaf turgor pressure and the rate of photosynthesis of plant leaves decrease, which leads to the decrease of leaf area [9]. For morphological responses, *Prunus sargentii* and *Larix kaempferi* experienced a significant decrease in leaf size, respectively in leaf width and length under drought conditions [10]. Furthermore, *Maclura pomifera* [11], *Oryza sativa* [6], *Triticum aestivum* [12], *Lens culinaris* [13], *Dracocephalum moldavica* [14] all showed an obvious decrease in leaf area under drought stress. However, different plants have different responses to drought stress, such as sugarcane leaves showed marginal elongation under drought stress [5]. Another easily observed leaf morphology phenomenon is leaf rolling, for the loss of the potential pressure due to water loss from the upper epidermis of the leaf when plants are short of water. Under drought stress, the flag leaf of wheat would be severely rolling [15]. In a xerophytic environment, conifers have thick horny film, and their wilting and rolling motion can resist direct sunlight to improve their water retention [16,17].

Apart from leaves, plant roots, as organs that directly absorb water, also play a significant role in drought stress [18]. Developed roots can help plants to fully absorb and utilize the water stored in the soil so that plants can survive the drought period [19]. What is more, researches have shown that water is the main environmental factor affecting the development of plant roots [20]. Therefore, the morphological changes of plant roots in arid areas are particularly important. Root system configurations such as root hair, root branches, and root density can significantly affect the water deficiency of plants. Drought stress can inhibit the development of cotton seedlings, promote the elongation and thinning of fine roots, shorten the life of fine roots with different diameters, promote the elongation of root hairs, and accelerate their death [21]. *Cunninghamia lanceolate* can increase root complexity and elongation, reduce root branching angles, leading to steeper and deeper roots system to adapt to drought stress [22]. Maize treated by drought stress obtains more water from dry soil by reducing lateral root branch density, making axial root elongation and rooting depth larger [23]. Water also had a certain effect on the distribution of plant roots. Soybean, field pea, and chickpea were sensitive to the soil moisture content of biomass decreased more than the root, leading to a higher root/shoot ratio of soybean [24]. Drought can also affect the external morphology of plants in other ways, such as the average internode length of sugarcane increased by 39.02% after drought treatment in an early vegetative stage, and drought stress would destroy the full root structure [25]. In addition to the above characteristics, the root to stem ratio of plants also changes. The shoot and root biomass of soybean decreased significantly under drought stress. Under the condition of water restriction, the height, leaf size, and stem girth of maize plants decreased significantly [26].

2.2. Drought Stress, the Internal Structure, and Physical Property of Plants

In addition to the external form, the internal structure of plants also changed. There is a developed cuticle in the outer wall of the leaf epidermis. The cuticle is a kind of lipid membrane, which can reduce the loss of water to the atmosphere and acts as a barrier for plant water evaporation. The thick cuticle can improve the plant's energy reflection and reduce transpiration, thus enhancing the plant's drought resistance. The cuticular lipids content of *Arabidopsis thaliana* leaves increased significantly under water shortage treatment. The increase of epidermal wax per unit area under drought stress was mainly due to the increase of wax alkanes. Moreover, the water deficit increased the total amount of cutin monomers, changed the proportion of the cutin monomers amount, increased the thickness of the leaf cuticle, and the accumulation of osmium in the plant cuticle [27]. Tea leaves improve drought resistance through increasing wax coverage, cuticle thickness, and osmiophilicity [28]. In addition, plant leaves tend to increase mesophyll palisade tissue, decrease spongy tissue, increase the number of cell layers, but decrease the volume and shorten the intercellular space to adopt drought [29]. Stomatal development is another important index related to water stress. The drought process appeared to increase stomatal length, stomatal width, stomatal density, and stomatal opening. The reduced stomatal density of *Hordeum vulgare* leaves could increase its tolerance to water stress [30]. Apple cultivars, which exhibited significantly thicker cuticle, longer palisade cells, and thicker spongy parenchyma had superior drought tolerance [31]. In the observation of micromorphology of blackberry after drought treatment, it was found that with the extension of stress time, the morphology of leaf epidermis cells underwent a series of expansion changes. Moreover, the walls of the epidermal cells and spongy tissue cells of the leaves thickened with the duration of drought. Especially after treatment for a period of time, the spongy tissue cells were obviously compressed and filled with sclerenchyma [32]. What is more, the degree of lignification and channeling tissue on the epidermis had a great influence on the drought resistance of the plant. The study found that plants with water deficits had lower levels of lignin in their leaves than those with adequate water [33]. The xylem of the stems and roots of the stress-treated plants was thicker than that of the normal rapeseed plants. In addition, drought stress reduced the vessel's inner diameter and increased the number and inner diameter of root vessels [34]. Fresh and dry weights are also significantly reduced under water deficit conditions [35]. In the study of *Matthiola incana*, the relative water content did not change significantly with the increase of drought stress, but plant height, stem fresh weight, stem dry weight, root fresh weight, and root dry weight all decreased significantly [36]. Besides, water stress had significant effects on the essential oil content and essential oil composition of Rosemary. With the decrease of soil water content, stalk length, fresh weight, and fresh and dry weight of root decreased. At the same time, the content of essential oil also presents the trend of first rising and then falling [37].

3. Effects of Drought Stress on Plant Physiological and Biochemical Characteristics

When plants are subjected to drought stress, a series of changes will occur in their appearance, leading to a series of physiological and biochemical changes in plants. For example, the changes in photosynthesis, osmotic regulatory substances, drought-induced proteins, and antioxidant enzymes all reflect the different degrees of influence of plants under drought stress.

3.1. Photosynthetic Capacity

Photosynthesis is one of the main processes affected by water stress. Leaf photosynthetic products are the material basis of plant growth. The net photosynthetic rate directly reflects the material productivity per leaf area. Therefore, theoretically speaking, it is a reliable index to measure the biological production level of plants. The photosynthetic rate and transpiration rate decrease with the decrease of soil relative water content. Previous studies have shown that the decrease of photosynthetic rate under drought stress is the result of stomatal limitation and non-stomatal limitation. The stomatal limitation was the main factor of photosynthetic rate decrease under mild drought. However, under severe drought conditions, non-stomatal factors were the main reason for the decline of the photosynthetic rate. When water is deficient, it will lead to the decrease of photosynthesis directly through decreasing CO_2 availability resulted in diffusion limitations of the stomata and the mesophyll [38]. Stomatal closure limits leaf absorption of CO_2 and prevents transpiration water loss due to turgor pressure and/or reduced water potential. In a study by Victor Santos et al., they pointed out that photosynthesis in the canopy of the central Amazon forest decreased by 28% in the dry season and by 17% in the undergrowth, compared with that in other seasons in 2015. They further suggested that the reduction in photosynthesis was only related to the closure of stomata in trees in the canopy and undergrowth [39]. It was also found in wheat that drought decreased stomatal conductance, increased stomatal resistance, and decreased photosynthetic rate and transpiration rate [40]. However, with the increase of water deficit, non-stomatal factors began to play an important role. At this time, the potential photosynthetic CO_2 assimilation rate decreases, which cannot be eliminated by increasing the external CO_2 concentration. The decrease of photosynthesis, which is dominated by non-stomatal factors, is related to the decrease of activity or component content of many important processes related to photosynthesis. For example, Ribulose-1,5-bisphosphate (RuBP) content and activity, as well as apparent quantum yield, play a very important role in the photosynthetic assimilation process. In the study by Carmen Gimenez, it was found that there was an obvious S-shaped curve relationship between the photosynthetic rate and RuBP in sunflower leaves, suggested that the reduction of photosynthetic rate was to some extent restricted by RuBP content [41]. Dhammika's study on tobacco also confirmed that RuBP synthesis is limited under water stress due to inhibition of the activity of synthetic enzymes [42]. Another substance that is important for plant photosynthesis under water stress is the enzyme Ribulose-1,5-bisphosphate carboxylase/oxygenase (RuBisCo). The activity of the RuBisCo enzyme had no significant change or was less affected under mild water shortage, but decreased under severe drought. In addition, changes in photochemical and biochemical processes such as electron transfer rate decrease and photophosphorylation are also observed. The direct manifestation of these changes is the occurrence of "photoinhibition". With the increase of drought intensity, the net photosynthetic rate, transpiration rate, and stomatal conductance of cotton decreased [43]. Ma Ping et al. studied the effects of drought on photosynthesis in apples. Soil relative water content (SRWC) decreased from 87% to 24% within 15 days after the irrigation treatment was stopped, while leaf relative water content (LRWC), net photosynthetic rate (Pn), and stomatal conductance (GS) all showed a decreasing trend. Moreover, they noted that the photochemical reaction was only slightly downregulated under severe drought conditions. With the intensification of drought conditions, the activity of RuBisCo decreased significantly, and the actual efficiency of photosystem II (ΦPSII) decreased [44].

The chloroplast is the site of photosynthesis in green plant leaves, which mainly uses chlorophyll to absorb, transfer and transform light energy. Chlorophyll is continuously metabolized in plants, closely related to photosynthesis and yield formation of plants. As the most important and effective pigment in photosynthesis, chlorophyll can reflect the growth status of plants and the degree of stress. Chlorophyll content tends to decrease under drought stress and the ratio of chlorophyll "a", "b", and carotenoid was changed, thus, in turn, causes changes in photosynthetic function [45]. The reason for the decrease of chlorophyll content in leaves may be the degradation of chlorophyll directly caused by drought. Drought stress could significantly reduce the contents of chlorophyll a, chlorophyll b, and total chlorophyll in chickpea during vegetative growth and anthesis [46]. In the study of 13 durum wheat native varieties from Iran and Azerbaijan, it was found that different wheat varieties had different responses to drought stress. The chlorophyll level of susceptible wheat cultivars decreased significantly under drought stress, while the chlorophyll content of resistant wheat cultivars was still maintained [47]. In Chinese cork oak (*Quercus variabilis*) seedlings, chlorophyll a, chlorophyll b, carotenoids (Car), and total chlorophyll contents were significantly decreased at 40% and 20% field capacity, despite there was no significant change in Chl a/Chl b and Car/Chl ratios [48]. The total chlorophyll content and the ratio of Chl a/Chl b in oil palm were significantly decreased under water stress [49]. However, not all plants show reduced chlorophyll content under drought stress. Soheila pointed out that chlorophyll content in borage increased at lower irrigation levels, mainly due to lower leaf area index and more radiation interception [50]. Besides, drought makes it difficult for plants to absorb nutrient elements and causes symptoms of deficiency of elements, which is also manifested as decreased chlorophyll content.

Changes in plant pigments lead to the color of the plant changed into yellowish-brown when they suffer from drought. From the point of view of drought resistance, plants with high chlorophyll content generally have stronger drought resistance.

The pathways of CO_2 assimilation in photosynthesis can be divided into the C_4 pathway, C_3 pathway, and Crassulacean acid metabolism (CAM) pathway. Under drought stress, the C_4 pathway was significantly superior to the C_3 pathway. The leaves of C_4 plants have a typical Kranz wreath structure and water use efficiency (WUE) is significantly higher than that of the C_3 pathway. Under the condition of water shortage, the C_4 pathway can assimilate CO_2 to produce more organic matter, which is conducive to the plant to resist early drying. The stomata of CAM plants open at night, absorb CO_2, and form malic acid catalyzed by phosphoenolpyruvate carboxylase (PEPC), which is stored in vacuoles. Stomatal closure during the day, MAL decarboxylation gives off CO_2. Because CAM fixes CO_2 by stomatal opening at night, transpiration loss during the stomatal opening in the day is avoided, and the contradiction between stomatal transpiration and CO_2 absorption under drought stress is solved. Different species have different assimilation pathways and different environmental conditions can significantly change the carbon metabolism pathways of plants. That is to say, changes in growth and development level, growth conditions, nutritional status, and biological regulators can lead to a mutual transformation of CO_2 fixation pathways in plants. Increased ABA content in plants under drought conditions promotes the operation of the C_4 pathway. Similarly, the C_3 pathway can also be transformed into the CAM pathway [51]. Winter's studies on different varieties of orchids and *Mesembryanthemum crystallinum* L. (Aizoaceae) have shown that some species with highly flexible photosynthetic phenotypes have changes in assimilation pathways when the external environment changes. They operate in C_3 mode when not stressed, or in CAM mode when drought or salinity stressed [52–54]. Milton Garcia support that the C_3-CAM shift is present in the cactus seeding process. Ideas are put forward that there is a facultative component of CAM expression in the cactus. Shortly after germination, the expression of C_3 photosynthesis can promote plant growth when there is sufficient water. Facultative CAM components can accelerate the development of constitutive CAM and contribute to plant survival in water-deficient environments [55].

3.2. Osmotic Regulation Metabolism

Osmotic regulation is an important way for plants to reduce osmotic potential and resist adversity stress under water stress. When plants are subjected to drought stress, osmotic regulation can be realized in three ways, namely, the decrease of intracellular water, the decrease of cell volume, and the increase of cell contents. These three pathways coexist in plants, but not all plants have osmotic regulation. Osmotic regulation is generally considered to be the active regulation of cells to reduce osmotic potential by increasing solute. Its initial effect is to reduce the free energy of water bound inside the cell, maintain the difference of water potential inside and outside the cell, and enable the cell to absorb water under the condition of lower external water potential. Thus maintaining the turgor pressure required for cell growth [56]. Osmotic regulation can maintain stomatal conductance to moderate water deficit by maintaining turgor pressure. It helps to keep the content of CO_2 in mesophyll intercellular space at a high level so as to avoid or reduce the photosynthetic inhibition on photosynthetic organs. Osmotic regulation can maintain normal or minimize damage to biochemical, physiological, and morphological processes related to cell growth, stomatal opening, and photosynthesis during environmental stress. The osmotic regulating substances in plants mainly include organic osmotic regulating substances and inorganic ions entering from the external environment. Organic osmotic regulating substances, such as amine compounds (glycine betaine and polyamines), amino acid compounds (proline), and trehalose, fructan, mannitol, and other compounds, play a major role in regulating the osmotic type of cytoplasm. These substances are usually of small molecular weight, highly soluble, and have little toxicity to cells. They can maintain the normal osmotic pressure level, protect the protein activity and cell membrane structure, etc.

The osmotic regulation of inorganic ions is closely related to the ion pump. For example, the Na^+, K^+, H^+ pump can regulate the concentration of inorganics inside and outside the cell, thus changing the osmotic potential of the cell. At the same time, the change of inorganic ion concentration will cause a change in cell morphology and function. Suomin Wang et al. proposed that K^+ and free proline accumulation played an important role in drought adaptation of xerophytic plants and Na^+ accumulation is one of the most effective strategies for succulent xerophyte to adapt to drought [57]. At present, there are more studies on osmotic regulation substances such as proline (Pro), soluble sugar (SS), glycine betaine (GB), etc. Some studies have found that Pro accumulation is a protective measure taken by plants to resist drought stress [58]. When PEG concentration was 30%, Pro content in rice increased significantly [6]. Under drought stress conditions, osmotic regulation substance content increased, which was positively correlated with plant stress resistance. However, the variation range of osmotic regulation substance was different among different species. By decreasing soluble sugar, polysaccharide, and fructose contents and increasing proline, glucose, and trehalose contents, Lanzhou lily can improve its resistance to drought stress by changing the contents of osmotic regulation, and secondary metabolites [3]. The contents of soluble carbohydrates sucrose, glucose, and fructose in *Maclura pomifera* increased at the initial stage of drought stress but decreased after 22 days of severe drought stress. In addition, the affinity of osmotic substances proline and mannitol increased significantly under drought stress [11]. In the study of Farooq et al., the contents of proline, glycine betaine, total soluble carbohydrate, and sucrose were significantly increased due to drought stress in several pistachio genotypes [59].

As an osmotic regulating substance, proline (Pro) is preferentially stored in plant vacuoles. When the cell is subjected to osmotic stress, Pro is transported to the cytoplasm, and the osmotic potential is reduced by increasing the concentration of the cytoplasm so that the cell can still absorb extracellular water under the condition of low osmotic potential; thus, maintain the cell protoplasm and the external environment of osmotic balance [60]. Pro has a strong ability to hydrate, so it can also play a protective role in cell structure. In the event of plant injury, Pro interacts with proteins to form a hydrophobic skeleton to stabilize and protect biological macromolecules and cell membrane structures. Pro is also a variety of free radical scavengers [58]. Pro can reduce the oxygen damage caused by stress through chelating singlet oxygen and hydroxyl radical. Another way of Pro to remove ROS is to stimulate the activity of POD, catalase (CAT), superoxide dismutase (SOD), polyphenol oxidase (PPO), and other enzymes in plants. Under the stress of adversity, Pro can bind to proteins to form a protective film with water molecules on the surface of proteins. The formation of a protective membrane restrains the flow of water to the outside of the cell and reduces the loss of water. Moreover, the protective membrane has a good protective effect on proteins and other biological macromolecules, maintaining the high structure and activity of biological macromolecules. For denatured proteins under the stress of adversity, Pro can improve the hydrophilicity of denatured proteins after combining with it. It keeps the dissolved state of the denatured proteins so as to avoid the agglutination of the denatured proteins interfering with the metabolic activities of the cells. Therefore, Pro is an important osmotic regulating substance.

Glycine betaine is a water-soluble substance with amphoteric characteristics. As an effective non-toxic osmotic regulator, it can bind to both hydrophilic and hydrophobic regions of biological macromolecules such as enzymes. Drought stress can cause the accumulation of glycine betaine and it can improve the drought-resistant ability of plants [61], which have been proven in sunflower [62], wheat [63], barley [64], pepper [65], *Axonopus compressus* [66], etc. The application of glycine betaine can effectively improve the osmotic regulation ability, stomatal conductance, and carboxylation efficiency of CO_2 assimilation so as to promote photosynthesis [67]. In other words, under drought stress, glycine betaine can stabilize the structure and properties of biological macromolecules, such as the key enzymes of the dicarboxylic acid cycle, terminal oxidases, and the photosystem, etc., which have important physiological significance in maintaining normal respiration and photosynthesis of plants [58]. Soluble sugar (SS) is an important energy and carbon source in the organism and participates in many processes of plant life metabolism. The soluble sugar in general plants includes glucose, fructose, sucrose, and other carbohydrates. The accumulation of soluble sugar can reduce the water potential of cells and improve the ability of plants to absorb water and retain water. In addition, most osmotic regulators fail to protect proteins and biofilms with further water loss under severe drought stress. Only soluble sugars can take the place of water molecules and form hydrogen bonds with proteins to maintain the specific structure and function of proteins. Moreover, the increase of soluble carbohydrates between biofilms can avoid the direct collapse of the biofilm system.

However, osmotic regulation also has limitations. The improvement in drought resistance of plants is only temporary. Moreover, it has a very limited effect on plant drought tolerance. If drought stress is severe, the turgor pressure of plants cannot be maintained. The effects of drought are present even within the range of osmotic adjustment of water potential. Osmotic regulation can only alleviate drought damage of plants to a certain extent.

3.3. Drought-Induced Proteins

Drought-induced proteins are newly synthesized proteins in plants under drought stress, which play a protective role in plant adaptation to stress and can improve plant drought tolerance. Drought-induced proteins can be divided into two categories according to their functions: (1) functional proteins, which play a direct protective role in cells, mainly include ion channel proteins, LEA proteins, OSM proteins, and metabolic enzymes, etc. (2) Regulatory proteins, including protein kinases, phospholipase C, phospholipase D, G protein, calmodulin, transcription factors, and some signaling factors, are involved in signal transduction or gene expression regulation in water stress and play indirect protective roles. Three important drought-inducible proteins, LEA, AQP, and dehydrin, are highlighted below.

3.3.1. Late Embryogenesis Abundant Protein

Late embryogenesis abundant (LEA) protein is a dehydrating protective protein enriched in the late stage of seed embryo development. LEA protein is rich in lysine and glycine, most of which are between 10 and 30 kD, and a few of which are above 30 kD. LEA is a large family of proteins, with more than 50 of them found in *Arabidopsis* alone [68]. It is regulated by plant development stage, ABA and dehydration signal, etc., and can be expressed in many tissues and organs of plants, with high hydrophilicity and thermal stability. The ability to capture enough water into cells is closely related to the dehydration tolerance of plants and the protection of tissues from water stress. Most LEA proteins do not have a stable secondary structure, but they may acquire an α-helix structure after drying [69]. LEA protein can participate in the process of crop resistance to environmental stress and plays a key role in this process, which is closely related to its amino acid composition and structure. Most LEA proteins contain a high proportion of polar amino acids, which makes them highly hydrophilic. What is more, most LEA proteins contain some conserved sequences. These sequences can form high helical folding under stress conditions, and such structure may have a hydrophobic effect with the membrane system of some denatured proteins. By stabilizing lipid membrane or functional proteins, a large amount of water loss can be prevented, thus reducing the influence of the external environment on intracellular metabolism [70]. In addition, there is a dynamic equilibrium between random conformation and α-helix in the dissolved state of LEA protein, which is also one of the reasons that LEA protein can participate in the resistance of crops to environmental stress [71].

One of the important functions of LEA protein in response to stress such as drought is its ability to scavenge ROS. Plants will produce a large number of reactive oxygen free radicals under adverse conditions, which have strong oxidation properties and can damage cell membranes and proteins, etc. Therefore, the scavenging of reactive oxygen free radicals becomes an important protection mechanism of plants under adverse conditions. Hara et al. found that the dehydrin CuCOR19 in *Citrus reticulata* can scavenging hydroxyl radicals and hydrogen peroxide and reduce the damage of reactive oxygen free radicals to plants [72]. Dean et al. found that glycine, lysine, and histidine were vulnerable to free radical attacks, and the total contents of glycine, lysine, and histidine in citrus dehydrin CuCOR19 were up to over 40%. Therefore, LEA protein could consume part of free radicals through three amino acids and play a protective role on plants. Because the LEA protein has no obvious secondary structure, the oxidation of some amino acids is obviously not sufficient to completely destroy its function [73]. Lea protein can prevent the loss of water by binding to the lipid membrane through the α helix structure, so it has the function of binding to the membrane to stabilize the membrane. For example, in corn, dehydrin DHN1 can bind to vesicle membrane containing acidic phospholipids, and its helicity will be significantly enhanced, indicating that functional conformational changes of DHN1 have occurred at the membrane interface, which may be related to dehydrin maintaining the stability of vesicle membrane and other intimal structures under adversity [74]. Hara et al. also found that overexpression of *CuCOR19* in tobacco could inhibit lipid membrane peroxidation of tobacco [75]. Thalhammer et al. found that the Arabidopsis LEA proteins COR15A and COR15B could bind to lipid membranes under drought conditions and play a protective role in lipid membranes [76]. In addition to scavenging reactive oxygen free radicals and maintaining the stability of the intimal system, LEA protein can be used as a cryo-protectant and metal ion protectant to participate in a wide range of stress. It was found that most of the LEA proteins of the dehydrin family play an important role in the process of cold resistance in plants. The accumulation of dehydrin gene *WCS120* was significantly correlated with its survival rate during winter in wheat [77]. Overexpression of *DHNS* in *Arabidopsis thaliana* showed that the cold tolerance of *Arabidopsis thaliana* was improved [78]. There is no domain associated with metal ion binding in dehydrin, but the proportion of histidine in dehydrin is high. It is speculated that the high proportion of histidine in dehydrin makes dehydrin have the ability to bind metal ions. It was also found that His-X and His-X3-His (X is an arbitrary amino acid) structures existed in many dehydrators, and further study confirmed that the ability of these domains to bind metal ions was significantly higher than that of other amino acids [79]. However, in the study of *Ricinus communis*, it was found that there was no correlation between the ability of dehydrin to bind metal ions and histidine content, which indicated that dehydrin had a more complex mechanism of binding metal ions [80].

Similar to soluble sugar in cells, LEA proteins with high hydrophilicity bind a large number of water molecules, allowing plants to maintain normal metabolism without damaging cells even in the event of severe dehydration [81]. The facultative α-helix structure formed by LEA protein interacts with the cell membrane under dehydration conditions, making the cell membrane maintain a relatively stable state even under dehydration conditions, thus preventing water loss [71]. In addition, most of the *LEA* genes have ABA response elements in their promoter regions, so the increase of endogenous ABA content in plants under drought conditions can also lead to the increase of *LEA* gene expression [82]. Under drought stress, LEA protein is also a good enzyme protectant. For example, the *LEA* genes of *Boea hygrometrica*, *LEA1*, and *LEA2*, were transferred into tobacco (*Nicotiana tabacum* L.) to obtain transgenic tobacco. It has found that the transgenic tobacco superoxide dismutase (SOD), peroxidase (POD), and photosynthetic system II related enzyme activity was increased. Moreover, the water content of leaves also increased. Thus, the stability of its protein is enhanced [83]. Overexpression of some *LEA* genes can improve the drought resistance of plants. Studies have confirmed that the overexpression of wheat *LEA* gene *TaLEA3* into *Leymus chinensis* can improve drought resistance [84].

3.3.2. Dehydrin

Dehydrin, a member of the Lea-II family with a molecular weight of 9~200 kDa, is a drought-induced protein widely found in higher plants. It is produced during late embryogenesis and responds to low temperature and exogenous ABA, or typically accumulates in dehydration stressed plants under drought, salt, and extracellular freezing. Dehydrin is rich in glycine and lysine and lacks cysteine and tryptophan. It is highly hydrophilic. In addition, dehydrin is a heat-stable protein that remains stable in boiling water and is thought to play an important role in protecting cells from damage caused by cell dehydration. An important structural feature of dehydrin is that it has three conserved regions: K, S, and Y fragments. The K fragment consists of 15 amino acids (EKKGIMD-KIKEKLPG) and is rich in lysine. The K fragment is usually located at the C end of the protein sequence and can form the amphipathic α-helix, which is the important structural basis of its hydrophilicity [70]. The S fragment is composed of a series of serine residues, and phosphorylation of the S fragment has been shown to enable dehydrin to enter the nucleus guided by signal peptides [85]. The conserved sequence of fragment Y is (T/V) D (E/Q) YGNP, located at the N-terminal of the dehydrated protein. The fragment Y is homologous to the nucleic acid binding sites of some bacterial and plant molecular chaperones. In addition, dehydrin has some conservative less rich in polar amino acid of Φ fragments and approved a similar nuclear localization signal (NLS) sequence [86]. According to the number of K, S, and Y fragments, the plant dehydrin gene family can be divided into five subfamilies: K_n, SK_n, Y_nSK_n, Y_nK_n, and K_nS.

In an aqueous solution, dehydrin forms the largest amount of hydrogen bonds with neighboring water molecules, while the proportion of external hydrogen bonds is very low, which does not form the hydrophobic core required for folding protein. Therefore, the dehydrin protein presents an unstructured and disordered protein form without a fixed three-dimensional structure. However, when the microenvironment around the dehydrin protein changes, the conformation of the dehydrin protein also changes. In the dehydrated state, the K fragment forms an α-helix type conformation in which the negatively charged amino acids lie on one side of the helix, the hydrophobic amino acids on the other, and the positively charged amino acids lie on the polar nonpolar interface. The α-helix with both hydrophilic and hydrophobic properties can interact with the dehydrated surfaces of other proteins or biofilm surfaces [87]. Therefore, dehydrin plays a stabilizing role in protecting the membrane system. Stress often dehydrates plant cells, destroys the hydration protection system on the surface of membrane lipid bilayers, reduces the space between membrane lipid bilayers, and causes membrane fusion and severe destruction of membrane structure. The amphiphilic α-helix formed by the K fragment in the dehydration condition enables the dehydrin to participate in the hydrophilic and hydrophobic interactions. Due to its high hydration ability, dehydrin binds with membrane lipids to prevent excessive loss of water in cells, maintain the hydration protection system of membrane structure, prevent the decrease of membrane lipid bilayer spacing, and thus prevent membrane fusion and the destruction of biofilm structure [88]. Dehydrin also protects the protein. The amphipathic α-helix formed by the K fragment can bind the dehydrin to the hydrophobic point of the partially denatured protein, acting as a molecular chaperone to prevent the further denaturing of the protein. In addition, the middle fragment of dehydrin contains a large number of polar amino acid residues, which can produce synergistic effects with small polar molecules and low molecular weight substances (carbohydrates, amino acids, water molecules, etc.) in the nuclear matrix and cytoplasmic matrix, enhancing the protective effect of dehydrin on proteins [70].

3.3.3. Aquaporin

Aquaporin (AQP) is a class of intrinsic proteins in the plasma membrane or vacuolar membrane that specifically transport water, ranging from 26 kD to 30 kD, and belongs to the same family of major intrinsic protein (MIP) proteins as ion channels and glycerol channels. Based on the homology and structural characteristics of amino acid sequences, the AQPs family of plants is classified into four types: plasma membrane intrinsic proteins (PIPs); tonoplast intrinsic proteins (TIPs); nodulin 26-like intrinsic proteins (NIPs); small and basic intrinsic proteins (SIPs) [89]. Among them, PIPs are mainly located in the plasma membrane and can be divided into PIP1, PIP2, and PIP3 according to the homology difference between N-terminal and C-terminal sequences. TIPS are mainly distributed in the vacuolar membrane and can be divided into five groups according to different tissue location, namely α, β, γ, δ and ε, which are important aquaporins in plants. NOD26 is the first member of the NIP family found in plants, located on the symbiotic membrane of soybean and rhizobia [90]. According to the structural differences of ar/R of aquaporins and the specificity of transport substrates, NIPS is divided into three categories: NIPI, NIPI, and NIPIII. This subfamily can transport other substances except for water molecules [91]. SIPs are the smallest family of AQPs in plants, mainly located in the endoplasmic omentum, and can be divided into SIP1 and SIP2 according to the different NPA sequences in the N-terminal and B-ring [92].

The expression of AQP showed strong temporal and spatial specificity. AQP is highly expressed in tissues and organs that need a lot of water flow, such as root epidermis, outer cortex and endodermis cells, xylem parenchyma cells near xylem vessels, phloem associated cells, guard cells, etc. The physiological function of AQP is closely related to its expression period and location, and its functions cover a series of physiological processes such as seed maturation and germination, cell elongation, root growth, leaf extension and movement, petal expansion, pollen, and ovule development [93–98]. At the subcellular level, AQP is mainly distributed in membrane systems such as cell membrane, vacuole membrane, endoplasmic omentum membrane, chloroplast membrane, and mitochondrial membrane. It was also found that AQP was redistributed at the subcellular level in different tissue sites and in different environments [99,100]. AQP located in the cell membrane at the cellular and subcellular levels is mainly responsible for water absorption and effluent. AQP located on the invaginated plasma membrane contributes to water transport between the protoplast and the vacuole. AQP located in the vacuolar membrane plays a role in regulating turgor. The specific distribution of plant AQP indicates that strong water flow across cells occurs in this region. In general, at the cellular level, the plasma membrane intrinsic protein (PIP) is mainly responsible for water absorption and outflow, and the vacuolar membrane intrinsic protein—tonoplast intrinsic protein (TIP)—is responsible for regulating turgor pressure, thus maintaining the integrity of cells [101]. For the whole plant, the specific distribution of plant AQP indicates that there is strong water flow across cells in this region [102].

AQP plays an important role in water transport. During the transmembrane transport of water in plants, AQP promotes the transmembrane transport of water inside and outside of cells by reducing the resistance encountered in the transmembrane transport of water and accelerates the rate of water migration between cells along the gradient of water potential. This is an important function of AQP in the transmembrane transport of water between different intracellular regions. At the same time, AQP is also the main way of water in and out of the cell, balancing the water potential inside and outside the cell. For example, the AQP on the cell membrane of plant root cells can regulate 70%~90% of the water flowing through the root. Water is absorbed by the root system of the plant, which passes through the casparian strip into the vessels. The vascular system ensures that water is transported in large quantities through the plant. In many plants, AQP expression has been found in vascular bundles and adjacent tissues [103,104]. This suggests that plant AQP can accelerate water transport and facilitate water flow in and out of vascular bundles. In addition, plant AQP can maintain the water potential balance between xylem parenchyma cells and transpiration flow [105]. When the transpiration and water potential of the ducts are higher than that of parenchyma cells, the water will be stored in the vacuole through AQP transport. When the water potential of parenchyma cells is higher than the transpiration water potential of the ducts, AQP will transfer the stored water to the ducts. Water is transported across the plasma membrane and vacuole membrane of parenchyma cells through AQP. In addition to water molecules, aquaporin also transports other physiologically important neutral small molecules, such as CO_2, H_2O_2, glycerol, NH_3/NH_4^+, boron, silicon, and urea, which are involved in a series of important physiological processes in plants, such as photosynthesis, nutrient absorption, cell signal transduction, and stress response. The function of AQP determines its positive role in drought stress.

3.4. Reactive Oxygen Metabolism

3.4.1. Production and Basic Function of Reactive Oxygen Species

Oxygen is necessary for aerobic organisms to maintain their own life activities. When oxygen is not completely reduced in the metabolic process, a series of metabolites and their derivatives with more active chemical properties will be produced, called reactive oxygen species (ROS). ROS include superoxide radical O_2^-, H_2O_2, singlet oxygen $^1O_2$, hydroxyl radical $\cdot OH$, and organic oxygen radical $(RO\cdot, ROO\cdot)$, etc. [106]. Under normal conditions, the ROS produced in plants maintains a balance with its scavenging system. However, when plants are under drought stress, ROS production and clearance will be out of balance. Drought can cause the increase of reactive oxygen free radicals and make plant cells suffer oxidative stress. When ROS exceeds the capacity of the ROS scavenging system, it will cause the accumulation of ROS and oxidative damage. The production of these free radicals will lead to a variety of harmful cytological effects, such as biofilm lipid peroxidation, protein denaturation, DNA strand breakage, and blocked photosynthesis. Two types of protection systems, enzymatic and non-enzymatic, have been formed correspondingly in the process of long-term evolution in plants to maintain a moderate level of ROS.

ROS can be produced in plants through many metabolic pathways. For example, in the process of photosynthesis and respiration, plant mitochondria, chloroplasts and peroxisomes, and some other organelles or parts with high oxidation activity or strong electron transfer function can also produce ROS. Chloroplasts are the main source of ROS production in green plants [107]. When plants are in a water-deficient environment, the absorption efficiency of light energy decreases. The blocked fixation of carbon dioxide in plants results in a decrease in $NADP^+$ supply and a relative increase in the rate of photosynthetic electron transfer to O_2. Oxygen and so on are used as electron acceptors to form O_2^-. In turn, O_2^- can trigger a series of chain reactions to produce a large amount of ROS in plants [108]. Mitochondria are another important ROS-producing organelle. In the process of electron transfer in the respiratory chain, some electrons leak in the midway, making O2 form O_2^- [109]. ROS in plants can also be produced in the plasma membrane and plasmid. NADPH oxidase, pH-dependent cell wall peroxidase, oxalate oxidase, and amine oxidase on the plasma membrane are all sources of ROS. Besides, enzymes in the endoplasmic reticulum and other organelles, such as cyclooxygenase, peroxidase, and lipoxygenase, can produce ROS through a series of chemical reactions.

The ROS function has two sides. ROS can destroy plant biofilm systems. For example, $\cdot OH$ can directly induce the peroxidation decomposition of the unsaturated fatty acid chain in phospholipids, thus destroying of membrane structure. However, the peroxides and NO in ROS are mainly produced by NADPH oxidase, glutathione oxidase, and NO synthase, with low activity, so they cannot directly interact with lipids to induce lipid peroxidation (LPO). Pacher et al. showed that they can react quickly to produce peroxynitrite, which initiates the LPO reaction [110]. The forced destruction of membrane structure will lead to a series of biological dysfunction. ROS can also degrade biomacromolecules in plants. Almost all proteins or enzymes can be damaged by ROS oxidation. ROS can lead to decreased or loss of protein function, peptide chain breakage, protein cross-linking, the transformation of amino acid residues change, and changes in immunochemical properties, etc. [111]. The damage of ROS to protein is mainly through carbonylation and glycosylation. The oxygen-free radicals can interact with the sulfhydryl group of the active center of the enzyme to oxidize it into disulfide bonds, resulting in the inactivity of the enzyme. ROS can indirectly disrupt plant growth and development through the loss of enzyme activity. ROS can also interact with purines, pyrimidines, and deoxyribose in

DNA molecules to cause the breakage, degradation, and modification of single or double strands of DNA, thus damaging genetic material [112]. In addition to the toxic effects of plant damage, ROS in plants is also involved in the process of resisting external stress and regulating plant growth and development. Oxidative burst is an important process in which ROS is involved in plant defense response. When the pathogen infects the plant, the plant produces a large amount of ROS through oxidative burst, which directly kills the pathogen [113]. In addition to biotic stress, ROS also plays a key regulatory role in response to abiotic stress [114].

3.4.2. Reactive Oxygen Scavenging System

In order to protect plants from ROS damage, there are endogenous antioxidant protection systems, including non-enzymatic antioxidants and antioxidant enzymes. The synergistic effect of antioxidants and antioxidant enzymes makes the production and quenching of ROS in vivo in a dynamic balance, thus alleviating or mitigating stress damage and making plants adapt to drought stress. The non-enzymatic scavenging systems of ROS in plants mainly include ascorbate, reduced glutathione (GSH), vitamin E, mannitol, carotenoids, and flavonoids. These substances can react directly with ROS or appear as substrates of enzymes in the ROS scavenging mechanism. In addition, some small molecules such as vitamins are also involved in scavenging oxygen free radicals and preventing lipid peroxidation. It is an indispensable part of the body's anti-oxidation defense system. Enzymes involved in antioxidant protection in plants mainly include SOD, CAT, APX, DHAR, MDHAR, GR, and POD. The main function of SOD is to remove O_2^-, and can convert O_2^- to H_2O_2. SOD plays a key role in the enzyme system and is the first line of defense against ROS elimination system in plants. CAT and POD are mainly responsible for the removal of H_2O_2 in organisms. Besides, APX, GR, DHAR, and MDHAR are also very important H_2O_2 scavenging enzymes. Together, they form a second line of defense against ROS elimination systems in plants. GPX plays an important role in scavenging oxidative metabolism of lipids and alkyl peroxides, constituting the third line of defense against ROS scavenging.

SOD is one of the most important metal enzymes in the antioxidant enzyme system and plays a core role in the protective enzyme system. It alternately oxidizes and reduces the metals connected with the enzyme, and catalyzes the disproportionation reaction of O_2^- to generate O_2 and H_2O_2. Its activity is considered to be an important index of plant stress resistance. Generally speaking, SOD activity in plants under drought stress is positively correlated with an antioxidant capacity [115]. SOD activity increased under mild or short-term water stress but decreased under severe or long-term water stress. However, some studies believe that the change of SOD activity is complex. For example, with the increase of stress intensity, SOD activity always decreases, or first decreases and then increases, or remains unchanged. The above differences may be due to the fact that the response of plants to water deficit is initiated not by water deficit itself, but by the degree of water deficit perceived by plants. Plant SOD can be divided into three types: Mn-SOD, Cu/Zn-SOD, and Fe-SOD according to the metal atoms bound by SOD. Cu/Zn-SOD is composed of two subunits, each of which contains a Cu and a Zn, and is the most abundant one among the three superoxide dismutases. Each subunit of Mn-SOD and Fe-SOD contains only one metal ion. Mn-SOD and Fe-SOD have similar sequences and identical characteristic domains. Lower plants are dominated by Fe-SOD and Mn-SOD, while higher plants are dominated by Cu/Zn-SOD. Cu/Zn-SOD is mainly located in cytoplasm and chloroplasts, Mn-SOD is mainly located in mitochondria, and Fe-SOD is generally located in chloroplasts of some plants. In addition to cytoplasm, chloroplast, and mitochondria, SOD also exists in glyoxylate circulators and peroxisomes [116].

APx is one of the important components of the AsA-GSH redox pathway in plants. APx is about 30 kDa and generally exists in monomer form. Homodimer may also appear in some cAPx. It uses ascorbic acid (AsA) as an electron donor to catalyze the reaction between AsA and H_2O_2 to produce MD (monodehydroascorbate acid) and water. AsA, as both reactant and reaction product, can be recycled continuously, so that APx can be fully catalyzed to protect the chloroplast to maintain normal function. Four APx isozymes have been isolated: cytoplasmic isozyme cAPx, APx in chloroplasts, soluble sAPx in chloroplast stroma, and tAPx in membrane binding form in chloroplast thylakoids. In addition, a kind of peroxide object binding APx was also found. tAPx and sAPx exist in similar molar ratios in chloroplasts. Cytoplasmic cAPx and chloroplast APx have different electron donors and different internal sequences.

CAT is a heme-containing tetramer enzyme found in all plant cells that rapidly breaks down H_2O_2 into H_2O and O_2. CAT mainly exists in peroxisomes in cells and is responsible for scavenging H_2O_2 produced in peroxisomes. CAT is also found in glyoxylic acid circulators and its function is mainly to remove H_2O_2 produced by photorespiration or fatty acid β-oxidation reaction [117]. Since H_2O_2 can be directly diffused across the membrane, H_2O_2 generated by other parts can also be diffused into peroxisomes and decomposed by CAT. In synergy with SOD, H_2O_2 can remove potentially harmful O_2^- and H_2O_2 in plants, thus minimizing the formation of $\cdot$OH. CAT is not directly involved in the decomposition process of H_2O_2. Its scavenging mechanism is that the heme iron of the enzyme reacts with H_2O_2 to generate an iron peroxide active body, which then oxidizes 1 molecule of H_2O_2.

The non-enzymatic ROS scavenging system in plants mainly includes ascorbate, reduced glutathione (GSH), vitamin E, mannitol, carotenoids, and flavonoids, which can react directly with ROS or act as enzyme substrates in the ROS scavenging mechanism. In addition, as an indispensable part of the body's anti-oxidation defense system, some small molecules such as vitamins also participate in the removal of oxygen free radicals, preventing lipid peroxidation. For example, some cysteine-rich small molecular proteins in plants, such as metallothionein (MT) [118] and gibberellin-induced protein (GIP) [119], can also degrade H_2O_2. Overexpression of these antioxidant proteins can significantly reduce the content of H_2O_2 in plants after abiotic stress treatment, thus improving the stress resistance of transgenic plants.

Actually, there are two types of glutathione: reduced glutathione (GSH) and oxidized glutathione (GSSG). Among them, reduced glutathione (GSH) is commonly known as glutathione, which can scavenge free radicals in cells that have toxic effects. GSH is a mercapto tripeptide compound formed by the polymerization of glutamic acid, cysteine, and glycine, in which the mercapto group as the active group is easy to combine with some substances, such as free radicals and heavy metals to play a detoxification effect. In the biosynthesis of glutathione, GSH biosynthesis catalyzed by glutamate-cysteine ligase (GCL) and glutathione synthetase (GS) plays a crucial role in maintaining homeostasis and preventing redox damage [120]. For example, when a small amount of H_2O_2 is generated inside the cell, GSH reduces H_2O_2 to H_2O under the action of GPx, and its own is oxidized to GSSG. Under the action of glutathione reductase, GSSG receives H to reduce to GSH, so that the scavenging reaction of free radicals in the body can be carried out continuously, thus stabilizing the membrane structure.

Ascorbic acid (AsA), also known as vitamin C, is a kind of abundant small molecule antioxidant substance commonly found in plants [121]. AsA can act as an important antioxidant and enzyme cofactor in plants, regulating photosynthesis, photooxidation, cell division, and playing an important role in plant signal transduction [122,123]. In plants, AsA content was positively correlated with plant stress resistance. The content of AsA varies greatly among different tissues of plants. For example, Smirnoff has suggested that AsA is present in chloroplast stroma in significantly higher concentrations than in other tissues [124]. As an important antioxidant in plants, AsA can directly or indirectly reduce the amount of ROS. AsA can directly remove ROS including O_2^-, $^1O_2$. Indirectly, AsA can reduce α-tocopherol and act as an electron donor for APx to remove H_2O_2, thus achieving the ROS scavenging purpose [125]. In addition, AsA also plays an important role in photoprotection as a cofactor in the lutein cycle, thereby protecting organisms and their normal metabolism from damage caused by oxidative stress [126]. More importantly, because the end product of the AsA oxidation reaction is non-toxic DHA or 2, 3-DKG, the free radical reaction chain can be terminated.

The ascorbate-glutathione cycle (AsA-GSH) is the main pathway of AsA and GSH regeneration. In this cycle, AsA acts as an electron donor for ascorbate peroxidase (APx) to remove H_2O_2. Monodehydroascorbate (MDHA) generated by oxidation can be reduced by MDHAR, and can also disproportionate to generate AsA and dehydroascorbate (DHA). DHAR uses GSH as an electron donor to reduce DHA to AsA, and the oxidized glutathione (GSSG) generated can be reduced to GSH again by GR, so as to complete the process of scavenging ROS, such as H_2O_2 and regenerating AsA and GSH. The AsA-GSH cycle plays an important role in the antioxidant protection of plants under drought stress. A study on the response of AsA-GSH circulatory metabolic enzymes in *Coffea canephora* to drought stress showed that APX, GR, and DHAR activities increased under drought stress, but MDHAR activity had no significant change [127]. By studying sunflower and sorghum, Jingxian et al. found that there were differences in the response of AsA-GSH circulatory metabolic enzymes to drought stress in different plant organelles. They proposed that the chloroplast AsA-GSH cycle was the main method to remove H_2O_2 in sunflower under drought stress, while the cytoplasmic AsA-GSH cycle was the main method to remove H_2O_2 in sorghum [128].

4. Drought Stress Signal Transduction in Plants

The signal transduction process of plants from sensing environmental stimuli to responding to them generally includes three parts: (1) the sensory transduction and response of sensory cells to environmental stimuli, namely the original signal sensory transduction process, producing intercellular messenger; (2) the intercellular messenger is transmitted between cells or tissues, and finally acts on the receptor cell site; (3) the transduction and response of acceptor cells to intercellular messengers lead to physiological, biochemical, and functional changes in the acceptor tissues, which are ultimately reflected in the response of plants to environmental stimuli or adversity [129].

4.1. Plant Drought Stress Signal

The decrease of soil water content caused the change of leaf water status, and then affected the physiological function of plants. Leaf water potential reflects plant water status and is related to specific stress degrees. The decrease of leaf water potential and turgor pressure affected the synthesis, transportation, and distribution of plant hormones, such as ABA and cytokinin. Changes in turgor pressure caused by cell water loss may be the reason for cell perception of water stress, which is also known as the hydraulic signal of plant drought stress [130]. Besides hydraulic signal, the electrical signal also plays an important role in plant signal transduction under drought stress. Fromm et al. proposed through their study on maize in dry soil that electrical signals play an important role in the communication between roots and shoots of water-deficient plants [131]. What is more, when plants feel the initial drought signal, the osmotic stress signal is converted into an intracellular chemical signal by the membrane receptor, which triggers the downstream effector to produce the second messenger. Then the signal is amplified gradually through the cascade transmission of the signal. In the process of signal transduction of dry early stress, the second messengers involved in signal transduction mainly included plant hormone signals, Ca^{2+}, IP3, phosphatidic acid, and ROS signals.

Plant hormones are a kind of chemical signal molecules that regulate plant growth. They often play a regulatory role in a low concentration. They can transmit cell signals in different parts of plants and among cells so that the remote transmission of plant signals can be realized. When soil water content decreases, some physiologically active substances act as chemical signals, and their content increases, which is called a positive signal. For example, under drought stress, the content of IAA, ABA, and ethylene increases. In contrast, a decrease in a biologically active substance is called a negative signal, such as cytokinins.

ABA is a small molecule lipophilic plant hormone, which is a crucial signal molecule in plant water stress. As a kind of plant hormone, ABA can control plant growth, inhibit seed germination and promote aging. In addition to regulating plant growth and development, ABA is also involved in regulating plant responses to various external stresses, embodied in content increasing greatly when the plant is in drought, high salt, low temperature, and other adversities. Moreover, ABA plays a pivotal role in the information connection between the aboveground and underground parts of plants. When plants are under drought stress, ABA produced in the rhizosphere can be used as a positive signal to regulate the physiological activities of aboveground parts. When plants are under water stress, root cells are the first to experience environmental changes and produce ABA, which transmits the signal to other organs and tissues of plants through vascular bundles, causing senescence of leaves and stomatal closure, so as to reduce water loss. ABA can be transported from the underground part to the aboveground part through the xylem, leading to increased ABA content in the leaves. In fact, ABA induces a wide range of downstream signaling factor responses, including kinases, phosphatases, G-proteins, and proteins in the ubiquitin pathway.

ABA has multiple receptors, such as ABAR/CHLH, •GCR2, •GTG1/2, and PYR/PYL/RCAR. These receptor proteins have the activity of protein kinases, which can be activated by binding ABA molecules to change the protein structure, and then activate or inhibit the activity of downstream signaling proteins to transmit signals between cells. Research on ABA receptors is still ongoing, and the exact function of the different receptors remains questionable. ABAR/CHLH is a magnesium ion chelatase H subunit located in plant cyto-plastids/chloroplasts. It not only catalyzes the synthesis of chlorophyll in cells but also participates in the reverse signal transfer between plastids/chloroplasts and the nucleus under stress conditions [132,133]. GCR2 protein is a G protein coupled receptor located in the plasma membrane of the cell. The C-terminal of GCR2 protein can interact with the A subunit of G protein (GPA1) to form a complex. The specific binding of ABA and GCR2 protein induces the release of G protein. The G protein is then separated into $G\alpha$ and $G\beta\gamma$ dimer, and the signal response of ABA is regulated by the downstream effector of GCR2 protein [134]. G protein, consisting of $G\alpha$, $G\beta$, and $G\gamma$ subunits, plays an important role in response to plant hormone signaling by synergistic G-protein coupled receptors and their downstream effectors. GTG1/2 was first identified and named by Pandey et al. through bioinformatics analysis. In the ABA signal transduction pathway model with GTG1/2 as the receptor, GPA1–GTP promoted GTG–GTP to maintain a high level by inhibiting GTG1/2 protease activity, thus reducing the binding probability of GTG–GDP and ABA. On the contrary, the binding of GTGSGDP to ABA can lead to the configuration change and then initiate ABA signaling response, but the specific molecular mechanism has not been clarified. The PYR/PYL/RCAR protein binds to ABA molecules outside the cell membrane, which in turn binds and inhibits the phosphatase activity of the downstream protein phosphatase PP2C [135].

As an essential mineral element in plants, Ca^{2+} plays an important role in maintaining the stability of cell membrane and cell wall structure and participating in intracellular homeostasis and regulation of growth and development in terms of cell structure and physiological functions. Wang et al. found that extracellular Ca^{2+} can activate the increase of intracellular Ca^{2+} concentration through the calcium-sensing receptor (CAS) on the plasma membrane of guard cells of *Arabidopsis thaliana*, thus confirming the role of extracellular Ca^{2+} as the first messenger [136]. In addition, as mentioned above, in response to drought, plants synthesize the hormone ABA, which causes stomatal closure to reduce water loss. During stomatal closure, the concentration of Ca^{2+} in the cytoplasm increases, and Ca^{2+} acts as the second messenger in osmotic stress response [137]. Drought-induced transient increase of intracellular Ca^{2+} in guard cells promotes stomatal closure, maintains plant water, improves water use efficiency, and ultimately enhances plant adaptation to drought by interacting with or without ABA signaling pathways and downstream signal transduction mechanisms. In stomatal closure, the ABA-dependent Ca^{2+} signaling pathway is the main pathway. ABA activates plasma membrane calcium channels in various ways and stimulates intracellular calcium reservoirs to release Ca^{2+}. More Ca^{2+} will inhibit the inward potassium channel and further affect the anion channel. The phenomenon of anion outflow and depolarization will block the inward potassium channel and promote the outward potassium channel, leading to potassium ion outflow [138]. The guard cells are under low turgor pressure due to a large outflow of anions and potassium ions, making the stomata close gradually. IP3 and cyclic adenosine 5′-diphosphate ribose (cADPR) are also key second messengers in guard cells that can regulate Ca^{2+} concentration. IP3 and cADPR can release Ca^{2+} in guard cells and increase the concentration of Ca^{2+}, while ABA can rapidly increase IP3 and cADPR in guard cells. These three second messengers initiate calcium channels to transfer calcium ions into the cytoplasm and accumulate in large quantities, causing ion channels to interact with each other to produce a series of effects that promote stomatal closure [139,140]. Ca^{2+} transmits stress signals downstream by interacting with protein receptors. Major Ca^{2+} signal transduction pathways are involved in calcium-regulated kinase-mediated phosphorylation, including the regulation of downstream gene expression by Ca^{2+} regulating transcription factors and Ca^{2+} sensitive promoter elements [141]. Calcium-dependent protein kinases (CDPKs), calmodulin (CaM), and calcineurin B-like proteins (CBLs), which have been identified in plants, can recognize specific Ca^{2+} and rely on these calcium signals to transmit downstream to adapt to drought stress.

A certain amount of ROS produced under stress can be used as signal molecules to activate relevant active substances or defense systems, and mitigate the damage caused by abiotic stress [142]. Among ROS, H_2O_2 is mostly used as an important signal molecule for animal and plant cells to respond to various stresses because H_2O_2 is a very stable ROS with the longest half-life and strong diffusivity. Different plant organelles have different responses to cellular REDOX signals under drought stress. Although H_2O_2 is produced faster in peroxisomes and chloroplasts, mitochondria are the most vulnerable organelles to oxidative damage [143,144]. Increased mitochondrial production of H_2O_2 may be an important alarm signal, up-regulating the antioxidant defense system or triggering programmed cell death when oxidative stress intensifies. Studies have shown that H_2O_2 can regulate calcium mobilization, protein phosphorylation, and gene expression. Pei et al. found that H_2O_2 can regulate Ca^{2+} influx in protoplasts and increase of $[Ca^{2+}]$cyt in guard cells by activating Ca^{2+} channels in the plasma membrane of guard cells of *Arabidopsis thaliana*. In addition, they further proposed that ABA-induced H_2O_2 production and $H_2O_2{}^-$activated Ca^{2+} channels are important mechanisms of ABA-induced stomatal closure [145]. Mori et al. also reported an inevitable link between ROS signaling and stomatal closure in plants [146]. Yan et al. also reached the same conclusion: ABA can promote the production of ROS, and the ROS produced can act as signal molecules to regulate stomatal closure [147]. In addition, H_2O_2 also induces the phosphorylation of mitogen-activated protein kinase (MAPK), which is involved in multiple signal transduction cascades that regulate downstream gene expression [148].

4.2. Intracellular Transduction Pathways and Regulation Mechanisms of Plant Drought Stress Signals

Drought stress signal transduction can be divided into two pathways. The first pathway is the ROS-activated MAPK cascade pathway. MAPK cascade regulates antioxidant defense system and osmotic regulation system in plants. Furthermore, the damage caused by drought stress can be relieved by removing ROS and changing the osmotic potential of cells. The other pathway is Ca^{2+}-dependent stress signaling bypass mediated by calmodulin-dependent protein kinase (CDPK). Ca^{2+} signal is produced under drought stress, and Ca^{2+} signal further regulates the expression of plant protective proteins, such as LEA protein through CDPK, which is involved in the late response to drought stress, and ultimately enhances the drought resistance of plants.

Mitogen-activated protein kinases (MAPKs) are a class of important protein kinases involved in signal transduction, which play an extremely important role in plant growth, development, and stress response [149]. The MAPK cascade consists of three components: MAPK, MAPKK (MAPK kinase), and MAPKKK (MAPK kinase kinase). When the first member of this pathway, MAPKKK, is activated, the other two components undergo sequential phosphorylation and are activated in turn. The reason is that MAPKKK can double phosphorylate the serine (Ser) and serine/threonine (Ser/Thr) in MAPKK, thus activating it. The protein kinase of MAPK containing n conservative district and a very conservative TXY motif between the VII and the III subregion [150]. MAPKK initiates MAPKK by dual phosphorylation of threonine (T) and tyrosine (Y) residues at both ends of the X site [151]. As a result, MAPKK phosphorylates MAPKK and MAPKK phosphorylates MAPKK. Activated MAPKK can activate transcription factors and also cause cellular signaling responses through interactions with other proteins.

The full name of CDPK is calmodulin-dependent/calmodulin-independent protein kinase or calmodulin-like domain protein kinase. It belongs to Ser/Thr type protein kinases and is a large family encoded by multiple genes. Under the stimulation of external signals, plant cells showed changes in Ca^{2+} concentration and then activated CDPK. CDPK regulates downstream gene expression and product activity through the phosphorylation cascade. These products play an important role in the regulation of gene expression, enzyme metabolism, ion, and water transmembrane transport, and other microscopic aspects so that plants show macroscopic changes such as growth and development, stress resistance changes [152].

5. Drought Stress Signal Transduction in Plants

Generally, drought stress response genes can be divided into functional genes and regulatory genes. The products of functional genes directly resist environmental stress, such as aquaporin genes, osmoregulatory factors (such as sucrose, proline, and betaine) synthase genes, protective proteins (such as LEA protein, molecular chaperone, etc.) genes. The products of regulatory genes, such as protein kinase genes, protein phosphatase genes, phospholipid metabolism-related genes, and stress-related transcription factor genes, are involved in signal transduction and regulation of gene expression to indirectly respond to stress. These proteins act by participating in plant stress signal transduction pathways or by regulating the expression and activity of other effector molecules.

5.1. Functional Genes

5.1.1. Osmotic Adjustment Related Genes

According to the different pathways of proline accumulation, the related enzymes can be divided into three categories. The first category is the enzymes related to proline synthesis, including △-pyrroline-5-carboxylate synthetase (P5CS), pyrroline-5-carboxylate reductase (P5CR), and ornithine-δ-aminotransferase (δ-OAT). The second category is related to the degradation of proline enzymes, including proline dehydrogenase (ProDH) and △-pyrroline-5-carboxylate dehydrogenase (P5CDH). The third category is proline transport-related enzyme ProT. The synthesis sites of proline in plants are cytoplasm and chloroplast, and the synthesis pathways include glutamic acid (Glu) and ornithine (Orn) synthesis pathways [153]. Glutamic acid synthesis pathway mainly occurred under osmotic stress and nitrogen deficiency, while ornithine synthesis pathway existed in nitrogen abundant environment [154]. In the glutamic acid synthesis pathway, Glu is catalyzed by △-pyrroline-5-carboxylate synthetase (P5CS) to produce glutamic semialdehyde (GSA). Subsequently, GSA is automatically cycled to form pyrroline-5-carboxylic acid (P5C), which generates proline (Pro) under the action of pyrroline-5-carboxylate reductase (P5CR) [155,156]. Substrates and enzymes in the first step of the ornithine synthesis pathway are different from those in the glutamate pathway. The substrate was ornithine (Orn) and the enzyme was ornithine-δ-aminotransferase (δ-OAT). The substrates and products under the two pathways mainly include Glu, Orn, GSA, P5C, and Pro. The enzymes required for the reaction include P5CS, P5CR, and δ-OAT. Kishor et al. transferred the *P5CS* and *P5CR* genes into tobacco. It was found that although the mRNA levels of both were increased, the proline level of *P5CR* transgenic tobacco was not significantly increased, while the proline level of *P5CS* transgenic tobacco was significantly increased [157]. La Rosa et al. obtained the same result that when soybean *P5CR* gene was overexpressed in tobacco, the activity of P5CR was increased five times, but the level of proline in transgenic tobacco was not significantly increased [158]. These results indicated that the increase of proline was more affected by *P5CS* than by *P5CR*. Therefore, the P5CS enzyme is the rate-limiting enzyme of proline metabolism and determines the synthesis of proline. Sharma et al. found that Arabidopsis *P5CS1* mutants underproduce proline during stress [159]. Baocheng et al. introduced P5CS cDNA from moth bean (*Vigna aconitifolia* L.) into rice (*Oryza sativa* L.) genome. The transgenic plants showed overproduction of the P5CS enzyme and accumulation of proline [160]. Similarly, in transgenic *AtP5CS* tobacco, its proline content was significantly increased, and its osmotic regulation ability was enhanced [161]. The same effect was also shown in potato [162], sugarcane [163], soybean [164], etc. δ-OAT is another key enzyme in proline synthesis and its activity was significantly enhanced under drought conditions [165]. Overexpression of *δ-OAT* in plants can significantly increase proline content in tobacco, rice, etc. [166,167]. In addition, the degradation of proline occurs in mitochondria and is the reversal of the synthesis pathway of glutamic acid. Proline is first oxidized by proline dehydrogenase (ProDH) to P5C, which is reduced to glutamic acid by $\triangle$-pyrroline-5-carboxylate dehydrogenase (P5CDH) [168]. Studies have shown that Arabidopsis proline dehydrogenase (*PDH1*) mutants block Pro catabolism and found that plants maintain growth through active Pro catabolism under low water potential [159]. What is more, proline transport requires the participation of ProT. This transporter belongs to the amino acid/auxin permease (AAAP) gene family in plants and is a typical Na^+-dependent sub-amino acid transporter. The transporter is directly absorbed by proline coupling along with the Na^+-electrochemical gradient, which requires the participation of Na^+-K-ATPase and belongs to active transport [169]. However, many studies have proved that the alteration of ProT expression cannot change proline accumulation in a directed way. In *Arabidopsis thaliana* plants overexpressing *HvProT*, the proline content in the aboveground part decreased while that in the root increased [170].

The synthesis of glycine betaine (GB) in plants, mainly accomplished by the enzymatic reaction, has been elucidated in many studies. Choline, as the initiator of GB synthesis, is obtained through the methylation of three adenosine-methionine-dependent phospho-ethanolamine (PE) catalyzed by the cytoplasmic enzyme phospho-ethanolamine N-methyltransferase (PEAMT) [171]. The PEAMT enzyme has two tandem methyltransferase domains at the N terminal and C terminal. The N-terminal methyltransferase domain methylate PE to phosphate-monomethyl-ethanolamine (P-MME), and the C-terminal methyltransferase domain methylate P-MME to phosphate-dimethylethanolamine (P-DME), and P-DME to phosphocholine (PC) [172]. PC is then converted to choline in different ways. McNeil et al. found a different transformation pathway for PC in spinach and tobacco, the former by direct dephosphorylation to choline, and the latter by first containing PC in phosphatidylcholine and then metabolizing it to choline [173]. Next, betaine is synthesized by a two-step oxidation reaction. The first step was to oxygenate choline into betaine aldehyde with the help of a ferredoxin-dependent choline monooxygenase (CMO). The CMO catalyzed step is the rate-limiting step in GB biosynthesis [174]. The second step is NAD^+-dependent betaine aldehyde dehydrogenase (BADH) catalyzed the oxidation of betaine aldehyde into betaine [175,176]. CMO is a ferredoxin-dependent rate-limiting enzyme encoded by a single gene. CMO has Rieske-type [2Fe-2s] active site and is the only matrix enzyme with the Rieske iron-sulfur center, usually localized in the chloroplast or other subcellular compartments [177]. Under normal conditions, CMO activity is low and unstable. Since the reduced ferredoxin is produced by photosynthetic electron transport, the CMO activity in plants can be improved to a certain extent under light induction. The CMO plays a balancing and speed-limiting role in this process. Due to the toxic effect of betaine aldehyde on plant cells in this step, CMO should not only synthesize enough betaine aldehyde for further synthesis of betaine but also limit the excessive accumulation of betaine aldehyde in the plant. The catalytic enzyme BADH is a dimer encoded by a single chain nuclear gene with two alleles. It is composed of two monomers of equal molecular weight. It belongs to the superfamily of aldehyde dehydrogenases and also has nonspecific effects on other aldehyde substrates [178]. BADH is dependent on both NAD^+ and $NADP^+$, but in plants, BADH shows higher activity in the presence of NAD^+ [179]. The BADH of monocotyledons may be located in microsomes, while that of dicotyledons may be located in the chloroplast stroma. BADH has two isozymes (BADH I and BADH II), in which BADHII plays a more important role [180]. BADH, as the most important catalytic enzyme in the synthesis of betaine, has low activity under normal conditions. However, under the stress conditions of low temperature, drought, and high salinity, the isozyme activity of BADH was significantly increased, which resulted in the synthesis of a large amount of betaine, indicating that the activity of BADH was induced by stress. With advances in genomics and proteomics as well as genetic engineering techniques, some plant species have been engineered using genes from the GB biosynthetic pathway that confer tolerance to abiotic stresses. Most of the plants that have been genetically engineered to produce GB are naturally non-GB accumulative plants [181]. Shen et al. isolated and identified the *CMO* gene from spinach and transferred it into tobacco, and found that salt tolerance and drought tolerance of transgenic tobacco were also significantly improved [182]. Similarly, other studies have also shown that *CMO* transgenic rice and tobacco can significantly improve their tolerance to salt and drought stress [183,184]. Ishitani et al. isolated and cloned the *BADH* gene from barley and transferred it into tobacco, which improved the drought tolerance of tobacco to a certain extent [185]. Fan et al. transferred the *SoBADH* gene from spinach into the sweet potato and found that the transgenic plants showed stronger BADH activity and eventually showed increased tolerance to abiotic stress [186]. Li et al. transferred the *SoBADH* gene into tomatoes to produce transgenic plants with higher levels of betaine and greater stress resistance [187].

The metabolism of soluble sugar in plants is very complex. Taking sucrose as an example, FBPase (fructose-1, 6-bisphosphatase) and sucrose phosphate synthase (SPS) are important rate-limiting enzymes in the sucrose synthesis pathway. The enzyme FBPase, one of the key enzymes in the gluconeogenesis pathway, catalyzes the hydrolysis of fructose-1 6- diphosphate (FDP) to fructose -6- phosphate (F6P). The catalytic product of FBPase in the cytoplasm is sucrose, while the catalytic product of FBPase in the chloroplast is starch. Cho et al. constructed FBPase overexpressed Arabidopsis lines and found that the soluble sugar content of the transgenic plants was significantly increased [188]. On the contrary, decreasing the activity of FBPase in potato cytoplasm by antisense technique resulted in a decrease in sucrose synthesis rate [189]. SPS catalyzed the synthesis of sucrose-6-phosphate using uridine diphosphate glucose (UDPG) as the donor and fructose 6-phosphate as the receptor. Sucrose 6-phosphate is dephosphorylated and hydrolyzed to form sucrose and phosphate ions under the action of sucrose phosphate phosphorylase (SPP). This reaction is basically irreversible. However, SPS and SPP exist in the plant body in the form of complex, so SPS catalysis of sucrose production is actually irreversible. Therefore, SPS is a key enzyme controlling sucrose synthesis in plants [190]. Park et al. transferred the Arabidopsis *AtSPS1* gene into tobacco and found that the sucrose content of transgenic tobacco increased, accompanied by plant height growth, stem diameter thickening, and fiber lengthening [191]. Moreover, previous studies have confirmed that the SPS activity and sucrose content of transgenic plants obtained by introducing *ZmSPS1* into tomato [192], potato [193], and Arabidopsis [194] were significantly increased.

5.1.2. Drought-Induced Protein Genes

LEA protein is a protein that is highly expressed in late embryonic development. It plays a crucial role in plant response and resistance to drought, mainly by capturing water, stabilizing and protecting the structure and function of proteins and membranes, and protecting cells from water stress as a molecular chaperone and hydrophilic solute [195]. Sivamani introduced the ABA-responsive gene *HVA1* (a member of group 3 LEA protein genes) into spring wheat (*Triticum aestivum* L.), and found that the transgenic wheat had significantly higher water use efficiency and better growth characteristics under water deficit condition than the control wheat [196]. Under drought stress, seed germination rate, seedling fresh weight, and root length of *CmLEA-S* (a melon Y3SK2-type *LEA* gene) transgenic plants were significantly higher than those of wild-type plants. They also had less wilting and yellowing, more proline, less MDA, and stronger APX and CAT activities [197]. Luo et al. constructed *Capsicum annuum* L. plants with the expression of *CaDHN5* (a dehydrin gene) downregulated by VIGS (Virus-induced Gene Silencing) and Arabidopsis plants with transgenic overexpression of *CaDHN5*. It was found that *CaDHN5* was positively correlated with the expression of manganese superoxide dismutase (*MnSOD*) and peroxidase (*POD*) genes [198]. Under drought stress, seed germination rate and survival rate of *OeSRC1* (a Ks-type dehydrin gene) transgenic tobacco plants were higher than those of wild-type tobacco plants, and they accumulated more free proline, but electrolyte leakage did not change significantly [199].

Plant aquaporin (AQP) is a membrane channel located in the plasma membrane and intracellular module, which can promote the transport of water, small neutral molecules, and gases across biofilm [200]. Aquaporin belongs to the MIP family of proteins that regulate cellular water movement and maintain water relationships in plants, especially under drought stress. As mentioned earlier, AQP can be divided into PIPs, TIPs, NIPs, and SIPs, as well as the genes that encode them. Among them, plasma membrane intrinsic proteins (PIPs) and tonoplast intrinsic proteins (TIPs) mediate the main pathways of intracellular water transport, maintain intracellular and intercellular water relations under stress, and are involved in many processes of the drought stress response. Zhang etc. found that rose water channel protein RhPIP2;1 can influence plant growth and stress reaction by interacting with the membrane MYB protein RhPTM [201]. Overexpression of *CrPIP2;3* in *Arabidopsis thaliana* (a PIP2 gene from rose) can promote the survival and recovery of transgenic plants under drought stress by regulating water homeostasis, thus affecting drought tolerance of plants [202]. The seed germination rate, seed yield, seed vigor, and root length of transgenic *Arabidopsis thaliana* lines overexpressing *JcPIP2;7* (a plasma membrane intrinsic protein gene) and *JcTIP1;3* (a tonoplast intrinsic protein gene) under mannitol condition were significantly higher than those of the control [203]. Peng et al. tested the effect of the ginseng *PgTIP1* gene by transgenic it into Arabidopsis plants and showed that it altered root morphology and leaf water channel activity, thereby altering drought tolerance [204]. The overexpression of *CsTIP2;1* in Arabidopsis plants increased the expansion of mesophyll cells, midrib aquiferous parenchyma abundance, H_2O_2 detoxification, and stomatal conductance, and then significantly improved the water and oxidation state, photosynthetic capacity, transpiration rate, and water use efficiency of leaves under the condition of continuous dry soil [205].

5.2. Regulatory Genes

Regulatory genes are genes that regulate stress signal transduction and functional gene expression. The regulatory genes of drought stress response can be divided into the following categories. The first is the transcription factors related to the regulation of stress gene expression, including bZIP, MYB, MYC, EREBP/APZ, CBFI (CRT/DRE binding factor), DREB1A (DRE binding), etc. These transcription factors can be strongly induced by water stress and their expression can further regulate the expression of various functional genes. The second type of protein kinases is related to the sensing and transduction of stress signals, such as receptor protein kinases, ribosomal protein kinases, transcription regulatory protein kinases, etc. These kinases usually play the role of stress signal cascading amplification. Among them, the most important are the three key kinases included in the MAPK cascade: MAPK, MAPKK, and MAPKKK. The third type is related to the second messenger generation and transduction of enzymes, such as phospholipase D, phospholipase C. Phospholipase C catalyzes the hydrolysis of PIP into diesterphthalein glycerol (DG) and inositol triphosphate (IP3). IP3 can induce the release of Ca^{2+} stored in the endoplasmic reticulum into the cytoplasm, and thus initiate the intracellular signal transduction process.

5.2.1. Signal Transduction Related Genes

The key step of ABA biosynthesis is to catalyze 9-cis-epoxycarotenoid dioxygenase (NCED) [206]. In *Arabidopsis thaliana*, drought tolerance is regulated by the *NCED* gene. The overexpression of the *AtNCED3* gene in Arabidopsis leads to the increase of endogenous ABA level, and drought and ABA promote gene transcription. Overexpression of this gene in plants resulted in a decrease in leaf respiration rate and an increase in drought resistance. The antisense inhibition of this gene made it sensitive to drought, suggesting that the expression of this gene plays a key role in ABA biosynthesis under drought stress [207]. Under drought stress, the increased activity of ABA synthase (such as *ZEP, NCED, LOS5/ABA3*, and *AAO*) in plant root cells produced a large amount of ABA, which was transported to leaf cells through transpiration flow. ABA is perceived by ABA receptors on guard cells and is transported across the membrane by intracellular second messengers [calcium messenger, proton messenger, inositol triphosphate (IP3), etc.]. Thus, a variety of ion channels and enzymes related to physiological and biochemical reactions are activated to regulate stomatal movement and eventually lead to stomatal closure. Other studies have shown that under drought conditions, ABA promotes open stomatal closure and inhibits closed stomatal opening in isolation. During stomatal closure, ABA, H_2O_2, and NO may all act on the MAPK signaling pathway. In the future, tomato-derived *LeNCED1* was transferred into tobacco (with tetracycline as control). When tobacco leaves were treated with tetracycline, the increase of ABA content in the leaves induced *NCED* transcription, but there was no significant difference in the tomato transformed with *LeNCED1* under the strong promoter CaMV35S [208].

Calmodulin, calmodulin-like proteins, calmodulin B-like proteins, and calcium-dependent protein kinases (CDPKs) are the four major families of calcium-binding proteins in plants. As a Ca^{2+} signal sensor, CDPK is closely related to the further transmission of cellular Ca^{2+} signal. Because the N-terminal serine/threonine-protein kinase domain of CDPKs can be fused with the carboxy-terminal calmodulin-like domain containing the EF-hand calcium-binding site, CDPKs are independent of exogenous calmodulin interactions but can be directly activated by Ca^{2+} binding [209,210]. Although most *CDPK* genes are commonly expressed in organisms, some *CDPK* genes are expressed only in specific tissues or are induced by hormonal, biological, or abiotic stress conditions. Salt stress, drought stress, and other abiotic stress can significantly improve the transcription level of CDPK [211,212]. Urao et al. cloned two *CDPK* genes, named *AtCDPK1* and *AtCDPK2*, from *Arabidopsis thaliana*. The expression of these two genes can be induced by drought, suggesting that these two genes are involved in osmotic stress signal transduction [213]. The protein kinases *AtCDPK10* and *AtCDPK30* expressed in maize protoplasts can activate the promoter of the *HVA1* gene induced by drought and high salt stress, and the mutant without the *CDPK* region is not responsive to various stresses and ABA. Therefore, it is speculated that *AtCDPK10* and *AtCDPK30* are the positive regulators of the plant stress signal transduction pathway [214]. Moreover, Saijo et al. found that overexpression of *OsCDPK7* in rice enhanced drought stress resistance of rice [215].

The MAP protein kinase genes isolated from *Arabidopsis thaliana* were induced by drought, high salinity, and low-temperature stress, including *AtMPK3, AtMPK4, AtMPK6, AtMEK1*, and *AtMEKK1*. Studies have shown that the MAP kinase cascade system is not only regulated by phosphorylation and dephosphorylation at the protein level but also induced by environmental stress signals at the transcriptional level. Mizoguchi et al. found that *AtMEKK1* is involved in the MAP kinase cascade signaling of drought, high salinity, low temperature, and traumatic stress in Arabidopsis. The cascade pathway consists of At-MEKK1 (MAPKK kinase), AtMEK1 (MAPKK kinase), and AtMPK4 (MAPKK kinase) [216]. It has also been reported that drought or high salinity also activates SIMK (stress-induced MAPK) in *Medicago sativa* cells and SIPK (salicylic acid-induced protein kinase) in tobacco cells. Chitlaru et al. found that hypertonic stress could rapidly activate a protein kinase, and confirmed that the protein kinase belonged to MEK1 [217]. Xiong et al. found that the *OsMAPK5* gene in rice was induced by a variety of biological and abiotic stresses, and overexpression of this gene in rice could enhance drought resistance, salt resistance and low-temperature tolerance of transgenic rice [218].

5.2.2. Transcription Factor Genes

In the process of signal transduction under drought stress, transcription factors (TF) regulate and reduce the damage to plants from multiple levels by activating multiple pathways, which plays a crucial role in the growth and development of plants under stress [219]. Among them, the transcription factor gene families related to drought stress mainly include *HD-Zip/bZIP, AP2/ERF, NAC, MYB,* and *WRKY*. However, Different transcriptional factors play different transcriptional regulatory roles under drought conditions, depending on plant species and strain, development stage, and drought treatment intensity. Gong et al. pointed out that the 43 transcription factor genes in drought response of tomato drought-resistant lines mainly came from 5 families with the most abundant expression changes, which were *WRKY, NAC, BHLH, AP2/EREBP,* and *HSF* in turn, while *MYB, bZIP,* and *CCAAT* families had less abundant expression [220]. Different from tomato, the highest abundance of the 261 transcription factor genes in rice were *MYB* (35 members) and *AP2/EREBP* (28 members), followed by 21 *bHLH*, 11 *HSF*, 27 *NAC*, and 15 *WRKY*. Moreover, drought-resistant rice cultivars could activate more upregulated transcription factors than non-drought-resistant rice cultivars. For example, the number of upregulated transcription factors in the *AP2* family of rice drought-resistant variety was 35 more than that of rice non-drought-resistant variety after 18 days of drought [221].

The HD-Zip transcription factors belong to a homeobox protein encoding 60 conserved amino acid homeodomains (HD), which consists of six families, namely HD-Zip, KNOX, PHD, BELL, WOX, and ZF-HD [222]. Among them, homeodomain-leucine zipper (HD-zip) is a plant-specific transcription factor, which consists of DNA-homologous domain and additional Leu zipper (Zip) components [223]. The former binds specifically to DNA, while the latter mediates the formation of protein dimer, a transcription factor involved in regulating plant growth and development under normal growth conditions and environmental stress [224]. Based on sequence conservatism, structural characteristics, function, and other characteristics, HD-Zip transcription factors can be divided into four subfamilies (HD-Zip I~ HD-Zip IV). Different subfamily members have different biological functions, some are involved in the cross-interaction of multiple hormonal pathways, and some interact with key genes and downstream genes of hormonal pathways [222]. Atalou et al. proposed that the expression of subfamily I and II genes of the HD-Zip family of transcription factors were induced by drought stress. These two genes participate in the hormone signaling pathway, regulate the expansion, division, and differentiation of plant cells by interacting with the hormone pathway genes and downstream genes, and thus improve the drought resistance of plants [225]. Expression analysis by Deng et al. showed that CpHB-7 negatively regulates the expression of ABA-responsive genes, which also explains the reduced sensitivity of transgenic plants with ectopic *CpHB-7* to ABA during seed germination and stomatal closure [226]. *Arabidopsis thaliana* with overexpression of HD-Zip I subfamily gene Hahb-4 showed strong tolerance to water stress and insensitivity to ethylene because the overexpression of *Hahb-4* gene inhibited the expression of ethylene synthesis genes *ACO, SAM,* and downstream ethylene signaling genes *ERF2* and *ERF5* [227]. Fan et al. silenced RhHB1, which encodes a homeodomain-leucine zipper I γ-clade transcription factor in rose flowers, resulting in an increased content of JA-Ile and a decreased tolerance to dehydration. It has also been shown that RhHB1 can inhibit the expression of lipoxygenase 4 (*RhLOX4*) by directly binding to the promoter of *RhLOX4*. In other words, the JA feedback loop mediated by the *RhHB1/RhLOX4* regulatory module provides dehydration tolerance by fine-tuning the level of bioactive JA [228].

AP2/ERF transcription factors play an important role in plant stress resistance and previous studies have shown that they can participate in the process of drought stress resistance in plants through different pathways. AP2/ERF can regulate drought stress response by affecting the synthesis of plant hormones. Cheng et al. proposed that as the upstream component of jasmonic acid and ethylene signals, ERF1 can integrate JA, ET, and abscisic acid signals through stress-specific gene regulation, and play a positive role in drought tolerance [229]. Wan et al. found a drought-induced upregulated ERF transcription factor gene *OsDERF1*, and the overexpression of *OsDERF1* in rice reduced the tolerance of rice to drought stress at the seedling stage. It has been demonstrated that *Os*DERF1 can directly bind to the GCC boxes in the promoter regions of negative regulatory factors *OsAP2-39* and *OsERF3* and activate their expression. However, *Os*AP2-39 and *Os*ERF3 can bind to the GCC box of *ACS* and *ACO* promoter of ethylene synthesis genes and inhibit the expression of these genes, thus inhibiting the synthesis of ethylene. Therefore, the reduction of ethylene content by overexpression of *OsDERF1* is one of the important reasons for the decrease of drought tolerance in rice [230]. Zhang et al. found that overexpression of *JERF1* can improve drought tolerance of transgenic rice and that *JERF1* can activate expression of *OsABA2* and *Os03G0810800*, two key enzymes of ABA synthesis, and increase ABA content. These results suggest that *JERF1* may regulate drought response through the ABA pathway. Moreover, AP2/EREBP can also respond to drought stress by affecting metabolite synthesis in plants. By overexpressing *DREB1A* in *Arabidopsis thaliana*, Maruyama et al. found that the contents of starch degrading enzyme, sucrose metabolizing enzyme, and sugar alcohol synthase changed, affecting the content changes of monosaccharide, disaccharide, and trisaccharide, thus enhancing the drought resistance of transgenic plants [231]. In upland rice, *OsERF71*-overexpressing lines, different from *OsERF71* interference lines, were found to enhance drought resistance by increasing the expression of *OsP5CS1* and *OsP5CS2* regulating proline synthesis [232]. The overexpression of *FaDREB2* in *Broussonetia papyrifera* can increase the content of soluble sugar and proline in vivo, and thus enhance the tolerance [233]. It was also found in rice that when the rice gene *JERF3* was overexpressed, the accumulation of sugar and proline in rice could be increased to resist drought [234]. Similarly, overexpression of *GmERF3* could improve the drought resistance of tobacco by increasing soluble sugar and proline content, respectively [235]. In addition, some people have pointed out that AP2/EREBP protein is also involved in ROS clearance. Through GUS activity test and SOD activity detection, Wu et al. found that JERF3 can bind to the GCC box of *NtSOD*, thereby activating the expression of *NtSOD*, improving the activity of SOD, enhancing the ability of ROS scavenging, and improving the tolerance of tobacco to osmotic stress [236]. What is more, AP2/ERF transcription factor can also regulate drought resistance of plants by participating in the regulation of wax synthesis. Wang et al. found that OsWR1 physically interacts with the DRE and GCC boxes in the promoter of wax-related genes *OsLACS2* and *OsFAE1'-L*, which can directly regulate the expression of these genes, thereby altering long-chain fatty acids and alkanes to regulate wax synthesis. Therefore, the drought resistance of overexpression of *OsWR1* was significantly improved [237].

MYB is one of the largest transcription factor families in plants. It is widely involved in the regulation of secondary metabolism, response to hormones and environment, the guidance of cell differentiation and morphogenesis, and also plays a key role in resistance to drought and other abiotic stresses [238]. The N-terminal of MYB transcription factor is a conserved helix-turn-helix (HTH) protein DNA binding domain consisting of 52 amino acids, which directly determines the accuracy of binding to target genes and can bind to cis components, such as GCC box, DRE, ABRE, W box, etc. The C-terminal is the transcriptional initiating region, which determines the transcriptional activity of a transcription factor and its interaction with other genes or components to manipulate the expression efficiency of downstream genes [239]. According to the structure of the DNA binding domain, the MYB transcription factor family can be divided into 1R-MYB, 2R-MYB, 3R-MYB, and 4R-MYB subfamilies. In Arabidopsis plants overexpressing *OsMYB3R-2*, expression of genes dehydration-responsive element-binding protein 2A, *COR15a*, and *RCI2A* was significantly increased, leading to enhanced abiotic stress resistance [240]. The *Arabidopsis thaliana* overexpressing *GaMYB85* had higher free proline and chlorophyll content, showed higher seed germination rate under mannitol treatment, and higher drought resistance efficiency than the wild type under water shortage conditions, most probably via an ABA-induced pathway. Furthermore, the ectopic expression of *GaMYB85* resulted in increased transcription levels of stress-related markers such as *RD22*, *ADH1*, *RD29A*, *P5CS*, and *ABI5* [241]. *GbMYB5* gene silencing decreased the proline content and antioxidant enzyme activity increased the malondialdehyde (MDA) content and decreased the tolerance of cotton to drought stress. However, in tobacco lines overexpressing *GbMYB5*, proline content and antioxidant enzyme activity increased, while MDA content decreased. The expression levels of the antioxidant genes *SOD*, *CAT*, and *GST*, polyamine biosynthesis genes *ADC1* and *SAMDC*, and the late embryogenesis abundant protein-encoding gene ERD10D and dry-responsive genes *NCED3*, *BG*, and *RD26* were significantly increased in tobacco overexpressing *GbMYB5* [242]. Wang et al. constructed *GmMYB84* overexpressing soybeans, which has longer primary root length, greater proline, and ROS contents, higher antioxidant enzyme activities [peroxidase (POD), catalase (CAT), and superoxide dismutase (SOD)], lower dehydration rate, and reduced malondialdehyde (MDA) content. In addition, they found that some ROS-related genes of the transgenic plants were upregulated under abiotic stress, and *GmMYB84* could directly bind to the promoters of *GmBOHB-1* and *GmBOHB-2* genes through electrophoretic mobility shift assay and luciferase reporter analysis [243]. Chen found that MdMYB46 can directly bind to lignin biosynthesis-related gene promoter to promote secondary cell wall biosynthesis and lignin deposition, and can also directly activate stress response signals to improve salt and osmotic stress tolerance of apple [244]. Geng also found that MdMYB88 and MdMYB124 could regulate root xylem development and regulate cellulose and lignin accumulation in response to drought by directly binding to *MdDVND6* and *MdMYB46* promoter under drought conditions [245].

The WRKY protein family, named for its highly conserved WRKYGQK DNA domain, is a zinc finger-type transcription regulator, which is a unique transcription factor in plants. In addition to the presence of at least one highly conserved WRKYGQK sequence and zinc finger structure, the WRKY domain also specifically interacts with the (T) (T) TGAC (C/T) sequence (W box) of the target gene promoter [246]. W-boxes are found in the promoters of many genes related to plant defense response and even in the self-promoters of some WRKY transcription factor genes. Therefore, WRKY transcription factors may regulate the expression of downstream functional genes or other regulatory genes through binding with W-box, thus participating in the regulation process of various physiological activities in plants. Overexpression of *TaWRKY10* in tobacco enhanced drought resistance, which was characterized by higher proline and soluble sugar content, lower ROS, and MDA content, and increased germination rate, root length, survival rate, and relative water content under stress conditions. This is because TaWRKY10 plays a positive role in drought stress by regulating osmotic balance, scavenging ROS and transcription of stress-related genes [247]. Moreover, Yan et al. found that GhWRKY17 regulates plant sensitivity to drought by reducing ABA levels, and regulates the expression of ROS scavenging genes, such as *APX*, *CAT*, and *SOD*. In other words, GhWRKY17 responds to drought and salt stress by regulating the ABA signaling pathway, and ROS production in plant cells [248]. Similarly, GhWRKY68 responds to drought and salt stress by regulating ABA signaling and cellular ROS, too [249]. In addition, WRKY transcription factors also can participate in the process of stress resistance by regulating the expression of other transcription factors. Wei et al. proposed that two ERF family genes *NtERF5* and *NtEREBP-1* in transgenic plants overexpressing *TcWRKY53* were negatively induced, suggesting that *TcWRKY53* may regulate osmotic stress responses through interaction with ERF transcription factors rather than direct regulation of functional genes [250].

The NAC family of transcription factors is a class of plant-specific transcription factors with a variety of biological functions, which is characterized by highly conserved and specific NAC domains in the N-terminal of proteins. NAC plays an important role in plant resistance to drought stress by directly or by regulating the expression of genes involved in drought response. Fujita et al. indicated that RD26, as a dehydration-induced NAC protein, plays a transcriptional role in ABA-induced gene expression in plants under abiotic stress [251]. Moreover, Yong et al. showed that LlNAC2 was involved in DREB/CBF-COR and ABA signaling pathways to regulate stress tolerance in lily [252]. *Arabidopsis thaliana* with overexpression of *PwNAC2* exhibited greater drought tolerance by scavenging ROS, reducing membrane damage, slowing water loss, and increasing stomatal closure. In addition, the ABA or CBF pathway marker genes transgenic with the *PwNAC2* gene were significantly increased in Arabidopsis, suggesting that PwNAC2 enhanced plant tolerance to abiotic stress through multiple signaling pathways [253]. Jiang et al. proposed that RhNAC$_3$, as a positive regulator, could improve the dehydration tolerance of rose petals mainly by regulating osmotic regulation-related genes [254]. In transgenic *Arabidopsis thaliana*, overexpression of *VvNAC17* enhanced drought resistance and upregulated expression of ABA and stress-related genes such as *ABI5, AREB1, COR15A, COR47, P5CS, RD22,* and *RD29A* [255]. However, excessive expression of stress-related genes may have negative effects on plant growth and development. Nakashima found that transgenic plants overexpressing *OsNAC6* improved drought, high salinity, and blast resistance, but resulted in dwarfing and low yield.

6. Conclusions

As mentioned above, in the past studies, the changes of plant external morphology and internal biochemical properties under drought stress have been described in detail. We have also gained a good understanding of signal transduction networks and molecular regulatory mechanisms in plants. Nevertheless, our current research is still incomplete and there are still many scientific problems to be solved. For example, ABA signaling networks are poorly described. In addition, although some famous abiotic stress-related gene families, such as AP2/ERF, MYB, NAC, etc., have been extensively studied by predecessors, there are still many unknown mechanisms in the large molecular regulation. Plants respond to water scarcity in different ways, and this is a complex process that we still need to work on unraveling. The research on the strategies of plants to cope with drought stress can help us to better use scientific means to improve the adaptability of plants to water shortage environment and increase the yield of crops to play a more important role. Therefore, this review provides valuable background knowledge and theoretical basis for selective breeding, cross breeding, and molecular breeding of agricultural and forestry crops in the future by systematically analyzing and summarizing the mechanisms of plant response to drought.

Author Contributions: This review was mainly organized by X.Y., M.L., Y.W. (Yufei Wang), Y.W. (Yiran Wang), and Z.L., who were the contributors for collection of the materials and reversion of the manuscript. S.C. conceived the study and revised the manuscript. All authors have read and agreed to the published version of the manuscript.

Funding: This review was funded by the Fundamental Research Funds for the Central Universities (2572019CG08), the National Natural Science Foundation of China, grant number 31870659 and Heilongjiang Touyan Innovation Team Program (Tree Genetics and Breeding Innovation Team).

Institutional Review Board Statement: Not applicable.

Informed Consent Statement: Not applicable.

Conflicts of Interest: The authors declare no conflict of interest.

References

1. Khan, M.A.; Iqbal, M.; Akram, M.; Ahmad, M.; Hassan, M.W.; Jamil, M. Recent advances in molecular tool development for drought tolerance breeding in cereal crops: A review. *Zemdirb. Agric.* **2013**, *100*, 325–334. [CrossRef]
2. Escalona, J.M.; Flexas, J.; Medrano, H. Stomatal and non-stomatal limitations of photosynthesis under water stress in field-grown grapevines. *Aust. J. Plant Physiol.* **1999**, *27*, 421–433. [CrossRef]
3. Li, W.; Wang, Y.; Zhang, Y.; Wang, R.; Xie, Z. Impacts of drought stress on the morphology, physiology, and sugar content of Lanzhou lily (*Lilium davidii* var. unicolor). *Acta Physiol. Plant.* **2020**, *42*, 127. [CrossRef]

4. Anjum, S.A.; Ashraf, U.; Tanveer, M.; Khan, I.; Hussain, S.; Shahzad, B.; Zohaib, A.; Abbas, F.; Saleem, M.F.; Ali, I.; et al. Drought Induced Changes in Growth, Osmolyte Accumulation and Antioxidant Metabolism of Three Maize Hybrids. *Front. Plant Sci.* **2017**, *8*, 69. [CrossRef] [PubMed]

5. Misra, V.; Solomon, S.; Mall, A.K.; Prajapati, C.P.; Hashem, A.; Abd Allah, E.F.; Ansari, M.I. Morphological assessment of water stressed sugarcane: A comparison of waterlogged and drought affected crop. *Saudi J. Biol. Sci.* **2020**, *27*, 1228–1236. [CrossRef]

6. Patmi, Y.S.; Pitoyo, A.; Solichatun; Sutarno. Effect of drought stress on morphological, anatomical, and physiological characteristics of Cempo Ireng Cultivar Mutant Rice (*Oryza sativa* l.) strain 51 irradiated by gamma-ray. *J. Phys. Conf. Ser.* **2020**, *1436*, 012015. [CrossRef]

7. Werner, C.; Correia, O.; Beyschlag, W. Two different strategies of Mediterranean macchia plants to avoid photoinhibitory damage by excessive radiation levels during summer drought. *Acta Oecologica* **1999**, *20*, 15–23. [CrossRef]

8. Taiz, L.; Zeiger, E. *Plant Physiology and Development*; Sinauer Associates: Sunderland, MA, USA, 2015.

9. Rucker, K.S.; Kvien, C.K.; Holbrook, C.C.; Hook, J.E. Identification of Peanut Genotypes with Improved Drought Avoidance Traits 1. *Peanut Sci.* **1995**, *22*, 14–18. [CrossRef]

10. Bhusal, N.; Lee, M.; Han, R.; Han, A.; Kim, H. Responses to drought stress in *Prunus sargentii* and *Larix kaempferi* seedlings using morphological and physiological parameters. *For. Ecol. Manag.* **2020**, *465*. [CrossRef]

11. Khaleghi, A.; Naderi, R.; Brunetti, C.; Maserti, B.E.; Babalar, M. Morphological, physiochemical and antioxidant responses of *Maclura pomifera* to drought stress. *Entific Rep.* **2019**, *9*, 19250. [CrossRef]

12. Hosseini, F.; Mosaddeghi, M.R.; Dexter, A.R. Effect of the fungus *Piriformospora indica* on physiological characteristics and root morphology of wheat under combined drought and mechanical stresses. *Plant Physiol. Biochem.* **2017**, *118*, 107–112. [CrossRef] [PubMed]

13. Mishra, B.K.; Srivastava, J.P.; Lal, J.P. Drought Resistance in Lentil (*Lens culinaris* Medik.) in Relation to Morphological, Physiological Parameters and Phenological Developments. *Int. J. Curr. Microbiol. Appl. Sci.* **2018**, *7*, 2288–2304. [CrossRef]

14. Asl, K.K.; Ghorbanpour, M.; Khameneh, M.M.; Hatami, M. Influence of Drought Stress, Biofertilizers and Zeolite on Morphological Traits and Essential Oil Constituents in *Dracocephalum moldavica* L. *J. Med. Plants* **2018**, *17*, 91–112.

15. Willick, I.R.; Lahlali, R.; Vijayan, P.; Muir, D.; Karunakaran, C.; Tanino, K.K. Wheat flag leaf epicuticular wax morphology and composition in response to moderate drought stress are revealed by SEM, FTIR-ATR and synchrotron X-ray spectroscopy. *Physiol. Plant.* **2018**, *162*, 316–332. [CrossRef]

16. Rueda, M.; Godoy, O.; Hawkins, B.A. Spatial and evolutionary parallelism between shade and drought tolerance explains the distributions of conifers in the conterminous United States. *Glob. Ecol. Biogeogr.* **2017**, *26*, 31–42. [CrossRef]

17. Picotte, J.J.; Rhode, J.M.; Cruzan, M.B. Leaf morphological responses to variation in water availability for plants in the *Piriqueta caroliniana* complex. *Plant Ecol.* **2008**, *200*, 267–275. [CrossRef]

18. Lobet, G.; Draye, X. Novel scanning procedure enabling the vectorization of entire rhizotron-grown root systems. *Plant Methods* **2013**, *9*, 1. [CrossRef]

19. Gowda, V.R.P.; Henry, A.; Yamauchi, A.; Shashidhar, H.E.; Serraj, R. Root biology and genetic improvement for drought avoidance in rice. *Field Crop. Res.* **2011**, *122*, 1–13. [CrossRef]

20. Cao, X.; Chen, C.; Zhang, D.; Shu, B.; Xiao, J.; Xia, R. Influence of nutrient deficiency on root architecture and root hair morphology of trifoliate orange (*Poncirus trifoliata* L. Raf.) seedlings under sand culture. *Sci. Hortic.* **2013**, *162*, 100–105. [CrossRef]

21. Xiao, S.; Liu, L.; Zhang, Y.; Sun, H.; Zhang, K.; Bai, Z.; Dong, H.; Li, C. Fine root and root hair morphology of cotton under drought stress revealed with RhizoPot. *J. Agron. Crop Sci.* **2020**, *206*, 679–693. [CrossRef]

22. Yang, Z.; Zhou, B.; Chen, Q.; Ge, X.; Shi, Y. Effects of drought on root architecture and non-structural carbohydrate of *Cunninghamia lanceolata*. *Acta Ecol. Sin.* **2018**, *38*, 6729–6740.

23. Zhan, A.; Schneider, H.; Lynch, J.P. Reduced Lateral Root Branching Density Improves Drought Tolerance in Maize. *Plant Physiol.* **2015**, *168*, 1603–1615. [CrossRef]

24. Du, Y.; Zhao, Q.; Chen, L.; Yao, X.; Zhang, W.; Zhang, B.; Xie, F. Effect of drought stress on sugar metabolism in leaves and roots of soybean seedlings. *Plant Physiol. Biochem.* **2020**, *146*, 1–12. [CrossRef]

25. Benjamin, J.G.; Nielsen, D.C. Water deficit effects on root distribution of soybean, field pea and chickpea. *Field Crop. Res.* **2006**, *97*, 248–253. [CrossRef]

26. Bismillah Khan, M.; Hussain, M.; Raza, A.; Farooq, S.; Jabran, K. Seed priming with $CaCl_2$ and ridge planting for improved drought resistance in maize. *Turk. J. Agric. For.* **2015**, *39*, 937–952. [CrossRef]

27. Kosma, D.K.; Bourdenx, B.; Bernard, A.; Parsons, E.P.; Lu, S.; Joubes, J.; Jenks, M.A. The Impact of Water Deficiency on Leaf Cuticle Lipids of Arabidopsis. *Plant Physiol.* **2009**, *151*, 1918–1929. [CrossRef]

28. Chen, M.; Zhu, X.; Zhang, Y.; Du, Z.; Chen, X.; Kong, X.; Sun, W.; Chen, C. Drought stress modify cuticle of tender tea leaf and mature leaf for transpiration barrier enhancement through common and distinct modes. *Sci. Rep.* **2020**, *10*, 6696. [CrossRef]

29. Chartzoulakis, K.; Patakas, A.; Kofidis, G.; Bosabalidis, A.; Nastou, A. Water stress affects leaf anatomy, gas exchange, water relations and growth of two avocado cultivars. *Sci. Hortic.* **2002**, *95*, 39–50. [CrossRef]

30. Hughes, J.; Hepworth, C.; Dutton, C.; Dunn, J.A.; Hunt, L.; Stephens, J.; Waugh, R.; Cameron, D.D.; Gray, J.E. Reducing Stomatal Density in Barley Improves Drought Tolerance without Impacting on Yield. *Plant Physiol.* **2017**, *174*, 776–787. [CrossRef]

31. Bai, T.; Li, Z.; Song, C.; Song, S.; Zheng, X. Contrasting Drought Tolerance in Two Apple Cultivars Associated with Difference in Leaf Morphology and Anatomy. *Am. J. Plant Sci.* **2019**, *10*, 709–722. [CrossRef]

32. Zhang, C.; Yang, H.; Wu, W.; Li, W. Effect of drought stress on physiological changes and leaf surface morphology in the blackberry. *Braz. J. Bot.* **2017**, *40*, 625–634. [CrossRef]
33. Vincent, D.; Lapierre, C.; Pollet, B.; Cornic, G.; Negroni, L.; Zivy, M. Water deficits affect caffeate O-methyltransferase, lignification, and related enzymes in maize leaves. A proteomic investigation. *Plant Physiol.* **2005**, *137*, 949–960. [CrossRef]
34. Yin, N.-W.; Li, J.-N.; Liu, X.; Lian, J.-P.; Fu, C.; Li, W.; Jiang, J.-Y.; Xue, Y.-F.; Wang, J.; Chai, Y.-R. Lignification Response and the Difference between Stem and Root of *Brassica napus* under Heat and Drought Compound Stress. *Acta Agron. Sin.* **2017**, *43*, 1689. [CrossRef]
35. Zhao, T.-J.; Sun, S.; Liu, Y.; Liu, J.-M.; Liu, Q.; Yan, Y.-B.; Zhou, H.-M. Regulating the Drought-responsive Element (DRE)-mediated Signaling Pathway by Synergic Functions of Trans-active and Trans-inactive DRE Binding Factors in *Brassica napus*. *J. Biol. Chem.* **2006**, *281*, 10752–10759. [CrossRef]
36. Jafari, S.; Hashemi Garmdareh, S.E.; Azadegan, B. Effects of drought stress on morphological, physiological, and biochemical characteristics of stock plant (*Matthiola incana* L.). *Sci. Hortic.* **2019**, *253*, 128–133. [CrossRef]
37. Bidgoli, R.D. Effect of drought stress on some morphological characteristics, quantity and quality of essential oil in Rosemary (*Rosmarinus officinalis* L.). *Adv. Med. Plant Res.* **2018**, *6*, 40–45. [CrossRef]
38. Flexas, J.; Bota, J.; Loreto, F.; Cornic, G.; Sharkey, T.D. Diffusive and Metabolic Limitations to Photosynthesis under Drought and Salinity in C3 Plants. *Plant Biol.* **2004**, *6*, 269–279. [CrossRef] [PubMed]
39. Santos, V.; Ferreira, M.J.; Rodrigues, J.; Garcia, M.N.; Ceron, J.V.B.; Nelson, B.W.; Saleska, S.R. Causes of reduced leaf-level photosynthesis during strong El Nino drought in a Central Amazon forest. *Glob. Chang. Biol.* **2018**, *24*, 4266–4279. [CrossRef] [PubMed]
40. Ahmed, M.; Stockle, C.O. *Quantification of Climate Variability, Adaptation and Mitigation for Agricultural Sustainability*; Springer International Publishing: Cham, Switzerland, 2017.
41. Gimenez, C.; Mitchell, V.J.; Lawlor, D.W. Regulation of Photosynthetic Rate of Two Sunflower Hybrids under Water Stress. *Plant Physiol.* **1992**, *98*, 516–524. [CrossRef]
42. Gunasekera, D.; Berkowitz, G.A. Use of Transgenic Plants with Ribulose-1,5-Bisphosphate Carboxylase/Oxygenase Antisense DNA to Evaluate the Rate Limitation of Photosynthesis under Water Stress. *Plant Physiol.* **1993**, *103*, 629–635. [CrossRef]
43. Deeba, F.; Pandey, A.K.; Ranjan, S.; Mishra, A.; Singh, R.; Sharma, Y.K.; Shirke, P.A.; Pandey, V. Physiological and proteomic responses of cotton (*Gossypium herbaceum* L.) to drought stress. *Plant Physiol. Biochem.* **2012**, *53*, 6–18. [CrossRef]
44. Ma, P.; Bai, T.-h.; Ma, F.-w. Effects of progressive drought on photosynthesis and partitioning of absorbed light in apple trees. *J. Integr. Agric.* **2015**, *14*, 681–690. [CrossRef]
45. Farooq, M.; Wahid, A.; Kobayashi, N.; Fujita, D.; Basra, S.M.A. Plant drought stress: Effects, mechanisms and management. *Agron. Sustain. Dev.* **2009**, *29*, 185–212. [CrossRef]
46. Mafakheri, A.; Siosemardeh, A.; Bahramnejad, B.; Struik, P.C.; Sohrabi, Y. Effect of drought stress on yield, proline and chlorophyll contents in three chickpea cultivars. *Aust. J. Crop Sci.* **2010**, *4*, 580–585.
47. Zaefyzadeh, M.; Quliyev, R.A.; Babayeva, S.M.; Abbasov, M.A. The Effect of the Interaction between Genotypes and Drought Stress on the Superoxide Dismutase and Chlorophyll Content in Durum Wheat Landraces. *Turk. J. Biol.* **2009**, *33*, 1–7.
48. Wu, M.; Zhang, W.H.; Ma, C.; Zhou, J.Y. Changes in morphological, physiological, and biochemical responses to different levels of drought stress in Chinese cork oak (*Quercus variabilis* Bl.) seedlings. *Russ. J. Plant Physiol.* **2013**, *60*, 681–692. [CrossRef]
49. Azzeme, A.M.; Abdullah, S.N.A.; Aziz, M.A.; Wahab, P.E.M. Oil palm leaves and roots differ in physiological response, antioxidant enzyme activities and expression of stress-responsive genes upon exposure to drought stress. *Acta Physiol. Plant.* **2016**, *38*, 1–12. [CrossRef]
50. Dastborhan, S.; Ghassemi-Golezani, K. Influence of seed priming and water stress on selected physiological traits of borage. *Folia Hortic.* **2015**, *27*, 151–159. [CrossRef]
51. Huber, W.; Sankhla, N. *C4 Pathway and Regulation of the Balance Between C4 and C3 Metabolism*; Springer: Berlin/Heidelberg, Germany, 1976.
52. Silvera, K.; Santiago, L.S.; Winter, K. Distribution of crassulacean acid metabolism in orchids of Panama: Evidence of selection for weak and strong modes. *Funct. Plant Biol.* **2005**, *32*, 397–407. [CrossRef] [PubMed]
53. Winter, K.; Holtum, J.A.M. The effects of salinity, crassulacean acid metabolism and plant age on the carbon isotope composition of *Mesembryanthemum crystallinum* L., a halophytic C3-CAM species. *Planta* **2005**, *222*, 201–209. [CrossRef] [PubMed]
54. Winter, K.; Holtum, J.A. Environment or development? Lifetime net CO2 exchange and control of the expression of Crassulacean acid metabolism in *Mesembryanthemum crystallinum*. *Plant Physiol.* **2007**, *143*, 98–107. [CrossRef]
55. Winter, K.; Garcia, M.; Holtum, J.A.M. Drought-stress-induced up-regulation of CAM in seedlings of a tropical cactus, *Opuntia elatior*, operating predominantly in the C3 mode. *J. Exp. Bot.* **2011**, *62*, 4037–4042. [CrossRef]
56. Osakabe, Y.; Osakabe, K.; Shinozaki, K.; Tran, L.S. Response of plants to water stress. *Front. Plant Sci.* **2014**, *5*, 86. [CrossRef]
57. Wang, S.; Wan, C.; Wang, Y.; Chen, H.; Zhou, Z.; Fu, H.; Sosebee, R. The characteristics of Na+, K+ and free proline distribution in several drought-resistant plants of the Alxa Desert, China. *J. Arid Environ.* **2004**, *56*, 525–539. [CrossRef]
58. Ashraf, M.; Foolad, M.R. Roles of glycine betaine and proline in improving plant abiotic stress resistance. *Environ. Exp. Bot.* **2007**, *59*, 206–216. [CrossRef]
59. Khoyerdi, F.F.; Shamshiri, M.H.; Estaji, A. Changes in some physiological and osmotic parameters of several pistachio genotypes under drought stress. *Sci. Hortic.* **2016**, *198*, 44–51. [CrossRef]

60. Szabados, L.; Savoure, A. Proline: A multifunctional amino acid. *Trends Plant Sci* **2010**, *15*, 89–97. [CrossRef]
61. Jones, R.W.; Storey, R. Salt Stress and Comparative Physiology in the Gramineae. II. Glycinebetaine and Proline Accumulation in Two Salt- and Water-Stressed Barley Cultivars. *Funct. Plant Biol.* **1978**, *5*, 817–829. [CrossRef]
62. Hussain, M.; Malik, M.A.; Farooq, M.; Ashraf, M.Y.; Cheema, M.A. Improving drought tolerance by exogenous application of glycinebetaine and salicylic acid in sunflower. *J. Agron. Crop Sci.* **2008**, *194*, 193–199. [CrossRef]
63. Raza, M.A.S.; Saleem, M.F.; Shah, G.M.; Khan, I.H.; Raza, A. Exogenous application of glycinebetaine and potassium for improving water relations and grain yield of wheat under drought. *J. Soil Sci. Plant Nutr.* **2014**, *14*, 348–364. [CrossRef]
64. Wang, N.; Cao, F.; Richmond, M.E.A.; Qiu, C.; Wu, F. Foliar application of betaine improves water-deficit stress tolerance in barley (*Hordeum vulgare* L.). *Plant Growth Regul.* **2019**, *89*, 109–118. [CrossRef]
65. Korkmaz, A.; Değer, Ö.; Kocaçınar, F. Alleviation of water stress effects on pepper seedlings by foliar application of glycinebetaine. *N. Z. J. Crop Hortic. Sci.* **2015**, *43*, 18–31. [CrossRef]
66. Nawaz, M.; Wang, Z. Abscisic Acid and Glycine Betaine Mediated Tolerance Mechanisms under Drought Stress and Recovery in *Axonopus compressus*: A New Insight. *Sci. Rep.* **2020**, *10*, 6942. [CrossRef]
67. Ma, X.L.; Wang, Y.J.; Xie, S.L.; Wang, C.; Wang, W. Glycinebetaine application ameliorates negative effects of drought stress in tobacco. *Russ. J. Plant Physiol.* **2007**, *54*, 472. [CrossRef]
68. Bies-Ethève, N.; Gaubier-Comella, P.; Debures, A.; Lasserre, E.; Jobet, E.; Raynal, M.; Cooke, R.; Delseny, M. Inventory, evolution and expression profiling diversity of the LEA (late embryogenesis abundant) protein gene family in *Arabidopsis thaliana*. *Plant Mol. Biol.* **2008**, *67*, 107–124. [CrossRef] [PubMed]
69. Bremer, A.; Wolff, M.; Thalhammer, A.; Hincha, D.K. Folding of intrinsically disordered plant LEA proteins is driven by glycerol-induced crowding and the presence of membranes. *Febs J.* **2017**, *284*, 919–936. [CrossRef]
70. Close, T.J. Dehydrins: Emergence of a biochemical role of a family of plant dehydration proteins. *Physiol. Plant.* **2010**, *97*, 795–803. [CrossRef]
71. Soulages, J.L.; Kim, K.; Arrese, E.L.; Walters, C.; Cushman, J.C. Conformation of a group 2 late embryogenesis abundant protein from soybean. Evidence of poly (L-proline)-type II structure. *Plant Physiol.* **2003**, *131*, 963–975. [CrossRef] [PubMed]
72. Hara, M.; Fujinaga, M.; Kuboi, T. Radical scavenging activity and oxidative modification of citrus dehydrin. *Plant Physiol. Biochem.* **2004**, *42*, 657–662. [CrossRef] [PubMed]
73. Dean, R.T.; Fu, S.; Stocker, R.; Davies, M.J. Biochemistry and pathology of radical-mediated protein oxidation. *Biochem. J.* **1997**, *324*, 1–18. [CrossRef]
74. Koag, M.C.; Fenton, R.D.; Wilkens, S.; Close, T.J. The binding of maize DHN1 to lipid vesicles. Gain of structure and lipid specificity. *Plant Physiol.* **2003**, *131*, 309–316. [CrossRef]
75. Hara, M.; Terashima, S.; Kuboi, F.T. Enhancement of cold tolerance and inhibition of lipid peroxidation by citrus dehydrin in transgenic tobacco. *Planta* **2003**, *217*, 290–298. [CrossRef]
76. Thalhammer, A.; Hundertmark, M.; Popova, A.V.; Seckler, R.; Hincha, D.K. Interaction of two intrinsically disordered plant stress proteins (COR15A and COR15B) with lipid membranes in the dry state. *Bba Biomembr.* **2010**, *1798*, 1812–1820. [CrossRef] [PubMed]
77. Vítámvás, P.; Kosová, K.; Prá Ilová, P.; Prášil, I.T. Accumulation of WCS120 protein in wheat cultivars grown at 9 °C or 17 °C in relation to their winter survival. *Plant Breed.* **2010**, *129*, 611–616. [CrossRef]
78. Puhakainen, T.; Hess, M.W.; Mäkelä, P.; Svensson, J.; Palva, E.T. Overexpression of Multiple Dehydrin Genes Enhances Tolerance to Freezing Stress in Arabidopsis. *Plant Mol. Biol.* **2004**, *54*, 743–753. [CrossRef] [PubMed]
79. Chaga, G.S. Twenty-five years of immobilized metal ion affinity chromatography: Past, present and future. *J. Biochem. Biophys. Methods* **2001**, *49*, 313–334. [CrossRef]
80. Kruger, C.; Berkowitz, O.; Stephan, U.W.; Hell, R. A metal-binding member of the late embryogenesis abundant protein family transports iron in the phloem of *Ricinus communis* L. *J. Biol. Chem.* **2002**, *277*, 25062–25069. [CrossRef]
81. Ingram, J.; Bartels, D. The Molecular Basis Of Dehydration Tolerance In Plants. *Annu. Rev. Plant Biol.* **1996**, *47*, 377–403. [CrossRef] [PubMed]
82. Hundertmark, M.; Hincha, D.K. LEA (Late Embryogenesis Abundant) proteins and their encoding genes in *Arabidopsis thaliana*. *BMC Genom.* **2008**, *9*, 118. [CrossRef]
83. Liu, X.; Wang, Z.; Wang, L.; Wu, R.; Phillips, J.; Deng, X. LEA 4 group genes from the resurrection plant Boea hygrometrica confer dehydration tolerance in transgenic tobacco. *Plant Sci.* **2009**, *176*, 90–98. [CrossRef]
84. Wang, L.; Li, X.; Chen, S.; Liu, G. Enhanced drought tolerance in transgenic *Leymus chinensis* plants with constitutively expressed wheat TaLEA3. *Biotechnol. Lett.* **2009**, *31*, 313–319. [CrossRef] [PubMed]
85. Alsheikh, M.K. Ion Binding Properties of the Dehydrin ERD14 Are Dependent upon Phosphorylation. *J. Biol. Chem.* **2003**, *278*, 40882–40889. [CrossRef] [PubMed]
86. Monroy, A.F.; Castonguay, Y.; Laberge, S.; Sarhan, F.; Dhindsa, L.P.V.A. A new cold-induced alfalfa gene is associated with enhanced hardening at subzero temperature. *Plant Physiol.* **1993**, *102*, 873–879. [CrossRef] [PubMed]
87. Hara, M.; Terashima, S.; Kuboi, T. Characterization and cryoprotective activity of cold-responsive dehydrin from *Citrus unshiu*. *J. Plant Physiol.* **2001**, *158*, 1333–1339. [CrossRef]
88. Allagulova, C.R.; Gimalov, F.R.; Shakirova, F.M.; Vakhitov, V.A. The Plant Dehydrins: Structure and Putative Functions. *Biochem. Biokhimiia* **2003**, *68*, 945–951. [CrossRef]

89. Johanson, U.; Karlsson, M.; Johansson, I.; Gustavsson, S.; Kjellbom, P.J. The complete set of genes encoding major intrinsic proteins in Arabidopsis provides a framework for a new nomenclature for major intrinsic proteins in plants. *Plant Physiol.* **2001**, *126*, 1358–1369. [CrossRef] [PubMed]

90. Rougé, P.; Barre, A. A molecular modeling approach defines a new group of Nodulin 26-like aquaporins in plants. *Biochem. Biophys. Res. Commun.* **2008**, *367*, 60–66. [CrossRef]

91. Mitani-Ueno, N.; Yamaji, N.; Zhao, F.J.; Jian, F.M. The aromatic/arginine selectivity filter of NIP aquaporins plays a critical role in substrate selectivity for silicon, boron, and arsenic. *J. Exp. Bot.* **2011**, *62*, 4391–4398. [CrossRef] [PubMed]

92. Ishikawa, F.; Suga, S.; Uemura, T.; Sato, M.H.; Maeshima, M. Novel type aquaporin SIPs are mainly localized to the ER membrane and show cell-specific expression in *Arabidopsis thaliana*. *FEBS Lett.* **2005**, *579*, 5814–5820. [CrossRef]

93. Frigerio, L. Mapping of Tonoplast Intrinsic Proteins in Maturing and Germinating Arabidopsis Seeds Reveals Dual Localization of Embryonic TIPs to the Tonoplast and Plasma Membrane. *Mol. Plant* **2011**, *4*, 180–189.

94. Muto, Y.; Segami, S.; Hayashi, H.; Sakurai, J.; Murai-Hatano, M.; Hattori, Y.; Ashikari, M.; Maeshima, M. Vacuolar proton pumps and aquaporins involved in rapid internode elongation of deepwater rice. *Biosci. Biotechnol. Biochem.* **2011**, *75*, 114–122. [CrossRef] [PubMed]

95. Javot, H. Role of a Single Aquaporin Isoform in Root Water Uptake. *Plant Cell* **2003**, *15*, 509–522. [CrossRef]

96. Heinen, R.B.; Ye, Q.; François, C. Role of aquaporins in leaf physiology. *J. Exp. Bot.* **2009**, 2971–2985. [CrossRef] [PubMed]

97. Bots, M. Aquaporins of the PIP2 Class Are Required for Efficient Anther Dehiscence in Tobacco. *Plant Physiol.* **2005**, *137*, 1049–1056. [CrossRef] [PubMed]

98. Liu, D.; Tu, L.; Wang, L.; Li, Y.; Zhu, L.; Zhang, X. Characterization and expression of plasma and tonoplast membrane aquaporins in elongating cotton fibers. *Plant Cell Rep.* **2008**, *27*, 1385–1394. [CrossRef]

99. Wudick, M.M.; Luu, D.T.; Maurel, C. A look inside: Localization patterns and functions of intracellular plant aquaporins. *New Phytol.* **2010**, *184*, 289–302. [CrossRef]

100. Maurel, C.; Santoni, V.; Luu, D.T.; Wudick, M.M.; Verdoucq, L. The cellular dynamics of plant aquaporin expression and functions. *Curr. Opin. Plant Biol.* **2009**, *12*, 690–698. [CrossRef]

101. Fotiadis, D.; Jeno, P.; Mini, T.; Wirtz, S.; Muller, S.A.; Fraysse, L.; Kjellbom, P.; Engel, A. Structural Characterization of Two Aquaporins Isolated from Native Spinach Leaf Plasma Membranes. *J. Biol. Chem.* **2001**, *276*, 1707–1714. [CrossRef] [PubMed]

102. Otto, B.; Kaldenhoff, R. Cell-specific expression of the mercury-insensitive plasma-membrane aquaporin NtAQP1 from *Nicotiana tabacum*. *Planta* **2000**, *211*, 167–172. [CrossRef]

103. Yamada, S. A family of transcripts encoding water channel proteins: Tissue-specific expression in the common ice plant. *Plant Cell* **1995**, *7*, 1129–1142.

104. Kaldenhoff, R. The blue light-responsive AthH2 gene of *Arabidopsis thaliana* is primarily expressed in expanding as well as in differentiating cells and encodes a putative channel protein of the plasmalemma. *Plant J. Cell Mol. Biol.* **2010**, *7*, 87–95. [CrossRef] [PubMed]

105. Netting, A.G. pH, abscisic acid and the integration of metabolism in plants under stressed and non-stressed conditions: Cellular responses to stress and their implication for plant water relations. *J. Exp. Bot.* **2000**, *51*, 147–158. [CrossRef] [PubMed]

106. Mignolet-Spruyt, L.; Xu, E.; Idänheimo, N.; Hoeberichts, F.A.; Mühlenbock, P.; Brosché, M.; Van Breusegem, F.; Kangasjärvi, J. Spreading the news: Subcellular and organellar reactive oxygen species production and signalling. *J. Exp. Bot.* **2016**, *67*, 3831–3844. [CrossRef] [PubMed]

107. Dietz, K.-J. Thiol-Based Peroxidases and Ascorbate Peroxidases: Why Plants Rely on Multiple Peroxidase Systems in the Photosynthesizing Chloroplast? *Mol. Cells* **2016**, *39*, 20–25. [PubMed]

108. Pospía il, P. Molecular mechanisms of production and scavenging of reactive oxygen species by photosystem II. *Biochim. Biophys. Acta* **2012**, *1817*, 218–231. [CrossRef] [PubMed]

109. Dröse, S.; Brandt, U. Molecular mechanisms of superoxide production by the mitochondrial respiratory chain. *Adv. Exp. Med. Biol.* **2012**, *748*, 145–169.

110. Pacher, P.; Beckman, J.S.; Liaudet, L. Nitric Oxide and Peroxynitrite in Health and Disease. *Physiol. Rev.* **2007**, *87*, 315–424. [CrossRef] [PubMed]

111. Stadtman, E.R.; Moskovitz, J.; Levine, R.L. Oxidation of methionine residues of proteins: Biological consequences. *Antioxid Redox Signal* **2003**, *5*, 577–582. [CrossRef] [PubMed]

112. Marnett, L.J. Oxyradicals and DNA damage. *Carcinogenesis* **2000**, *21*, 361–370. [CrossRef] [PubMed]

113. Wojtaszek, P. Oxidative burst: An early plant response to pathogen infection. *Biochem. J.* **1997**, *322*, 681–692. [CrossRef]

114. Mittler, R. Oxidative stress, antioxidants and stress tolerance. *Trends Plant Sci.* **2002**, *7*, 405–410. [CrossRef]

115. Scandalios, J.G. Oxygen Stress and Superoxide Dismutases. *Plant Physiol.* **1993**, *101*, 7–12. [CrossRef]

116. Slooten, L. Factors Affecting the Enhancement of Oxidative Stress Tolerance in Transgenic Tobacco Overexpressing Manganese Superoxide Dismutase in the Chloroplasts. *Plant Physiol.* **1995**, *107*, 737–750. [CrossRef] [PubMed]

117. Willekens, H.; Chamnongpol, S.; Davey, M.; Schraudner, M.; Langebartels, C.; Van Montagu, M.; Inzé, D.; Camp, W. Catalase is a sink for H2O2 and is indispensable for stress defense in C3 plants. *Embo J.* **1997**, *16*, 4806–4816. [CrossRef] [PubMed]

118. Hassinen, V.H.; Tervahauta, A.I.; Schat, H.; K Renlampi, S.O. Plant metallothioneins–metal chelators with ROS scavenging activity? *Plant Biol.* **2011**, *13*, 225–232. [CrossRef] [PubMed]

119. Wigoda, N.; Ben-Nissan, G.; Granot, D.; Schwartz, A.; Weiss, D. The gibberellin-induced, cysteine-rich protein GIP2 from Petunia hybrida exhibits in planta antioxidant activity. *Plant J.* **2010**, *48*, 796–805. [CrossRef] [PubMed]
120. Musgrave, W.B.; Yi, H.; Kline, D.; Cameron, J.C.; Wignes, J.; Dey, S.; Pakrasi, H.B.; Jez, J.M. Probing the origins of glutathione biosynthesis through biochemical analysis of glutamate-cysteine ligase and glutathione synthetase from a model photosynthetic prokaryote. *Biochem. J.* **2013**, *450*, 63–72. [CrossRef]
121. Noctor, G.; Foyer, C.H. Ascorbate and glutathione: Keeping Active Oxygen Under Control. *Annu. Rev. Plant Physiol. Plant Mol. Biol.* **1998**, *49*, 249–279. [CrossRef]
122. Barth, C. The Timing of Senescence and Response to Pathogens Is Altered in the Ascorbate-Deficient Arabidopsis Mutant vitamin c-1. *Plant Physiol.* **2004**, *134*, 1784–1792. [CrossRef]
123. Veljovic-Jovanovic, S.D.; Pignocchi, C.; Noctor, G.; Foyer, C.H. Low ascorbic acid in the vtc-1 mutant of Arabidopsis is associated with decreased growth and intracellular redistribution of the antioxidant system. *Plant Physiol.* **2001**, *127*, 426–435. [CrossRef]
124. Smirnoff, N. Ascorbate biosynthesis and function in photoprotection. *Philos. Trans. R. Soc. Lond.* **2000**, *355*, 1455–1464. [CrossRef]
125. Liebler, D.C.; Kling, D.S.; Reed, D.J. Antioxidant protection of phospholipid bilayers by alpha-tocopherol. Control of alpha-tocopherol status and lipid peroxidation by ascorbic acid and glutathione. *J. Biol. Chem.* **1986**, *261*, 12114–12119. [CrossRef]
126. Davey, M.W.; Montagu, M.V. Plant L-ascorbic acid: Chemistry, function, metabolism, biovailability and effects of processing. *J. Sci. Food Agric.* **2000**, *80*, 825–860. [CrossRef]
127. Pinheiro, H.A.; Damatta, F.M.; Chaves, A.R.M.; Fontes, E.P.B.; Loureiro, M.E. Drought tolerance in relation to protection against oxidative stress in clones of *Coffea canephora* subjected to long-term drought. *Plant Sci.* **2004**, *167*, 1307–1314. [CrossRef]
128. Zhang, J.; Kirkham, M.B. Enzymatic responses of the ascorbate-glutathione cycle to drought in sorghum and sunflower plants. *Plant Sci.* **1996**, *113*, 139–147. [CrossRef]
129. Davies, W.J.; Zhang, J.H. Root Signals and the Regulation of Growth and Development of Plants in Drying Soil. *Annu. Rev. Plant Physiol. Plant Mol. Biol.* **1991**, *42*, 55–76. [CrossRef]
130. Chazen, O.; Neumann, P.M. Hydraulic Signals from the Roots and Rapid Cell-Wall Hardening in Growing Maize (*Zea mays* L.) Leaves Are Primary Responses to Polyethylene Glycol-Induced Water Deficits. *Plant Physiol.* **1994**, *104*, 1385–1392. [CrossRef]
131. Fromm, J.; Fei, H. Electrical signaling and gas exchange in maize plants of drying soil—ScienceDirect. *Plant Sci.* **1998**, *132*, 203–213. [CrossRef]
132. Walker, J.C.; Willows, D.R. Mechanism and regulation of Mg-chelatase. *Biochem. J.* **1997**, *327*, 321–333. [CrossRef]
133. Mochizuki, N.; Brusslan, J.A.; Larkin, R. Arabidopsis genomes uncoupled 5 (GUN5) mutant reveals the involvement of Mg-chelatase H subunit in plastid-to-nucleus signal transduction. *Proc. Natl. Acad. Sci. USA* **2001**, *98*, 2053–2058. [CrossRef]
134. Liu, X.; Yue, Y.; Li, B.; Nie, Y.; Li, W.; Wu, W.H.; Ma, L. A G protein-coupled receptor is a plasma membrane receptor for the plant hormone abscisic acid. *Science* **2007**, *315*, 1712–1716. [CrossRef]
135. Yin, P.; Fan, H.; Hao, Q.; Yuan, X.; Wu, D.; Pang, Y.; Yan, C.; Li, W.; Wang, J.; Yan, N. Structural insights into the mechanism of abscisic acid signaling by PYL proteins. *Nat. Struct. Mol. Biol.* **2009**, *16*, 1230–1237. [CrossRef] [PubMed]
136. Wang, W.H.; Yi, X.Q.; Han, A.D.; Liu, T.W.; Chen, J.; Wu, F.H.; Dong, X.J.; He, J.X.; Pei, Z.M.; Zheng, H.L. Calcium-sensing receptor regulates stomatal closure through hydrogen peroxide and nitric oxide in response to extracellular calcium in Arabidopsis. *J. Exp. Bot.* **2012**, *63*, 177–190. [CrossRef] [PubMed]
137. Case, R.M.; Eisner, D.; Gurney, A.; Jones, O.; Muallem, S.; Verkhratsky, A. Evolution of calcium homeostasis: From birth of the first cell to an omnipresent signalling system. *Cell Calcium* **2007**, *42*, 345–350. [CrossRef]
138. Kim, T.H.; Böhmer, M.; Hu, H.; Nishimura, N.; Schroeder, J.I. Guard Cell Signal Transduction Network: Advances in Understanding Abscisic Acid, CO_2, and Ca^{2+} Signaling. *Annu. Rev. Plant Biol.* **2010**, *61*, 561–591. [CrossRef]
139. Li, S.; Assmann, S.M.; Albert, R. Predicting essential components of signal transduction networks: A dynamic model of guard cell abscisic acid signaling. *Plos Biol.* **2006**, *4*, e312. [CrossRef]
140. McAdam, S.A.M.; Brodribb, T.J. Separating active and passive influences on stomatal control of transpiration. *Plant Physiol.* **2014**, *164*, 1578–1586. [CrossRef]
141. Kudla, J.; Batistic, O.; Hashimoto, K. Calcium Signals: The Lead Currency of Plant Information Processing. *Plant Cell* **2010**, *22*, 541–563. [CrossRef]
142. Miller, G.; Shulaev, V.; Mittler, R. Reactive oxygen signaling and abiotic stress. *Physiol. Plant.* **2008**, *133*, 481–489. [CrossRef]
143. Kohli, S.K.; Khanna, K.; Bhardwaj, R.; Abd Allah, E.F.; Ahmad, P.; Corpas, F.J. Assessment of Subcellular ROS and NO Metabolism in Higher Plants: Multifunctional Signaling Molecules. *Antioxidants* **2019**, *8*, 641. [CrossRef] [PubMed]
144. Singh, A.; Kumar, A.; Yadav, S.; Singh, I.K. Reactive oxygen species-mediated signaling during abiotic stress. *Plant Gene* **2019**, *18*, 100173. [CrossRef]
145. Pei, Z.M.; Murata, Y.; Benning, G.; Thomine, S.; Klüsener, B.; Allen, G.J.; Grill, E.; Schroeder, J.I. Calcium channels activated by hydrogen peroxide mediate abscisic acid signalling in guard cells. *Nature* **2000**, *406*, 731–734. [CrossRef] [PubMed]
146. Mori, I.C.; Schroeder, J.I. Reactive oxygen species activation of plant Ca^{2+} channels. A signaling mechanism in polar growth, hormone transduction, stress signaling, and hypothetically mechanotransduction. *Plant Physiol.* **2004**, *135*, 702–708. [CrossRef] [PubMed]
147. Yan, J.; Tsuichihara, N.; Etoh, T.; Iwai, S. Reactive oxygen species and nitric oxide are involved in ABA inhibition of stomatal opening. *Plant Cell Environ.* **2007**, *30*, 1320–1325. [CrossRef]

148. Hossain, M.A.; Bhattacharjee, S.; Armin, S.-M.; Qian, P.; Xin, W.; Li, H.-Y.; Burritt, D.J.; Fujita, M.; Tran, L.-S.P. Hydrogen peroxide priming modulates abiotic oxidative stress tolerance: Insights from ROS detoxification and scavenging. *Front. Plant Sci.* **2015**, *6*, 420. [CrossRef] [PubMed]

149. Colcombet, J.; Hirt, H. Arabidopsis MAPKs: A complex signalling network involved in multiple biological processes. *Biochem. J.* **2008**, *413*, 217–226. [CrossRef] [PubMed]

150. Hirt, H. Multiple roles of MAP kinases in plant signal transduction. *Trends Plant Sci. Trends Plant Sci* **1997**, *2*, 11–15. [CrossRef]

151. Payne, D.M.; Rossomando, A.J.; Martino, P.; Erickson, A.K.; Sturgill, T.W. Identification of the regulatory phosphorylation sites in pp42/mitogen-activated protein kinase (MAP kinase). *Embo J.* **1991**, *10*, 885–892. [CrossRef] [PubMed]

152. Tuteja, N.; Mahajan, S. Calcium signaling network in plants: An overview. *Plant Signal. Behav.* **2007**, *2*, 79–85. [CrossRef]

153. Delauney, A.J.; Verma, D.P.S. Proline biosynthesis and osmoregulation in plants. *Plant J. Cell Mol. Biol.* **1993**, *4*, 215–223. [CrossRef]

154. Hua, X.J.; van de Cotte, B.; Van Montagu, M.; Verbruggen, N. Developmental regulation of pyrroline-5-carboxylate reductase gene expression in Arabidopsis. *Plant Physiol.* **1997**, *114*, 1215–1224. [CrossRef]

155. Savouré, A.; Jaoua, S.; Hua, X.-J.; Ardiles, W.; Van Montagu, M.; Verbruggen, N. Isolation, characterization, and chromosomal location of a gene encoding the " 1-pyrroline-5-carboxylate synthetase in *Arabidopsis thaliana*. *FEBS Lett.* **1995**, *372*, 13–19. [CrossRef]

156. Hu, C.A.; Delauney, A.J.; Verma, D.P. A bifunctional enzyme (delta 1-pyrroline-5-carboxylate synthetase) catalyzes the first two steps in proline biosynthesis in plants. *Proc. Natl. Acad. Sci. USA* **1992**, *89*, 9354–9358. [CrossRef]

157. Kishor, P.; Hong, Z.; Miao, G.H.; Hu, C.; Verma, D. Overexpression of [delta]-Pyrroline-5-Carboxylate Synthetase Increases Proline Production and Confers Osmotolerance in Transgenic Plants. *Plant Physiol.* **1995**, *108*, 1387–1394. [CrossRef]

158. Larosa, P.C.; Rhodes, D.; Rhodes, J.C.; Bressan, R.A.; Csonka, L.N. Elevated Accumulation of Proline in NaCl-Adapted Tobacco Cells Is Not Due to Altered Delta-Pyrroline-5-Carboxylate Reductase. *Plant Physiol* **1991**, *96*, 245–250. [CrossRef]

159. Sharma, S.; Villamor, J.G.; Verslues, P.E. Essential role of tissue-specific proline synthesis and catabolism in growth and redox balance at low water potential. *Plant Physiol.* **2011**, *157*, 292–304. [CrossRef]

160. Zhu, B.; Su, J.; Chang, M.; Verma, D.P.S.; Fan, Y.-L.; Wu, R. Overexpression of a Δ1-pyrroline-5-carboxylate synthetase gene and analysis of tolerance to water- and salt-stress in transgenic rice. *Plant Sci.* **1998**, *139*, 41–48. [CrossRef]

161. Yamchi, A.; Rastgar Jazii, F.; Mousavi, A.; Karkhane, A.A.; Renu. Proline Accumulation in Transgenic Tobacco as a Result of Expression of Arabidopsis " 1-Pyrroline-5-carboxylate synthetase (P5CS) During Osmotic Stress. *J. Plant Biochem. Biotechnol.* **2007**, *16*, 9–15. [CrossRef]

162. Hmida-Sayari, A.; Gargouri-Bouzid, R.; Bidani, A.; Jaoua, L.; Savouré, A.; Jaoua, S. Overexpression of Δ1-pyrroline-5-carboxylate synthetase increases proline production and confers salt tolerance in transgenic potato plants. *Plant Sci.* **2005**, *169*, 746–752. [CrossRef]

163. Guerzoni, J.T.S.; Belintani, N.G.; Moreira, R.M.P.; Hoshino, A.A.; Domingues, D.S.; Filho, J.O.C.B.; Vieira, L.G.E. Stress-induced Δ1-pyrroline-5-carboxylate synthetase (P5CS) gene confers tolerance to salt stress in transgenic sugarcane. *Acta Physiol. Plant.* **2014**, *36*, 2309–2319. [CrossRef]

164. Zhang, G.-C.; Zhu, W.-L.; Gai, J.-Y.; Zhu, Y.-L.; Yang, L.-F. Enhanced salt tolerance of transgenic vegetable soybeans resulting from overexpression of a novel Δ1-pyrroline-5-carboxylate synthetase gene from *Solanum torvum* Swartz. *Hortic. Environ. Biotechnol.* **2015**, *56*, 94–104. [CrossRef]

165. Hervieu, F.; Dily, F.L.; Billard, J.-P.; Huault, C. Effects of water-stress on proline content and ornithine aminotransferase activity of radish cotyledons. *Phytochemistry* **1994**, *37*, 1227–1231. [CrossRef]

166. Roosens, N.H.; Bitar, F.A.; Loenders, K.; Angenon, G.; Jacobs, M. Overexpression of ornithine-δ-aminotransferase increases proline biosynthesis and confers osmotolerance in transgenic plants. *Mol. Breed.* **2002**, *9*, 73–80. [CrossRef]

167. Wu, L.; Fan, Z.; Guo, L.; Li, Y.; Zhang, W.; Qu, L.J.; Chen, Z. Over-expression of an Arabi-dopsisd δ-OAT gene enhances salt and drought tolerance in transgenic rice. *Chin. Sci. Bull.* **2003**, *48*, 2594–2600. [CrossRef]

168. Kiyosue, T.; Yoshiba, Y.; Yamaguchi-Shinozaki, K.; Shinozaki, K. A nuclear gene encoding mitochondrial proline dehydrogenase, an enzyme involved in proline metabolism, is upregulated by proline but downregulated by dehydration in Arabidopsis. *Plant Cell* **1996**, *8*, 1323–1335.

169. Rentsch, D.; Hirner, B.; Frommer, S.W.B. Salt Stress-Induced Proline Transporters and Salt Stress-Repressed Broad Specificity Amino Acid Permeases Identified by Suppression of a Yeast Amino Acid Permease-Targeting Mutant. *Plant Cell* **1996**, *8*, 1437–1446. [PubMed]

170. Ueda, A.; Shi, W.; Shimada, T.; Miyake, H.; Takabe, T. Altered expression of barley proline transporter causes different growth responses in Arabidopsis. *Planta* **2007**, *227*, 277–286. [CrossRef]

171. Nuccio, M.L.; McNeil, S.D.; Ziemak, M.J.; Hanson, A.D.; Jain, R.K.; Selvaraj, G. Choline import into chloroplasts limits glycine betaine synthesis in tobacco: Analysis of plants engineered with a chloroplastic or a cytosolic pathway. *Metab. Eng.* **2000**, *2*, 300–311. [CrossRef] [PubMed]

172. Brendza, K.M.; Haakenson, W.; Cahoon, R.E.; Hicks, L.M.; Palavalli, L.H.; Chiapelli, B.J.; McLaird, M.; McCarter, J.P.; Williams, D.J.; Hresko, M.C.; et al. Phosphoethanolamine N-methyltransferase (PMT-1) catalyses the first reaction of a new pathway for phosphocholine biosynthesis in *Caenorhabditis elegans*. *Biochem. J.* **2007**, *404*, 439–448. [CrossRef]

173. McNeil, S.D.; Nuccio, M.L.; Ziemak, M.J.; Hanson, A.D. Enhanced synthesis of choline and glycine betaine in transgenic tobacco plants that overexpress phosphoethanolamine N-methyltransferase. *Proc. Natl. Acad. Sci. USA* **2001**, *98*, 10001–10005. [CrossRef]

174. Bhuiyan, N.H.; Akira, H.; Nana, Y.; Vandna, R.; Takashi, H.; Teruhiro, T. Regulation of betaine synthesis by precursor supply and choline monooxygenase expression in Amaranthus tricolor. *J. Exp. Bot.* **2007**, 4203–4212. [CrossRef]

175. Chen, T.H.H.; Murata, N. Glycinebetaine protects plants against abiotic stress: Mechanisms and biotechnological applications. *PlantCell Environ.* **2011**, *34*, 1–20. [CrossRef]

176. Sakamoto, A.; Murata, N. The role of glycine betaine in the protection of plants from stress: Clues from transgenic plants. *Plant Cell Environ.* **2002**, *25*, 163–171. [CrossRef] [PubMed]

177. Rathinasabapathi, B.; Burnet, M.; Russell, B.L. Choline Monooxygenase, an Unusual Iron-Sulfur Enzyme Catalyzing the First Step of Glycine Betaine Synthesis in Plants: Prosthetic Group Characterization and cDNA Cloning. *Proc. Natl. Acad. Sci. USA* **1997**, *94*, 3454–3458. [CrossRef]

178. Rhodes, D.; Nadolska-Orczyk, A.; Rich, P.J. *Salinity, Osmolytes and Compatible Solutes*; Springer: Dordrecht, The Netherlands, 2002; pp. 181–204.

179. Fitzgerald, T.L.; Waters, D.L.E.; Henry, R.J. Betaine aldehyde dehydrogenase in plants. *Plant Biol.* **2009**, *11*, 119–130. [CrossRef]

180. Fujiwara, T.; Hori, K.; Ozaki, K.; Yokota, Y.; Mitsuya, S.; Ichiyanagi, T.; Hattori, T.; Takabe, T. Enzymatic characterization of peroxisomal and cytosolic betaine aldehyde dehydrogenases in barley. *Physiol. Plant.* **2008**, *134*, 22–30. [CrossRef] [PubMed]

181. Wani, S.H.; Singh, N.B.; Haribhushan, A.; Mir, J.I. Compatible solute engineering in plants for abiotic stress tolerance—Role of glycine betaine. *Curr. Genom.* **2013**, *14*, 157–165. [CrossRef]

182. Shen, Y.-G.; Du, B.-X.; Zhang, W.-K.; Zhang, J.-S.; Chen, S.-Y. AhCMO, regulated by stresses in *Atriplex hortensis*, can improve drought tolerance in transgenic tobacco. *Tag. Theor. Appl. Genet. Theor. Und Angew. Genet.* **2002**, *105*, 815–821. [CrossRef] [PubMed]

183. Mitsuya, S.; Kuwahara, J.; Ozaki, K.; Saeki, E.; Fujiwara, T.; Takabe, T. Isolation and characterization of a novel peroxisomal choline monooxygenase in barley. *Planta* **2011**, *234*, 1215–1226. [CrossRef]

184. Wu, S.; Su, Q.; An, L.J. Isolation of choline monooxygenase (CMO) gene from salicornia europaea and enhanced salt tolerance of transgenic tobacco with CMO genes. *Indian J. Geo-Mar. Sci.* **2010**, *47*, 298–305.

185. Ishitani, M.; Nakamura, T.; Han, S.Y.; Takabe, T. Expression of the betaine aldehyde dehydrogenase gene in barley in response to osmotic stress and abscisic acid. *Plant Mol. Biol.* **1995**, *27*, 307–315. [CrossRef]

186. Fan, W.; Zhang, M.; Zhang, H.; Zhang, P. Improved Tolerance to Various Abiotic Stresses in Transgenic Sweet Potato (*Ipomoea batatas*) Expressing Spinach Betaine Aldehyde Dehydrogenase. *PLoS ONE* **2012**, *7*, e37344. [CrossRef]

187. Li, M.; Li, Z.; Li, S.; Guo, S. Genetic Engineering of Glycine Betaine Biosynthesis Reduces Heat-Enhanced Photoinhibition by Enhancing Antioxidative Defense and Alleviating Lipid Peroxidation in Tomato. *Plant Mol. Biol. Rep.* **2014**, *32*, 42–51. [CrossRef]

188. Cho, M.H.; Jang, A.; Bhoo, S.H.; Jeon, J.S.; Hahn, T.R. Manipulation of triose phosphate/phosphate translocator and cytosolic fructose-1,6-bisphosphatase, the key components in photosynthetic sucrose synthesis, enhances the source capacity of transgenic Arabidopsis plants. *Photosynth. Res.* **2012**, *111*, 261–268. [CrossRef] [PubMed]

189. Zrenner, R.; Krause, K.e.; Apel, P.; Sonnewald, U. Reduction of the cytosolic fructose-1,6-bisphosphatase in transgenic potato plants limits photosynthetic sucrose biosynthesis with no impact on plant growth and tuber yield. *Plant J.* **2010**, *9*, 671–681. [CrossRef] [PubMed]

190. Huber, S.C.; Huber, J.L. Role and regulation of sucrose-phosphate synthase in higher plants. *Annu. Rev. Plant Physiol. Plant Mol. Biol.* **1996**, *47*, 431–444. [CrossRef]

191. Park, J.Y.; Canam, T.; Kang, K.Y.; Ellis, D.D.; Mansfield, S.D. Over-expression of an arabidopsis family A sucrose phosphate synthase (SPS) gene alters plant growth and fibre development. *Transgenic Res.* **2008**, *17*, 181–192. [CrossRef]

192. Nguyen-Quoc, B.; N'Tchobo, H.; Foyer, C.H.; Yelle, S. Overexpression of sucrose phosphate synthase increases sucrose unloading in transformed tomato fruit. *J. Exp. Bot.* **1999**, *50*, 785–791. [CrossRef]

193. Ishimaru, K.; Hirotsu, N.; Kashiwagi, T.; Madoka, Y.; Nagasuga, K.; Ono, K.; Ohsugi, R. Overexpression of a Maize SPS Gene Improves Yield Characters of Potato under Field Conditions. *Plant Prod. Sci.* **2008**, *11*, 104–107. [CrossRef]

194. Strand, Å.; Foyer, C.H.; Gustafsson, P.; Gardeström, P.; Hurry, V. Altering flux through the sucrose biosynthesis pathway in transgenic *Arabidopsis thaliana* modifies photosynthetic acclimation at low temperatures and the development of freezing tolerance. *Plant Cell Environ.* **2010**, *26*, 523–535. [CrossRef]

195. Hand, S.C.; Menze, M.A.; Toner, M.; Boswell, L.; Moore, D. LEA proteins during water stress: Not just for plants anymore. *Annu. Rev. Physiol.* **2011**, *73*, 115–134. [CrossRef]

196. Sivamani, E.; Bahieldin, A.; Wraith, J.M.; Al-Niemi, T.; Dyer, W.E.; Ho, T.D.; Qu, R. Improved biomass productivity and water use efficiency under water deficit conditions in transgenic wheat constitutively expressing the barley HVA1 gene. *Plant Sci. Int. J. Exp. Plant Biol.* **2000**, *155*, 1–9. [CrossRef]

197. Poku, S.A.; Chukwurah, P.N.; Aung, H.H.; Nakamura, I. Over-Expression of a Melon Y3SK2-Type LEA Gene Confers Drought and Salt Tolerance in Transgenic Tobacco Plants. *Plants* **2020**, *9*, 1749. [CrossRef]

198. Luo, D.; Hou, X.; Zhang, Y.; Meng, Y.; Zhang, H.; Liu, S.; Wang, X.; Chen, R. CaDHN5, a Dehydrin Gene from Pepper, Plays an Important Role in Salt and Osmotic Stress Responses. *Int. J. Mol. Sci.* **2019**, *20*, 1989. [CrossRef] [PubMed]

199. Poku, S.A.; Seçgin, Z.; Kavas, M. Overexpression of Ks-type dehydrins gene OeSRC1 from *Olea europaea* increases salt and drought tolerance in tobacco plants. *Mol. Biol. Rep.* **2019**, *46*, 5745–5757. [CrossRef] [PubMed]

200. Kaldenhoff, R.; Fischer, M. Aquaporins in plants. *Acta Physiol.* **2006**, *187*, 169–176. [CrossRef]

201. Zhang, S.; Feng, M.; Chen, W.; Zhou, X.; Lu, J.; Wang, Y.; Li, Y.; Jiang, C.Z. In rose, transcription factor PTM balances growth and drought survival via PIP2;1 aquaporin. *Nat. Plants* **2019**, *5*, 290–299. [CrossRef] [PubMed]

202. Zheng, J.; Lin, R.; Pu, L.; Wang, Z.; Jian, S. Ectopic Expression of CrPIP2;3, a Plasma Membrane Intrinsic Protein Gene from the Halophyte *Canavalia rosea*, Enhances Drought and Salt-Alkali Stress Tolerance in Arabidopsis. *Int. J. Mol. Sci.* **2021**, *22*, 565. [CrossRef]

203. Khan, K.; Agarwal, P.; Shanware, A.; Sane, V.A. Heterologous Expression of Two Jatropha Aquaporins Imparts Drought and Salt Tolerance and Improves Seed Viability in Transgenic *Arabidopsis thaliana*. *PLoS ONE* **2015**, *10*, e0128866. [CrossRef]

204. Peng, Y.; Lin, W.; Cai, W.; Arora, R. Overexpression of a Panax ginseng tonoplast aquaporin alters salt tolerance, drought tolerance and cold acclimation ability in transgenic Arabidopsis plants. *Planta* **2007**, *226*, 729–740. [CrossRef] [PubMed]

205. Martins, C.P.S.; Neves, D.M.; Cidade, L.C.; Mendes, A.F.S.; Silva, D.C.; Almeida, A.-A.F.; Coelho-Filho, M.A.; Gesteira, A.S.; Soares-Filho, W.S.; Costa, M.G.C. Expression of the citrus CsTIP2;1 gene improves tobacco plant growth, antioxidant capacity and physiological adaptation under stress conditions. *Planta* **2017**, *245*, 951–963. [CrossRef]

206. Seo, M. Complex regulation of ABA biosynthesis in plants. *Trends Plant Sci.* **2002**, *7*, 41–48. [CrossRef]

207. Iuchi, S.; Kobayashi, M.; Taji, T.; Naramoto, M.; Seki, M.; Kato, T.; Tabata, S.; Kakubari, Y.; Yamaguchi-Shinozaki, K.; Shinozaki, K. Regulation of drought tolerance by gene manipulation of 9-cis-epoxycarotenoid, a key in abscisic acid biosynthesis in Arabidopsis. *Plant J. Cell Mol. Biol.* **2001**, *27*, 325. [CrossRef]

208. Awan, S.Z.; Chandler, J.O.; Harrison, P.J.; Sergeant, M.J.; Bugg, T.D.H.; Thompson, A.J. Promotion of Germination Using Hydroxamic Acid Inhibitors of 9-cis-Epoxycarotenoid Dioxygenase. *Front. Plant Sci.* **2017**, *8*, 357. [CrossRef] [PubMed]

209. Cheng, H.S. Calcium signaling through protein kinases. The Arabidopsis calcium-dependent protein kinase gene family. *Plant Physiol.* **2002**, *129*, 469. [CrossRef] [PubMed]

210. Harmon, A.C.; Gribskov, M.; Gubrium, E.; Harper, J.F. The CDPK superfamily of protein kinases. *New Phytol.* **2001**, *151*, 175–183. [CrossRef]

211. Botella, J.R.; Arteca, J.M.; Somodevilla, M.; Arteca, R.N. Calcium-dependent protein kinase gene expression in response to physical and chemical stimuli in mungbean (*Vigna radiata*). *Plant Mol. Biol.* **1996**, *30*, 1129–1137. [CrossRef]

212. Patharkar, O.R.; Cushman, J.C. A stress-induced calcium-dependent protein kinase from *Mesembryanthemum crystallinum* phosphorylates a two-component pseudo-response regulator. *Plant J. Cell Mol. Biol.* **2000**, *24*, 679–691. [CrossRef]

213. Urao, T.; Katagiri, T.; Mizoguchi, T.; Yamaguchi-Shinozaki, K.; Hayashida, N.; Shinozaki, K. Two genes that encode Ca(2+)-dependent protein kinases are induced by drought and high-salt stresses in *Arabidopsis thaliana*. *Mol. Gen. Genet. MGG* **1994**, *244*, 331–340. [CrossRef]

214. Sheen, J. Ca2+-Dependent Protein Kinases and Stress Signal Transduction in Plants. *Science* **1997**, *274*, 1900–1902. [CrossRef]

215. Saijo, Y. A Ca2+-Dependent Protein Kinase that Endows Rice Plants with Cold- and Salt-Stress Tolerance Functions in Vascular Bundles. *Plant Cell Physiol.* **2001**, *42*, 1228–1233. [CrossRef]

216. Mizoguchi, T.; Irie, K.; Hirayama, T.; Hayashida, N.; Yamaguchi-Shinozaki, K.; Matsumoto, K.; Shinozaki, K. A gene encoding a mitogen-activated protein kinase kinase kinase is induced simultaneously with genes for a mitogen-activated protein kinase and an S6 ribosomal protein kinase by touch, cold, and water stress in *Arabidopsis thaliana*. *Proc. Natl. Acad. Sci. USA* **1996**, *93*, 765–769. [CrossRef]

217. Chitlaru, E.; Seger, R.; Pick, U. Activation of a 74 kDa plasma membrane protein kinase by hyperosmotic shocks in the halotolerant alga *Dunaliella salina*—ScienceDirect. *J. Plant Physiol.* **1997**, *151*, 429–436. [CrossRef]

218. Xiong, L. Disease Resistance and Abiotic Stress Tolerance in Rice Are Inversely Modulated by an Abscisic Acid Inducible Mitogen-Activated Protein Kinase. *Plant Cell* **2003**, *15*, 745–759. [CrossRef]

219. Shinozaki, K.; Yamaguchi-Shinozaki, K. Gene networks involved in drought stress response and tolerance. *J. Exp. Bot.* **2007**, *58*, 221. [CrossRef]

220. Gong, P.; Zhang, J.; Li, H.; Yang, C.; Zhang, C.; Zhang, X.; Khurram, Z.; Zhang, Y.; Wang, T.; Fei, Z. Transcriptional profiles of drought-responsive genes in modulating transcription signal transduction, and biochemical pathways in tomato. *J. Exp. Bot.* **2010**, *61*, 3563–3575. [CrossRef] [PubMed]

221. Degenkolbe, T.; Do, P.T.; Zuther, E.; Repsilber, D.; Walther, D.; Hincha, D.K.; K Hl, K.I. Expression profiling of rice cultivars differing in their tolerance to long-term drought stress. *Plant Mol. Biol.* **2009**, *69*, 133–153. [CrossRef]

222. Ariel, F.D.; Manavella, P.A.; Dezar, C.A.; Chan, R.L. The true story of the HD-Zip family. *Trends Plant Sci.* **2007**, *12*, 419–426. [CrossRef]

223. Nakashima, K.; Yamaguchi-Shinozaki, I.K. Transcriptional Regulatory Networks in Response to Abiotic Stresses in Arabidopsis and Grasses. *Plant Physiol.* **2009**, *149*, 88–95. [CrossRef] [PubMed]

224. Harris, J.C.; Hrmova, M.; Lopato, S.; Langridge, P. Modulation of plant growth by HD-Zip class I and II transcription factors in response to environmental stimuli. *New Phytol.* **2011**, *190*, 823–837. [CrossRef]

225. Agalou, A.; Purwantomo, S.; Övernäs, E.; Johannesson, H.; Zhu, X.; Estiati, A.; de Kam, R.J.; Engström, P.; Slamet-Loedin, I.H.; Zhu, Z.; et al. A genome-wide survey of HD-Zip genes in rice and analysis of drought-responsive family members. *Plant Mol. Biol.* **2008**, *66*, 87–103. [CrossRef]

226. Deng, X.; Phillips, J.; Bräutigam, A.; Engström, P.; Johannesson, H.; Ouwerkerk, P.B.F.; Ruberti, I.; Salinas, J.; Vera, P.; Iannacone, R.; et al. A homeodomain leucine zipper gene from *Craterostigma plantagineum* regulates abscisic acid responsive gene expression and physiological responses. *Plant Mol. Biol.* **2006**, *61*, 469–489. [CrossRef]

227. Manavella, P.A.; Arce, A.L.; Dezar, C.A.; Bitton, F.; Renou, J.-P.; Crespi, M.; Chan, R.L. Cross-talk between ethylene and drought signalling pathways is mediated by the sunflower Hahb-4 transcription factor. *Plant J. Cell Mol. Biol.* **2006**, *48*, 125–137. [CrossRef]
228. Fan, Y.; Liu, J.; Zou, J.; Zhang, X.; Jiang, L.; Liu, K.; Lü, P.; Gao, J.; Zhang, C. The RhHB1/RhLOX4 module affects the dehydration tolerance of rose flowers (*Rosa hybrida*) by fine-tuning jasmonic acid levels. *Hortic. Res.* **2020**, *7*, 74. [CrossRef]
229. Cheng, M.C.; Liao, P.M.; Kuo, W.W.; Lin, T.P. The Arabidopsis ETHYLENE RESPONSE FACTOR1 regulates abiotic stress-responsive gene expression by binding to different cis-acting elements in response to different stress signals. *Plant Physiol.* **2013**, *162*, 1566–1582. [CrossRef]
230. Wan, L.; Zhang, J.; Zhang, H.; Zhang, Z.; Quan, R.; Zhou, S.; Huang, R. Transcriptional activation of OsDERF1 in OsERF3 and OsAP2-39 negatively modulates ethylene synthesis and drought tolerance in rice. *PLoS ONE* **2011**, *6*, e25216. [CrossRef]
231. Maruyama, K.; Takeda, M.; Kidokoro, S.; Yamada, K.; Sakuma, Y.; Urano, K.; Fujita, M.; Yoshiwara, K.; Matsukura, S.; Morishita, Y.; et al. Metabolic pathways involved in cold acclimation identified by integrated analysis of metabolites and transcripts regulated by DREB1A and DREB2A. *Plant Physiol.* **2009**, *150*, 1972–1980. [CrossRef]
232. Li, J.; Guo, X.; Zhang, M.; Wang, X.; Zhao, Y.; Yin, Z.; Zhang, Z.; Wang, Y.; Xiong, H.; Zhang, H. OsERF71 confers drought tolerance via modulating ABA signaling and proline biosynthesis. *Plant Sci.* **2018**, *270*, 131–139. [CrossRef]
233. Li, M.-R.; Li, Y.; Li, H.-Q.; Wu, G.-J. Ectopic expression of FaDREB2 enhances osmotic tolerance in paper mulberry. *J. Integr. Plant Biol.* **2011**, *53*, 951–960. [CrossRef] [PubMed]
234. Zhang, H.; Liu, W.; Wan, L.; Li, F.; Dai, L.; Li, D.; Zhang, Z.; Huang, R. Functional analyses of ethylene response factor JERF3 with the aim of improving tolerance to drought and osmotic stress in transgenic rice. *Transgenic Res.* **2010**, *19*, 809–818. [CrossRef]
235. Zhang, G.; Chen, M.; Li, L.; Xu, Z.; Chen, X.; Guo, J.; Ma, Y. Overexpression of the soybean GmERF3 gene, an AP2/ERF type transcription factor for increased tolerances to salt, drought, and diseases in transgenic tobacco. *J. Exp. Bot.* **2009**, *60*, 3781–3796. [CrossRef] [PubMed]
236. Wu, L.; Zhang, Z.; Zhang, H.; Wang, X.-C.; Huang, R. Transcriptional modulation of ethylene response factor protein JERF3 in the oxidative stress response enhances tolerance of tobacco seedlings to salt, drought, and freezing. *Plant Physiol.* **2008**, *148*, 1953–1963. [CrossRef] [PubMed]
237. Wang, Y.; Wan, L.; Zhang, L.; Zhang, Z.; Zhang, H.; Quan, R.; Zhou, S.; Huang, R. An ethylene response factor OsWR1 responsive to drought stress transcriptionally activates wax synthesis related genes and increases wax production in rice. *Plant Mol. Biol.* **2012**, *78*, 275–288. [CrossRef]
238. Liu, J.; Osbourn, A.; Ma, P. MYB Transcription Factors as Regulators of Phenylpropanoid Metabolism in Plants. *Mol. Plant* **2015**, *8*, 689–708. [CrossRef]
239. Dubos, C.; Stracke, R.; Grotewold, E.; Weisshaar, B.; Martin, C.; Lepiniec, L.C. MYB transcription factors in Arabidopsis. *Trends Plant Sci.* **2010**, *15*, 573–581. [CrossRef]
240. Dai, X.; Xu, Y.; Ma, Q.; Xu, W.; Wang, T.; Xue, Y.; Chong, K. Overexpression of an R1R2R3 MYB gene, OsMYB3R-2, increases tolerance to freezing, drought, and salt stress in transgenic Arabidopsis. *Plant Physiol.* **2007**, *143*, 1739–1751. [CrossRef] [PubMed]
241. Butt, H.I.; Yang, Z.; Gong, Q.; Chen, E.; Wang, X.; Zhao, G.; Ge, X.; Zhang, X.; Li, F. GaMYB85, an R2R3 MYB gene, in transgenic Arabidopsis plays an important role in drought tolerance. *BMC Plant Biol.* **2017**, *17*, 142. [CrossRef]
242. Chen, T.; Li, W.; Hu, X.; Guo, J.; Liu, A.; Zhang, B. A Cotton MYB Transcription Factor, GbMYB5, is Positively Involved in Plant Adaptive Response to Drought Stress. *Plant Cell Physiol.* **2015**, *56*, 917–929. [CrossRef]
243. Wang, N.; Zhang, W.; Qin, M.; Li, S.; Qiao, M.; Liu, Z.; Xiang, F. Drought Tolerance Conferred in Soybean (*Glycine max.* L) by GmMYB84, a Novel R2R3-MYB Transcription Factor. *Plant Cell Physiol.* **2017**, *58*, 1764–1776. [CrossRef]
244. Chen, K.; Song, M.; Guo, Y.; Liu, L.; Xue, H.; Dai, H.; Zhang, Z. MdMYB46 could enhance salt and osmotic stress tolerance in apple by directly activating stress-responsive signals. *Plant Biotechnol. J.* **2019**, *17*, 2341–2355. [CrossRef]
245. Geng, D.; Chen, P.; Shen, X.; Zhang, Y.; Li, X.; Jiang, L.; Xie, Y.; Niu, C.; Zhang, J.; Huang, X.; et al. MdMYB88 and MdMYB124 Enhance Drought Tolerance by Modulating Root Vessels and Cell Walls in Apple. *Plant Physiol.* **2018**, *178*, 1296–1309. [CrossRef]
246. Eulgem, T.; Rushton, P.J.; Robatzek, S.; Somssich, I.E. The WRKY superfamily of plant transcription factors. *Trends Plant Sci.* **2000**, *5*, 199–206. [CrossRef]
247. Wang, C.; Deng, P.; Chen, L.; Wang, X.; Ma, H.; Hu, W.; Yao, N.; Feng, Y.; Chai, R.; Yang, G.; et al. A wheat WRKY transcription factor TaWRKY10 confers tolerance to multiple abiotic stresses in transgenic tobacco. *PLoS ONE* **2013**, *8*, e65120. [CrossRef] [PubMed]
248. Yan, H.; Jia, H.; Chen, X.; Hao, L.; An, H.; Guo, X. The cotton WRKY transcription factor GhWRKY17 functions in drought and salt stress in transgenic *Nicotiana benthamiana* through ABA signaling and the modulation of reactive oxygen species production. *Plant Cell Physiol.* **2014**, *55*, 2060–2076. [CrossRef] [PubMed]
249. Jia, H.; Wang, C.; Wang, F.; Liu, S.; Li, G.; Guo, X. GhWRKY68 reduces resistance to salt and drought in transgenic *Nicotiana benthamiana*. *PLoS ONE* **2015**, *10*, e0120646. [CrossRef]
250. Wei, W.; Zhang, Y.; Han, L.; Guan, Z.; Chai, T. A novel WRKY transcriptional factor from *Thlaspi caerulescens* negatively regulates the osmotic stress tolerance of transgenic tobacco. *Plant Cell Rep.* **2008**, *27*, 795–803. [CrossRef]
251. Fujita, M.; Fujita, Y.; Maruyama, K.; Seki, M.; Hiratsu, K.; Ohme-Takagi, M.; Tran, L.-S.P.; Yamaguchi-Shinozaki, K.; Shinozaki, K. A dehydration-induced NAC protein, RD26, is involved in a novel ABA-dependent stress-signaling pathway. *Plant J. Cell Mol. Biol.* **2004**, *39*, 863–876. [CrossRef] [PubMed]

252. Yong, Y.; Zhang, Y.; Lyu, Y. A Stress-Responsive NAC Transcription Factor from Tiger Lily (LlNAC2) Interacts with LlDREB1 and LlZHFD4 and Enhances Various Abiotic Stress Tolerance in Arabidopsis. *Int. J. Mol. Sci.* **2019**, *20*, 3225. [CrossRef]
253. Zhang, H.; Cui, X.; Guo, Y.; Luo, C.; Zhang, L. *Picea wilsonii* transcription factor NAC2 enhanced plant tolerance to abiotic stress and participated in RFCP1-regulated flowering time. *Plant Mol. Biol.* **2018**, *98*, 471–493. [CrossRef]
254. Jiang, X.; Zhang, C.; Lü, P.; Jiang, G.; Liu, X.; Dai, F.; Gao, J. RhNAC3, a stress-associated NAC transcription factor, has a role in dehydration tolerance through regulating osmotic stress-related genes in rose petals. *Plant Biotechnol. J.* **2014**, *12*, 38–48. [CrossRef]
255. Ju, Y.-L.; Yue, X.-F.; Min, Z.; Wang, X.-H.; Fang, Y.-L.; Zhang, J.-X. VvNAC17, a novel stress-responsive grapevine (*Vitis vinifera* L.) NAC transcription factor, increases sensitivity to abscisic acid and enhances salinity, freezing, and drought tolerance in transgenic Arabidopsis. *Plant Physiol. Biochem.* **2020**, *146*, 98–111. [CrossRef] [PubMed]

MDPI

St. Alban-Anlage 66

4052 Basel

Switzerland

Tel. +41 61 683 77 34

Fax +41 61 302 89 18

www.mdpi.com

Horticulturae Editorial Office

E-mail: horticulturae@mdpi.com

www.mdpi.com/journal/horticulturae